PRODUCTION OF QUALITY STEELS IN MINI STEEL PLANTS

Dr. S.K. Goel
Dr. M.L. Kapoor
Dr. V.N.S. Mathur
Dr. G.C. Kaushal

KHANNA PUBLISHERS
2-B, NATH MARKET, NAI SARAK,
DELHI - 110 006
Phones : 2391 23 80; 2722 41 79 Fax : 23980311

Published by :
Romesh Chander Khanna
for **KHANNA PUBLISHERS**
2-B, Nath Market, Nai Sarak,
Delhi - 110 006.

ISBN NO. : 81-7409-207-2

First Edition : 2005

Price : Rs. 165.00

Printed at :
Bright Printers, Turkman Gate, Delhi-110006.

Preface

In the present times of very rapid scientific and technological developments, it is rather imperative for shop floor engineers, metallurgists and executives of Mini steel plants to keep pace with modern developments with regard to the production of quality steels. It is because the quality requirements of customers for various grades of steels are becoming more and more stringent with the passage of time. Meeting the quality requirements is not simple without the understanding of technology and practices involved. There is a greater emphasis today on the production of alloy steels/forging quality steels/special steels etc. by the Mini Steel Plants in order to be competitive and to improve their chances of survival. To meet these objectives, most of the forward looking Mini Steel Plants are going in for newer technologies such as EBT, Electric Arc Furnaces, Secondary treatment units such as Ladle furnace, AOD, VOD or ESR process and use of DRI in EAF and Induction Furnaces. Also continuous casting units for almost all grades of steels have become an integral part of Mini Steel Plants as well as integrated steel plants today.

The present course material on the **"Production of quality steels in Mini Steel Plants"** has been solely designed to acquaint the concerned students/practicing Engineers/Metallurgist/ Executives of Mini Steel plants with the latest techniques involved in the production of quality steels using EAF – Secondary Steel making – Continuous casting route. An attempt has been made to present a proper mix of fundamentals together with the latest technological developments. The entire material is laid out in two parts. Part I is basically concerned with physical chemistry of electric steel making, raw materials, production of few typical grades of steels, casting practice and energy optimization and pollution control. Part II is concerned with the advances in Electric Arc Furnace steel making practice, continuous casting of steels, secondary steel making practices, non–metallic inclusions in steel and techno economics of stainless steel manufacture. It may be mentioned that different chapters in part I and Part II of this compilation have been contributed by different authors, who have expertise in respective areas. The objective is to bring the best out of individual authors and present it to the readers in a comprehensive manner.

A "List of References" has been provided in the end. These references, which contain textbooks/monographs/research papers etc. have been frequently consulted and quoted in the compilation of Part I and II of the present book. Our grateful thanks are due to all the authors/editors of books/monographs/research papers etc. contained in the "List of References". Also we would like to express our grateful thanks to our colleagues in the Deptt of Metallurgical Engg. for their kind help, encouragement and constructive criticism. Last but not the least, we would like to thank Dr. Kuldeep Singh, the then Professor and Director of Continuing Education Department, who was an enormous help during the conduct of specialist courses on "Mini Steel Plant Technology" and also later, when the entire course material was being re-cast in the form of present book.

Any suggestions by the readers meant to improve the contents of this book shall be most welcome and greatly appreciated by the authors.

Dr. S.K. Goel,
Editor

About the Authors

The present book is the result of contributions made by different authors having experience in specific areas.

Dr. M.L. Kapoor, Professor & Dy. Director (Retd.), IIT Roorkee and presently director, MIT, Moradabad, has expertise in Thermodynamics and Kinetics and has over 40 years of experience in research and teaching. He has worked as a consultant to various Mini steel plants, MECON, TISCO and SAIL.

Dr. S.K. Goel, Professor of Metallurgical Engg. and Emeritus Professor (Retd.), of IIT Roorkee who edited this book has wide variety of experience in industry and academic field alike. He has interest in Iron and Steel making and in Metal Casting Technology and has over 30 years of teaching and research experience.

Dr. V.N.S. Mathur, Professor of Metallurgical Engg. (Retd.) from IIT Roorkee, has experience in the field of Mineral Processing and engineering and has over 37 years of teaching and research experience.

Dr. G.C. Kaushal, Professor of Metallurgical Engg., IIT Roorkee, has interest in Metallurgical Thermodynamics and Kinetics and Steel making. Additionally, he has interest in Welding Science and technology and has vast teaching and research experience to his credit.

CONTENTS

Part — I

Part — II

PART–I

ONE

Physical Chemistry of Electric Steel Making

1.1. INTRODUCTION

The aim of this chapter is to provide a brief description of the physico-chemical principles underlying steel making processes. The interest in understanding of these principles has been stimulated by the challenge for the development of new steel making processes as well as to improve the efficiency of conventional methods to meet the stringent requirements of special and ordinary steels.

The complete physico-chemical description of steel making processes requires the consideration of thermodynamics, chemical kinetics and transport phenomenon. Thermodynamics of most of steel-making reactions can be adequately understood and the factors controlling the equilibrium distribution of various elements among gas, slag and metal can be assessed. One can also predict the temperature change in various phases from the knowledge of heat changes associated with the various reactions taking place during steel-making. In comparison to the effort made to understand and collect data on the thermodynamics of steel making reactions, very little work has been done on kinetic and transport aspects. Only recently some general pattern regarding these aspects has started emerging, because of the thrust on the formulation of mathematical models of various processes. In the following sections of this chapter we shall briefly deal with all the three above mentioned aspects of steel-making.

1.2. THERMODYNAMICS

Thermodynamics predicts the equilibrium state of a system or reaction under given conditions of temperature pressure and overall constitution. Although total equilibrium with respect to all the reactions is never reached in steel making, near equilibrium state for a number of reactions, particularly slag-metal reactions, does exist. So thermodynamics helps in calculating the distribution of elements among various phases for these reactions. Further when equilibrium has not been achieved, thermodynamic considerations lead one to find out the direction in which a reaction will proceed under given conditions. Thus an understanding of thermodynamics of steel-making processes becomes essential to affect an efficient control of the process. We shall first deal briefly with the basic principles of thermodynamics and then apply them to various phases and reactions which occur during steel making.

1.3. BASIC PRINCIPLES

Laws of thermodynamics lead to the conclusion that a system tends towards lowering of Gibb's free energy under the conditions of constant temperature and pressure. Further in a system which has more than one phase, the value of partial molar free energy of each of the components shall be same in both the phases. This partial molar free energy is also called chemical potential as a component tends to go from a state of higher to lower chemical potential. The chemical potential of component i, denoted by μ_i is given by the expression (at low pressures)

$$\mu_i = RT \ln p_i + C_i(T) \qquad ...(1.1)$$

where p_i is the vapour pressure of component i in a gas phase which is in equilibrium with the condensed phase and C_i (T) is a constant depending upon the nature of component i and temperature. It is this constant which makes the calculation of absolute value of chemical potential of a component impossible. Fortunately for our calculations we need only the change in the value of chemical potential when a component gets transferred from one state to the other. In general such a change is measured with respect to a certain assumed state of the component called the 'standard state' at which chemical potential of component i is denoted by $\mu_{i.}$ Thus according to Eq. (1.1), μ_i, chemical potential of component i in any state is given by the expression.

$$u_i = u_i + RT \ln p_i/p_i^o \qquad ...(1.2)$$

where p_i^o is the partial pressure of component i in its standard state. The ratio (p_i/p_i^o) is called the activity of component i and is denoted by the symbol a_i. Thus,

$$\mu_i = \mu_i^o + RT \ln a_i \qquad ...(1.3)$$

The above equation shows that the activity of a component in its standard state is unity. By making use of conventional experimental techniques the activities of various components in metallurgical solutions have been determined. These metallurgical systems can be classified into three categories namely, (*i*) Gas mixtures (*ii*) Metallic solution and (*iii*) Slags.

According to Dalton's law, the partial pressure of a component in a mixture of gases is proportional to its mole fraction. Thus :

$$p_i = N_i \, p \qquad ...(1.4)$$

where 'p' is total pressure of the gas phase. If this total pressure is equal to the pressure of the component in standard state, activity of component i is equal to its mole fraction, which is equal to the ratio of number of moles of component i and the total number of moles of all the components present in gas mixture.

The activities of various components in liquid and solid solutions are in general expressed by choosing the equilibrium vapour pressure of the pure component as standard state. Thus the activities of various components in solution shall be less than unity. As an example in Fig. 1.1 activity plots of three common elements, namely, Cu, Mn and Si in Iron are shown. This figure shows that in case of manganese, activity is equal to mole fraction. The solutions showing such a behaviour are called ideal solutions or Raoultian solutions as these solutions follow Raoult's law which states that

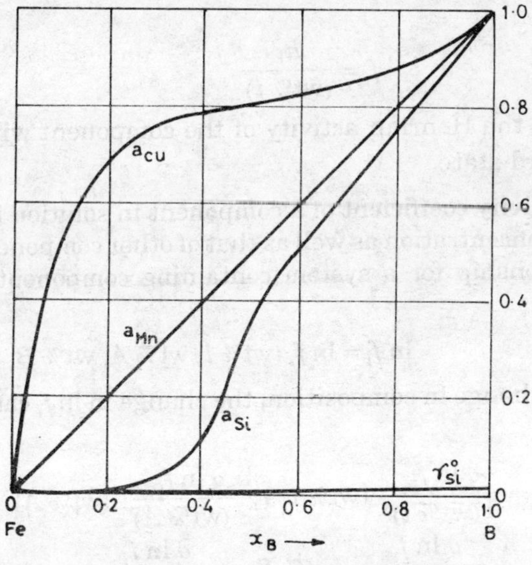

Fig. 1.1. The Raoultian activities of copper, manganese and silicon in solution in molten iron at 1600°C.

the vapour pressure of a component in solution is proportional to its mole fraction. Further Fig. 1.1 shows that the activity of silicon is less than that corresponding to ideal behaviour. Such solutions are said to show negative deviation from ideal behaviour. If the activity of a component is greater than that corresponding to Raultian behaviour as is the case with copper (ef. Fig. 1.1), the system is said to show positive deviation from ideal behaviour. The deviation from ideality is expressed in terms of a factor called Raoultian activity coefficient and denoted by γ_i and defined as

$$\gamma_i = \frac{a_i}{N_i} \qquad ...(1.5)$$

Thus for solutions showing positive deviation the activity coefficient is more than unity whereas for those showing negative deviation, it is less than unity. Further Fig. 1.1 shows that at low concentration of Cu and Si a proportionality exists between activity and mole fraction. This behaviour in solutions is quite general and is expressed as Henry's law which states that the activity of a component in dilute solution is proportional to its molefraction or weight percent. The constant of proportionality is known as Henry's law constant which is equal to γ_{si} for silicon as shown in Fig. 1.1. Strictly speaking this law is valid for infinitely dilute solution of a component and for all other concentration, the deviation from this law is expressed in terms of Henrian activity coefficient, f_i, defined as

$$f_i = \frac{h_i}{(\text{wt\%}\ i)} \qquad ...(1.6)$$

where h_i is the Henrian activity of the component with 1 wt% as the standard state.

The activity coefficient of a component in solution is a function of its own concentration as well as that of other components present. This relationship for a system containing components A, B, C is expressed as

$$\ln f_i = \ln f_i\ (\text{wt\%}\ i, \text{wt\%}\ A, \text{wt\%}\ B, \text{wt\%}\ C)$$

Thus with change in composition, the change in $\ln f_i$ can be written as

$$\ln f_i = \frac{\partial \ln f_i}{\partial\ (\text{wt\%}\ i)} \cdot (\text{wt\%}\ i) + \frac{\partial \ln f_i}{\partial\ (\text{wt\%}\ A)} \cdot (\text{wt\%}\ A)$$
$$+ \frac{\partial \ln f_i}{\partial\ (\text{wt\%}\ B)} \cdot (\text{wt\%}\ B) + \frac{\partial \ln f_i}{\partial\ (\text{wt\%}\ C)} \cdot (\text{wt\%}\ C) \qquad ...(1.7)$$

The partial derivatives in the above expression are constants and are called interaction parameters and are denoted by e_i^j, (where i refers to the component affected and j refers to the affecting component. Thus

$$e_i^i = \frac{\partial \ln f_i}{\partial (\text{wt\% } i)} \quad \text{is called self interaction parameter.}$$

Thus equation (1.7) can be rewritten as

$$\ln f_i = e_i^i (\text{wt\% } i) + e_i^A (\text{wt\% } A) + e_i^B (\text{wt\% } B) + e_i^C (\text{wt\% } C) \quad ...(1.8)$$

The interaction parameters for various components in a solvent are interrelated by the expression :

$$e_i^j = \frac{M_i}{M_j} \; e_j^i \quad ...(1.9)$$

The values of various interaction parameters have been determined experimentally and these can be used to calculate the activity of a component as illustrated by the following example.

Example 1. *Calculate the Henrian activity of sulphur at 1600°C in iron containing 0.1 wt% S, 1.3 wt% Si, 0.7 wt% Mn and 4.1 wt% C with the help of following data.*

$$e_s^S = -0.069,\; e_s^{Si} = +0.1445,\; e_s^{Mn} = -0.0575 \text{ and } e_s^C = +0.253$$

Solution. According to (1.8)

$$\ln f_s = -0.069 \times 0.1 + 0.1445 \times 1.3 - 0.0575 \times 0.7 + 0.253 \times 4.1$$
$$= 1.1845$$

Hence $f_s = 3.269$

Thus, h_s, Henrian activity of sulphur will be

$$h_s = 3.269 \times 0.1 = 0.3269$$

Once the chemical potential of various components are known in metal phase, the next step is to device means to bring their level to that prescribed by specifications. This is achieved by a number of chemical reactions involving the components. In general the result of these reactions is the exchange of concerned components between a metal and gas or slag phase. Therefore, the thermodynamic properties of both these phases should also be known. We have already discussed briefly the thermodynamics of gas phase, we shall briefly discuss the thermodynamics of slags.

The major components of slags are CaO, CaF_2, SiO_2, FeO, P_2O_5, MnO, Al_2O_3, Cr_2O_3 and MgO. These oxides either form a part of fluxes or come from the metal components on oxidation or from refractories. They can be divided into three categories according to their chemical behaviour. These categories are basic, acidic and amphoteric oxides. To which category an oxide will belong shall depend upon its capacity to part with its oxygen ion or react with it to form an ionic complex. Those oxides which can easily donate their oxygen ions are called basic oxides, whereas those which act as accepters of oxygen ions are called acidic oxides and those which have the capacity to accept or donate oxygen ions depending upon the condition of slag are called amphoteric oxides. For example, if we consider the reaction

$$2CaO + SiO_2 = Ca_2\,SiO_4 = 2Ca^{2+} + SiO_4^{4-} \qquad ...(1.10)$$

We find that CaO has donated its oxygen ions to silica leading to the formation of calcium orthosilicate which consists of Ca^{2+} and SiO_4^{4-} ions. Thus CaO is a basic oxide and SiO_2 an acidic oxide. The strength of acidic oxides can be judged from the intensity of their reaction *i.e.* the free energy of their reaction with a fixed base. Higher is the free energy of this reaction stronger is the acid. Similarly the strength of a base can be judged on the basis of the free energies of their reactions with a fixed acid. In case such a data is not available or one wants to assess qualitatively the nature of an oxide, following criteria are used.

(*i*) **Crystal Structure :** In general ionic crystals consist of two interlocking lattices, one for cations and other for anions, in such a way that each cation is surrounded as its nearest neighbour by anions and vice versa. The number of anions surrounding a cation or that of cation surrounding an anion is called the coordination number of cation or anion respectively. It has been observed that those oxides having a co-ordination number of eight or above for its cations, behave as basic oxides *e.g.* CaO, Na_2O, MnO. Those having co-ordination number of four or less behave as acidic oxides *e.g.*, SiO_2, P_2O_5, B_2O_3 and those having coordination number of six behave as amphoteric oxides *e.g.*, Al_2O_3.

(*ii*) **Cationic Field Strength :** It is defined as charge on the cation divided by the square of its radius. Higher is the field strength, stronger will be the electrostatic binding between cation and oxygen ion, hence the donation possibility will decrease *i.e.* more acidic will be the oxide. Cationic radii and field strength for some cations are as follows :

Cation	K^+	Na^+	Ca^{2+}	Mn^{2+}	Fe^{2+}	Mg^{2+}	Cr^{3+}	Fe^{3+}	Al^{3*}
Radii	1.33	0.95	0.99	0.8	0.75	0.65	0.64	0.6	0.5
Field strength	0.57	1.11	2.0	3.1	3.6	4.7	7.3	8.3	12
Ionic bond character	68	65	61	47	38	54	41	36	44

Cation	B^{3+}	Si^{4+}	P^{5+}
Radi	0.2	0.41	0.34
Field strength	75	23.8	43.3
Ionic bond character	31	36	28

(iii) **Degree of Ionic Bonding :** Higher is the percentage ionic character of an oxide bond more basic will be the oxide. Pauling has assigned a scale of electronegativities to elements whose difference between two elements is related to percentage ionic character. Higher is the difference in electronegativity more is the percentage of ionic character. The relationship between ionic character and electronegativity difference is shown in Fig. 1.2.

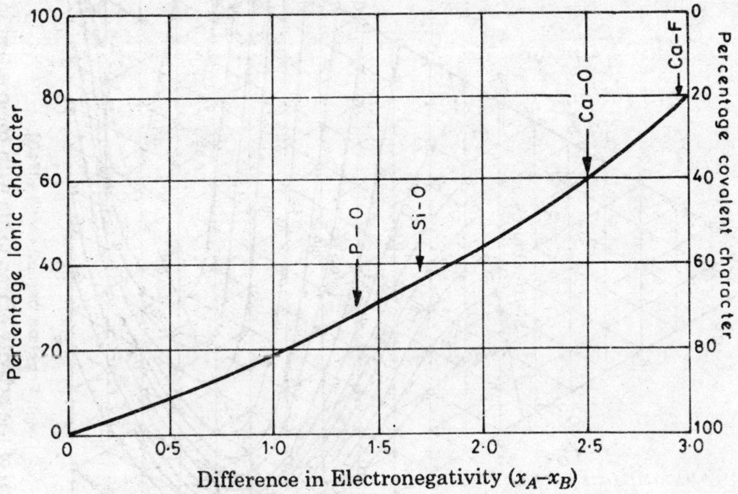

Fig. 1.2. Dependence of the degree of ionic bonding on the electronegativity difference between the ions involved.

Once oxides of different chemical character come together in a slag they react with one another to form various cationic and an ionic species. The activities of various components in slag phase, which ultimately determine the equilibrium distribution of various components between slag and metal, are controlled by the acid base reactions among various oxides. These activities can either be experimentally determined or roughly calculated taking into account the constitution of slag. An example of experimental determination of activities lies in the estimation of activity of FeO in all the important systems

$$(CaO + MgO) - FeO - (SiO_2 + P_2O_5) \qquad ...(1.11)$$

for steel making as shown in Fig. 1.3. This figure shows that a basic to acidic oxide ratio of 2 offers the highest activity of 'FeO' for any given iron oxide content. However this ratio is low as far as basic character of the slag is concerned hence one does work at higher base to acid ratios. It may be stated that the activities of all other components are also controlled by the acid base ratio. Most of these ratios called basicity indices or excess bases are

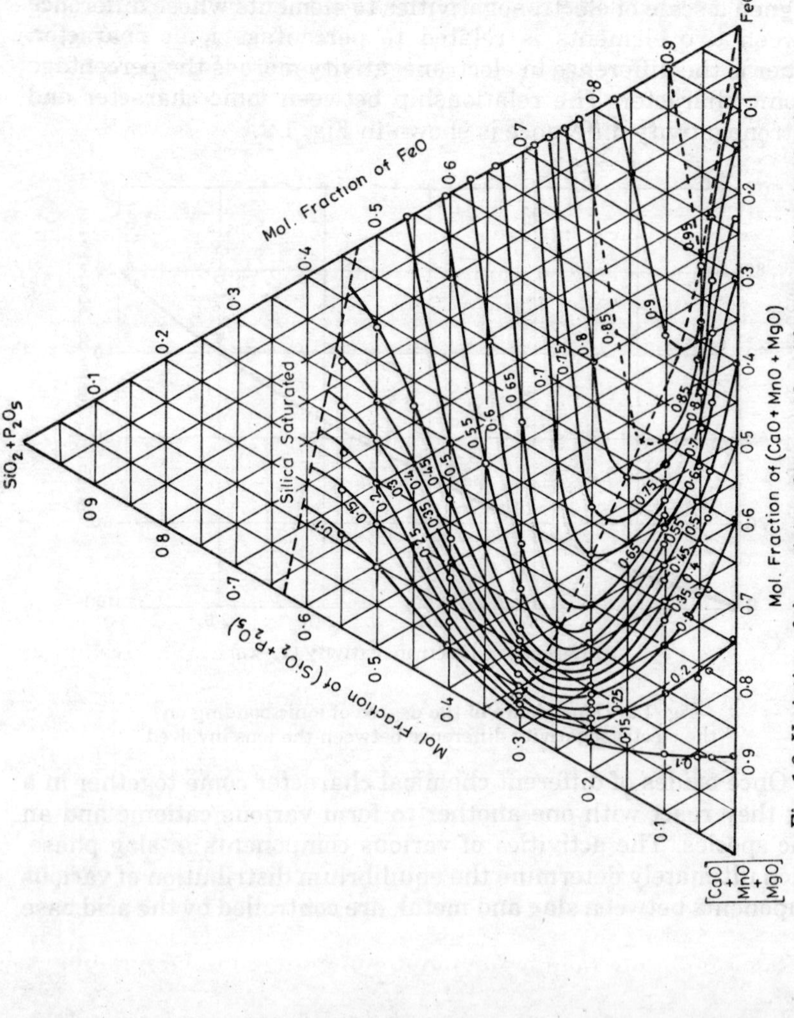

Fig. 1.3. Variation of the activity of iron oxide with composition in simple synthetic laboratory slags at steel-making temperatures.

based on the dominating influence of CaO and SiO_2. Some of these indices are as follows :

Vee Ratio $= \dfrac{\text{wt\% CaO}}{\text{wt\% SiO}}$

Modified 'V' Ratio $= \dfrac{\text{wt\% CaO}}{\text{wt\% SiO}_2 + \text{wt\% Al}_2\text{O}_3}$

Modified 'V' Ratio $= \dfrac{\text{wt\% CaO}}{\text{wt\% SiO}_2 + \text{wt\% P}_2\text{O}_5}$

Excess base B $= \text{wt\% CaO} - 1.86 \text{ wt\% SiO}_2 - 1.19 \text{ wt\% P}_2\text{O}_5$

Excess base $= RO - 2SiO_2 - 4P_2O_5 - 2Al_2O_3 - Fe_2O_3$

Basicity index $= \dfrac{CaO - 4P_2O_5}{SiO_2}$

and $= \dfrac{RO - 3\,P_2O_5}{SiO_2}$

Base-acid ratio $= \dfrac{RO}{SiO_2 + 2\,P_2O_5 + \frac{1}{3}Al_2O_3 + \frac{1}{2}Fe_2O_3}$

The weightage factors are only empirical and hence these factors work with success only in limited composition ranges.

Around second world war period, it was discovered that slags consist of ionic species. Further the explanation of slag-metal equilibrium became very difficult keeping in view the assumption of multitude of possible compounds formed and to take into account their dissociation constants. Thus the concept of a slag being completely ionised and containing a limited number of ionic species became increasingly attractive. These ionic theories have now been developed to a stage of being of great use in the understanding of slag-metal distribution behaviour in basic slag systems, but a satisfactory theory for acid slags remains to be propounded. According to these theories, the activity of a basic oxide MO is related to the ionic fractions by the equation

$$a_{MO} = N_M^{2+} N_O^{2-} \quad \text{...(1.12)}$$

where N_M^{2+} and N_O^{2-} is the cationic fraction of M^{2+} and N_O^{2-} is the anionic fraction of oxygen ions. For calculation of these fractions Temkin suggested the following ion formation scheme.

$$W\,SiO_2 + x\,P_2O_5 + y\,Al_2O_3 + z\,RO$$
$$= W\,SiO_4^{4-} + 2x\,PO_4^{3-} + 3y\,AlO_3^{3-} + (z - W - 2x - 3y)\,O^{2-} \quad \text{...(1.13)}$$

It was assumed that basic slags are those in which $(z - w - 2x - 3y)$ is positive. Any amount of CaF_2 present in slag was also assumed to be completely dissociated. To illustrate the use of above dissociation scheme let us consider a slag containing 0.55, 0.05, 0.15, 0.25 as mole fraction of CaO, MgO, FeO and SiO_2. Thus total RO for slag shall be (0.55 + 0.05 + 0.15) = 0.75. According to above scheme it will contain per mole of slag 0.55, 0.05, and 0.15 moles of Ca^{2+}, Mg^{2+} and Fe^{2+} cations and 0.25 and $(0.75 - 2 \times 0.25) = 0.25$ moles of oxygen ions. Thus

$$a_{CaO} = \frac{0.55}{0.55 + 0.05 + 0.15} \times \frac{0.25}{0.25 + 0.25}$$

$$= \frac{55}{75} \times \frac{25}{50} = 0.37$$

and

$$a_{FeO} = \frac{0.15}{0.55 + 0.05 + 0.15} \times \frac{0.25}{0.25 + 0.25}$$

$$= \frac{15}{75} \times \frac{25}{50} = 0.10$$

Thus one is able to calculate the activities of various species in both metal and slag phase. Exchange of components between two phases may be through a physical change or a chemical reaction belonging to reactions.

Let us now turn to the general treatment of chemical reaction. For this purpose consider the reaction

$$a\,A + b\,B = c\,C + d\,D \qquad \text{...(1.14)}$$

$$\Delta G = C\,\mu_c + d\,\mu_D - a\,\mu_A - b\,\mu_B$$

$$= C\,\mu_c^o + d\,\mu_D^o - a\,\mu_A^o - b\,\mu_B^o + RT \ln \frac{a_C^c \cdot a_D^d}{a_A^a \cdot a_B^b}$$

$$= G^o + RT \ln \frac{a_C^c \cdot a_D^d}{a_A^a \cdot a_B^b} \qquad \text{...(1.15)}$$

Reaction will proceed in forward direction if ΔG is negative, backward if ΔG positive and will be in a state of equilibrium if ΔG is equal to zero. Thus under this state Eq. (1.15) assumes the form

$$\frac{a_C^c \cdot a_D^d}{a_A^a \cdot a_B^b} = \exp\left[-\frac{\Delta G^\circ}{RT}\right] \qquad \text{...(1.16)}$$

In the above equation the left hand side contains the composition dependent functions, namely, the activities of components whereas the right hand side, composition independent function. The final composition of a system is determined by this relationship. The right hand side function is also termed as the equilibrium constant and is denoted by k. Thus from Eq. (1.16), one can write

$$k = \exp\left[-\frac{\Delta G^\circ}{RT}\right] \qquad ...(1.17)$$

or

$$\Delta G^\circ = -RT \ln k \qquad ...(1.18)$$

In steel making we are mainly concerned with the oxidation reduction reactions of the type

$$x\,M + \frac{y}{2}\,O_2 = M_x\,O_y \qquad ...(1.19)$$

where 'M' denotes a component present in the elemental form and M_xO_y is the chemical formula of the oxide formed on combination with oxygen. In order to compare the affinity for oxygen of various elements, the above reaction is written in such a way that it involves only one mole of oxygen, thus Eq. (1.19) can be rewritten as

$$\frac{2x}{y}\,M + O_2 = \frac{2}{y}\,M_x\,O_y \qquad ...(1.20)$$

Lower is the standard free energy, ΔG°, of the above reaction stronger is the affinity for oxygen of element 'M' and hence more stable is the oxide M_xO_y. In Fig. 1.4 the standard free energy of formation of various oxides is plotted as a function of composition. This diagram is known as Ellingham diagram and shows the relative stability of various oxides. For example the plot for the reaction $\frac{2}{3}\,Al + O_2 = Al_2O_3$ lies below that for the reaction $2Fe + O_2 = 2FeO$. Thus one can infer that aluminium has stronger affinity for oxygen than iron. In other words if 'FeO' is mixed with aluminium and heated to a temperature for starting the reaction, Al_2O_3 and Fe will be formed *i.e.* Aluminium will reduce 'FeO'. This reaction forms the basis of aluminothermic reduction which is applied to the production of a number of ferroalloys as well as to thermit welding.

One can also make use of Eq. (1.20) for the calculation of equilibrium constant of simple reactions. For example, for deoxidation of molten iron with aluminium the following reactions takes place

$$2\,[Al] + 3\,[O] = \left[\,Al_2O_3\,\right] \qquad ...(1.21)$$

The calculation of equilibrium constant of this reaction requires the splitting up of the above reaction into three, namely,

$$2\,Al\,(l) + \frac{3}{2}\,O_2 = Al_2\,O_3 \qquad ...(1.22)$$

$$Al\,(l) = [Al] \qquad ...(1.23)$$

$$\frac{1}{2}\,O_2 = [O] \qquad ...(1.24)$$

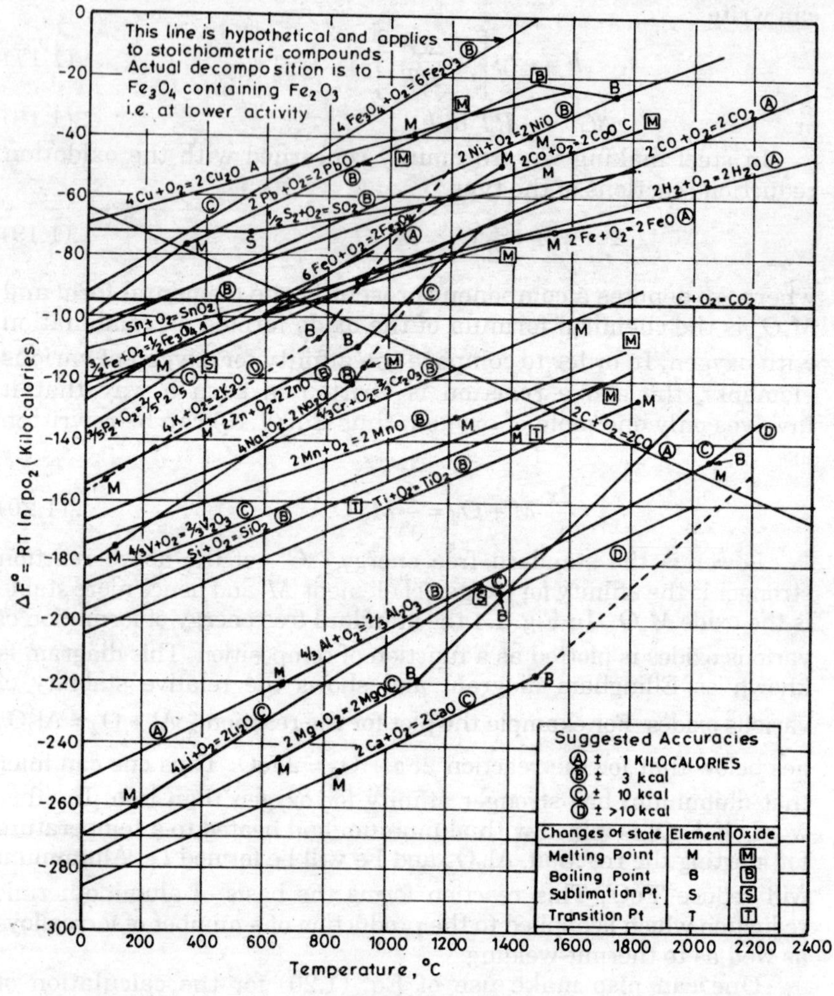

Fig. 1.4. Ellingham Diagram.

In reaction (1.21) - (1.24) square bracket refers to a component dissolved in metal. The standard free energy of reactions (1.22) - (1.24) at 1600°C are – 257500, – 24700, – 29300 cals respectively. The standard free energy of reaction (1.21) is obtained by subtraction from that of reaction (1.22), two times the free energy of reaction (1.23) and thrice that of (1.24). Thus one obtains for (1.21)

$$\Delta G^{\circ} = -\ 120400$$

Hence the equilibrium constant 'k' for reaction (1.21) will be

$$k = \exp\left[-\frac{\Delta G^\circ}{RT}\right] = \exp\left[\frac{120400}{2 \times 1873}\right] = \exp\,(32.14)$$
$$= 9.8 \times 10^{13}$$

The above example illustrates how one can find out the equilibrium constant for simple reactions. But in slag metal systems one is confronted with a number of simultaneous reactions. The equilibrium constant for such complex systems is influenced by that of each of the reactions. Flood, Forland and Grojotheim proposed a simple model for the calculation of equilibrium constant in these systems. This model is based on ionic theory of slags. According to this model, the equilibrium constant 'k^*' for a complex slag metal system involving a number of simultaneous reactions is given by the equation.

$$\log k^* = N_1^* \log k_1 + N_2^* \log k_2 + \ldots \qquad \ldots(1.25)$$

where N_1^*, N_2^* ...... are the weightage factors for reactions 1, 2 ...... which involve the exchange of the desired component between slag and metal. These weightage factors are called electrically equivalent fractions and can be calculated from the assumed ionic constitution of slag. The application of this equation shall be illustrated while considering some steel making reactions at a later stage.

1.4. CHEMICAL KINETICS

When the thermodynamically favourable conditions for a reaction to occur in a particular direction exist or are created, one requires knowing how fast will such a reaction occur. Such a study involves the study of the kinetics of the actual chemical reaction and allied transport processes which bring together the reactants and disburse the products. As steel-making involves transfer of components from one phase to the other, so such chemical reactions will take place only at specific reaction sites. Hence such reactions will involve the transport of reactants to reaction sites, chemical reaction at the sites and transport of products away from the reaction sites. A detailed treatment of these steps shall be presented at a later stage while discussing individual reactions. For the time being let us consider the general features of transport steps and chemical reactions.

In a transport process, the rate of transfer, J_i, of a component i per unit time area normal to the direction of transfer from position I in a system to position II is given by the expression

$$J_i = k_i\left[\,C_i^{\mathrm{I}} - C_i^{\mathrm{II}}\,\right] \qquad \ldots(1.26)$$

where C_i^I and C_i^{II} are the concentration of component i in terms of mass per unit volume at position I and II. k_i is called the mass transfer coefficient for component 'i'. It is directly proportional to the diffusion coefficient, D_i, of this component in quiescent bath whereas for stirred baths it is proportional to square root of D_i.

In chemical kinetics the rate of a reaction is expressed as

$$J_i = k_i\, C_i^n \qquad \text{...(1.27)}$$

where k_i is the specific reaction rate constant and n is the order of a reaction.

1.5. APPLICATION

Let us now consider the application of the above principles to various stages of steel making in electric furnaces which may be termed as :

(*i*) Melt down period
(*ii*) Oxidising period
(*iii*) Reducing period

During melt down period the scrap is exposed to heat and oxidising atmosphere. This leads to almost complete loss of Al, B, Be, Se, Ti and Zn present in scrap and partial loss of C, Cr, Pb, Mn, Si, V and W by oxidation. The oxides formed from these elements either leave the furnace in gas phase or go to slag phase as the case may be. The actual control of composition starts with the oxidising period, followed by reducing period.

1.5.1. Oxidising Period

During this period oxidising conditions are maintained in the furnace by either injecting oxygen or adding iron ore to the bath or maintaining oxidising conditions at the top of slag phase. During this period carbon, silicon, manganese, phosphorus and partially chromium (if present) and iron get oxidised and except carbon which forms a gas phase on oxidation, the components go to the slag phase. We shall consider the application of already discussed principles of thermodynamics and kinetics of these reactions turn by turn.

1.5.1.1. Carbon-oxygen Reaction

Both carbon and oxygen are present in iron in the form of dilute solution. Their solubilities follow the following thermodynamic equations.

$$\langle C \rangle_{gr} = [C]_{1\text{ wt\% in Iron}} \qquad \text{...(1.28)}$$

$$-\Delta G^\circ = 5400 - 10.1\, T. \qquad \text{...(1.28a)}$$

and

$$\frac{1}{2}\{O_2\} = [O]_{1\text{ Wt\% in Iron}} \qquad \text{...(1.29)}$$

$$\Delta G^\circ = -\,28220 - 0.57\,T \qquad ...(1.29a)$$

The square bracket in above equations refers to the component dissolved in iron. The maximum solubilities of carbon and oxygen individually in pure molten iron are given by the expressions :

$$\left[\text{wt\% C}\right]_{Sat} = 1.34 + 2.54 \times 10^{-3}\,(T - 298) \qquad ...(1.30)$$

$$\log\left[\text{wt\% O}\right]_{Sat} = -\frac{6320}{T} + 2.734 \qquad ...(1.31)$$

The henrian activity coefficients of carbon and oxygen in molten iron are given by the expressions

$$\log f_C^c = \frac{4350}{D}\left[1 + 4.10^{-4}\,(T - 1770)\right]\left[1 - N_{Fe}^2\right] \qquad ...(1.32)$$

$$\log f_o^o = -\,.20\left[\text{wt\% } O\right] \text{ at } 1600^\circ\text{C} \qquad ...(1.33)$$

In multicomponent dilute solutions the interaction parameters for various elements are as follows :

Element x	*S*	*Ni*	*Co*	*W*	*Cr*	*Mn*	*V*
e_C^x	0.09	0.012	0.012	– 0.03	– 0.024	—	– 0.038
e_O^x	1.0	0.006	0.007	0.0085	– 0.041	0.0	– 0.027

The free energy of carbon-oxygen reaction is given by the expression

$$\left[C\right]_{1\,wt\%} + \left[O\right]_{1\,wt\%} = \left\{CO\right\} \qquad ...(1.34)$$

$$\Delta G^\circ = -\,4860 - 9.8\,T \qquad ...(1.34a)$$

The equilibrium constant, k, for the above reaction at 1600°C can be written as

$$k = \frac{p_{CO}}{h_C \cdot h_O} = \exp\left[+\frac{4860 + 9.8 \times 1873}{2 \times 1873}\right] \approx 500 \qquad ...(1.35)$$

Hence $$h_C \cdot h_O \simeq 0.002\,p_{CO} \qquad ...(1.36)$$

This equation shows that (*i*) The amounts of dissolved carbon and oxygen shall depend upon the partial pressure of carbon-mono-oxide. Higher is this partial pressure, higher is the product of dissolved carbon and oxygen (*ii*) a decrease in the amount of dissolved carbon will lead to a simultaneous increase in dissolved equilibrium oxygen concentration. In Fig. 1.5 the relation between dissolved carbon and oxygen is shown as a function of p_{CO}. It is seen that, because of hyperbolic nature of this relationship, the amount of dissolved oxygen rises sharply at lower carbon contents, thus making it difficult to attain very low carbon level by simple oxidation in the furnace. Further Eq. (1.34) shows that the value of ΔG^o decreases with increase in temperature which will lead to

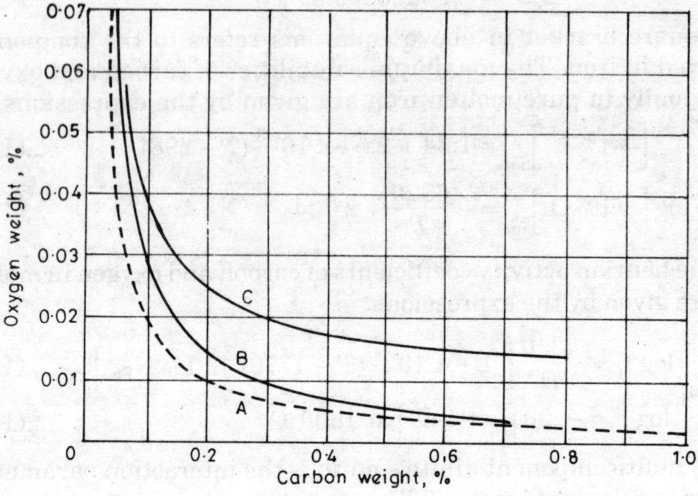

Fig. 1.5. Carbon-oxygen equilibrium in steel furnaces.
A. Equilibrium, 1600 C, Eq. 16-24a.
B. electric furnace, Marsh, J., Electric Furnace Proc. AIME, 1944, v. 16, p. 135.
C. Open hearth, Fetters, K., and. Chipman : Trans. AIME, 1940, v. 140, p. 170.

an increase in the carbon-oxygen concentration product for the same p_{CO} at high temperatures.

The amount of dissolved oxygen in bath bears a relationship with the 'FeO' content of the slag, according to the equation

$$\left[O \right]_{1\,wt\%} = \left(FeO \right) \qquad ...(1.37)$$

$$\Delta G^o = -30830 + 13.43\,T \qquad ...(1.37a)$$

Above equation shows that with the increase in oxygen content of the metal 'FeO' activity in slag will also increase. In other words if oxygen to metal bath is supplied by the slag phase, it should have a high 'FeO' activity. Such slags are called oxidising slags. In Fig. 1.3 the activity of 'FeO' is plotted as a function of composition of slag. As an application of this figure let us find out the equilibrium solubility of oxygen in metal which is in equilibrium with a slag containing 20% FeO, 58% CaO and 22% SiO_2. From Fig. 1.1 a_{FeO} for such a slag shall be 0.6 approximately. According to Eq. (1.37a)

$$k = \frac{a_{\text{FeO}}}{[\text{ wt\% O }]} = \exp\left(\frac{30830 - 13.43 \times 1873}{2 \times 1873} \right) = 4.55$$

Hence $$[\text{ wt\% O }] = \frac{0.6}{4.55} = 0.1320 = 1320 \text{ ppm.}$$

The actual value of [% C] × [% O] in electric furnace steel making is above the equilibrium value. This difference is due to the kinetic reasons and can be explained on the basis of mechanism of the reaction. Extensive work done on the kinetic studies of this reaction has shown that the CO bubbles are formed on the crevices in the refractory wall in contact with the metal. Thus the overall kinetics of carbon-oxygen reaction for the case when oxygen is transferred through the slag phase can be discussed in terms of the following steps.

(*i*) Pick up of oxygen by slag from the slag/gas interface. This pick up is considered to take place through the electrochemical reactions

$$\frac{1}{2}\{O_2\} = (O^{2-}) - 2e \qquad ...(1.38)$$

$$2\,Fe^{2+} = 2\,(Fe^{3+}) + 2e \qquad ...(1.39)$$

(*ii*) Transfer of oxygen from the gas/slag to slag/metal interface through slag phase. This transfer is controlled by the transport of Fe^{3+} through the slag phase, as oxygen ions are present in slag phase.

(*iii*) Transfer of oxygen from slag to metal phase at the slag metal interface through the reaction

$$2\,(Fe^{3+}) + 3\,(O^{2-}) = 2\,(Fe^{2+}) + 2\,(O^{2-}) + [O] \qquad ...(1.40)$$

During the slag formation stage, the above reaction is accompanied by simultaneous formation of 'FeO' in the slag according to equation

$$2\,(Fe^{3+}) + [Fe] = 3\,(Fe^{2+}) \qquad ...(1.41)$$

(*iv*) Transfer of oxygen from the slag/metal interface to the bulk of metal/phase.

(*v*) Transport of oxygen from bulk of metal phase to CO bubble/metal interface.

(*vi*) Transport of carbon from the bulk of metal phase to CO bubble/metal interface.

(*vii*) The reaction at bubble/metal interface leading to the formation of CO as follows

$$[C] + [O] = \{CO\} \qquad ...(1.42)$$

(*viii*) Detachment of the bubble from the pore and its rise through the metal and slag phase to the furnace atmosphere.

Because of high temperature prevailing in the furnace, one can conclude that the chemical reactions as given in steps (*i*), (*iii*) and (*vii*) shall be very fast and cannot be assumed to be rate controlling. Further as there will be enough pores at the bottom of the furnace and the rise of bubble shall also be fast, so one can safely assume step (*viii*) to be fast. Thus one can conclude that the transport steps will be the rate controlling ones. In case of the carbon concentration in the bath is high, it will be the transport of oxygen which will be the rate controlling. Experimental investigations have revealed that under the conditions when oxygen is transferred through the slag phase, it is the transfer of Fe^{3+} and hence that of oxygen through this phase (step 2) which is the slowest step in the entire process. Therefore, for faster removal of carbon, one tries to avoid this step. This can be accomplished in two ways, namely, either by the addition of iron ore to the metal bath or by the injection of oxygen to the bath. This reduces the problem to the analysis of two steps, namely, steps (*iv*) and (*v*). For both these steps, the rate equation will be

$$\dot{n}_{CO} = \dot{n}_O = k_o \cdot A_i \cdot \Delta C_o \quad ...(1.43)$$

where n_{CO} is the rate of evolution of CO. A_i is the interfacial area and k_o is the mass transfer coefficient for oxygen and ΔC_o is the difference in bath and equilibrium oxygen contents. Both A_i and k_o will be functions of the rate of evolution of CO and are thus indirectly the function of ΔC_O. Hence one can write

$$\dot{n}_{CO} = \text{constant.} (\Delta Co)^n \quad ...(1.44)$$

where $n > 1$. It is this equation which is considered to be followed for the kinetics of C – O reaction at high carbon contents. In case of low carbon (below 0.1%) contents, the transfer of carbon becomes the rate controlling. Hence, under these conditions

$$\dot{n}_{CO} = \text{constant.} (\Delta \text{Co})^n \quad ...(1.45)$$

The values of constants and exponents in eqns. (1.44) and (1.45) vary from furnace to furnace and can be calculated by measuring change in dissolved carbon and oxygen with time.

The rate of carbon boil in electric furnace steel making is kept to a value of about 0.3% per hour. This value is maintained to allow degassification by CO bubble purging. This aspect shall be considered in detail under the section on 'gases in steel'.

1.5.1.2. Manganese Reaction

Presence of Manganese in steel controls the adverse effects of sulphur and oxygen and improves the mechanical properties. Manganese is known to behave ideally in molten iron and thus one can write the following equation for dilute solution of maganese in iron

$$[\text{Mn}]_1 = [\text{Mn}]_{1\,wt\%} \quad ...(1.46)$$

$$\Delta G^o = -9.11\,T \quad ...(1.46a)$$

In general the alloying elements present in steel have zero interaction parameter with manganese except carbon which has a negative value for interaction parameter.

The extent of removal of manganese during steel making is controlled by the slag/metal reaction which according to ionic theory can be written as

$$(Fe^{2+}) + [Mn] = (Mn^{2+}) + [Fe] \quad ...(1.47)$$

Above equation states that the transfer of manganese from metal to slag takes place by a replacement reaction. The iron in slag may be present in association with any of the anions namely O^{2-}, SiO_4^{4-}, AlO_3^{3-} present in slag. Thus Eq. (1.47) represents a general scheme for a number of molecular reactions which may be written as

$$(FeO) + [Mn] = (MnO) + [Fe] \quad ...(1.48a)$$

$$\frac{1}{2}(Fe_2SiO_4) + [Mn] = \frac{1}{2}(Mn_2SiO_4) + [Fe] \quad ...(148b)$$

$$\frac{1}{3}(Fe_3(PO_4)_2) + [Mn] = \frac{1}{3}(Mn_3(PO_4)_2) + [Fe] \quad ...(1.48c)$$

According to Flood, Forland and Grjotheim model, the equilibrium constant for Eq. (1.47), is given by the expression

$$\log k = (N^*_{SiO_4^{4-}}) \log k^*_{SiO_4^{4-}} + (N^*_{O^{2-}}) \log k^*_{O^{2-}} + (N^*_{PO_4^{3-}}) \log k^*_{PO_4^{3-}} \quad ...(1.49)$$

$$= 2.9 \log N^*_{SiO_4^{4-}} + 2.4\, N^*_{O^{2-}} + 3.0\, N_{PO_4^{3-}} \text{ at } 1600°C \quad ...(1.50)$$

where $k^*_{O^{2-}}$, $k^*_{SiO_4^{4-}}$ and $k^*_{PO_4^{3-}}$ denote the equilibrium constants for Eq. (1.48*a*, *b* and *c*) respectively and N^*s, denote the electrically equivalent fraction of the anion given with subscript. It is defined as the fraction of anionic charge contributed by a particular anion to the total anionic charge of slag. Thus electrically equivalent fraction of oxygen ions in example will be given by

$$N^*_{O^{2-}} = \frac{2 \times 0.25}{2 \times 0.25 + 4 \times 0.25} = \frac{1}{3}$$

Manganese reaction is known to approach the state of equilibrium near the end of the heat. The rate of manganese transfer reaction is known to be controlled by the mass transfer in metal phase and is expressed as

$$\frac{d\,[\text{wt\% Mn}]}{dt} = -k_{Mn} A \left\{ [\text{wt\% Mn}] - [\text{wt\% Mn}]_i \right\} \quad ...(1.51)$$

where $[\text{wt\% Mn}]_i$ is the concentration of manganese on the metal side of slag metal interface. On integration the above equation yields,

$$\log \frac{[\text{wt\% Mn}]_t - [\text{wt\% Mn}]_e}{[\text{wt\% Mn}]_o - [\text{wt\% Mn}]_e} = -k' t \quad ...(1.52)$$

where '*k*' is the reaction rate constant. It varies with the fluid dynamic conditions in the bath. The subscripts '*t*' '*e*' and '*O*' in the above equation refer to value at any time "equilibrium value" and 'initial value.' Eq. (1.52) has been successfully used to calculate the concentration of manganese in the bath at any instant.

1.5.1.3. Silicon Reaction

Silicon shows very pronounced negative deviation from ideality in liquid iron. The free energy of solution of silicon in iron is expressed as

$$[\text{Si}] = [\text{Si}]_{1\,\text{wt\%}} \quad ...(1.53)$$

$$\Delta G = -28000 + 5.54\, T \quad ...(1.53a)$$

According to ionic theory, silicon transfer from metal to a basic slag takes place according to the equation

$$[\text{Si}] + 2\,[\text{O}] + 2\,(\text{O}^{2-}) = (\text{SiO}_4^{4-}) \quad ...(1.54)$$

$$K = \frac{N_{\text{SiO}_4^{4-}}}{N_{\text{O}^{2-}}^2\, [\text{wt\% Si}]^2\, [\text{wt\% O}]} \quad ...(1.54a)$$

The value of log *K* at 1600°C lies between 9.7 and 12.3 which shows that silicon is nearly completely oxidised during steel making. The rate of this reaction is quite high and therefore, equilibrium with respect to silicon is much before the completion of steel making process.

1.5.1.4. Phosphorus Reaction

Phosphorus is an undesirable element in most of the steels and its amount is kept below 0.04%. In some low carbon steels it is added upto 0.15% to add machinability.

The free energy of dissolution of phosphorus in iron is given by the expression

$$\text{P}_2 = 2\,[\text{P}]_{1\,\text{wt\%}} \quad ...(1.55)$$

$$\Delta G^\circ = -58500 - 9.1\, T \quad ...(1.55a)$$

The interaction parameter data for phosphorus is available only for few elements which is as follows :

The removal of phosphorus during steel making presents problems because of similarity of the free energies of formation of iron oxide and phosphorus pentaoxide as seen from Fig. 1.4. This figure shows that oxidation of phosphorus will be accompanied by the oxidation of nearly total amount of iron. This problem is overcome by the application of Van't Hoff's isotherm, which for the oxidation of phosphorus to phosphorus pentaoxide shall take the form

$$2\,[P] + 5\,[O] = (\,P_2\,O_5\,) \qquad ...(1.56)$$

$$\Delta G = \Delta G^\circ + RT \ln \frac{a_{P_2O_5}}{a_{[P]}^2\, a_{[O]}^5} \qquad ...(1.56a)$$

Eq. (1.56*a*) shows that '*G*' decreases as $a_{P_2O_5}$ decreases. Thus for removal of phosphorus one requires taking the help of such reaction for which free energy for reaction with phosphorus pentaoxide is negative. As phosphorus penta-oxide is acidic, so its reaction with basic oxide such as CaO, Na_2O etc. will be negative. Hence for removal of phosphorus one will have to work with basic slags. These slags will lower the activity of phosphorus penta-oxide and hence lower '*G*' in Eq. (1.56*a*) by clockwise rotation of $[P] - [P_2O_5]$ curve in Ellingham diagram as shown in Fig. (1.6). In general the *V*-ratio of slag for phosphorus removal is kept around 3. For assessing the activity coefficient in complex oxides, help is taken of the following empirical relationship.

$$\log \gamma_{P_2O_5} = -\,1.12\,[\,22\,N_{CaO} + 12\,N_{MgO} + 13\,N_{MnO} + 12\,N_{FeO} - 2\,N_{SiO_2}\,]$$

$$\log \gamma_{P_2O_5} = -\,\frac{44.600}{T} + 23.8 - 1.12\,\Sigma\,Ai\,Ni \qquad ...(1.57)$$

The extent to which $a_{P_2O_5}$ should be lowered by the steel making slag depends upon the condition at which carbon and phosphorus should be simultaneously removed. In terms of thermodynamics this means that at the temperature of oxidation, the free energy change for oxidation of phosphorus and that of carbon to carbon monoxide should be same when expressed on per mole of oxygen basis. Now the oxidation and equilibrium constants for these two reactions can be expressed as

$$\tfrac{4}{5}\,[P] + O_2 = \tfrac{2}{5}\,(P_2O_5) \qquad ...(1.58)$$

with $$K_p = \frac{(\,a_{P_2O_5}\,)^{2/5}}{(\,h_P\,)^{4/5}\,p_{O_2}}$$

and $$2\,[C] + \{O_2\} = 2\,\{CO\} \qquad ...(1.59)$$

with $$K_C = \frac{p_{CO}^2}{h_C^2\,p_{O_2}} \qquad ...(1.59a)$$

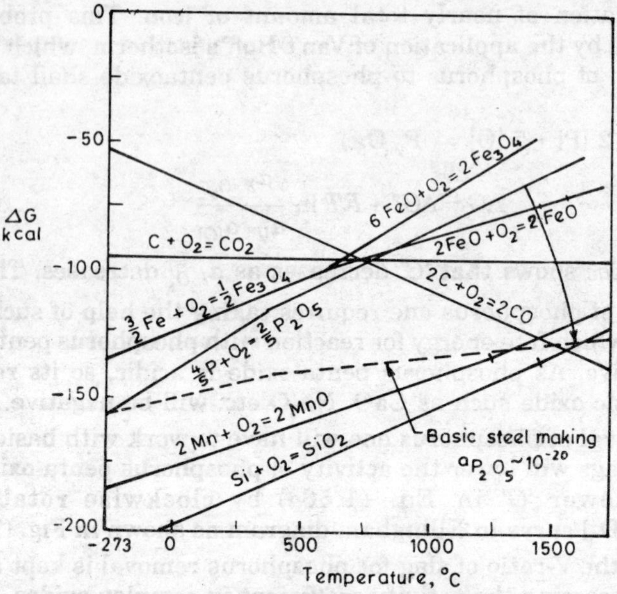

Fig. 1.6. Illustrating the free energies of formation for the oxides of the common elements removed during steel refining. The arrow indicates the increasing stability of P_2O_5 due to the use of basic steel-making slags.

Elimination of p_{O_2} form Eq. (1.58*a*) and (1.59*a*) yields

$$a_{P_2O_5} = \left[\frac{p_{CO}^2\, h_p^{4/5} \cdot K_P}{h_C^2 \cdot K_C}\right]^{5/2}$$

This equation can be used, taking the help of standard thermodynamic data, to find out the composition of slag to attain desired phosphorus level at a predetermined carbon or oxygen level in the bath. Though the distribution of phosphorus is controlled by the activities of concerned components in slag and metal, for industrial application one expresses the phosphorus distribution ratio as $\dfrac{(\text{wt\% } P_2O_5)}{[\text{wt\% } P]}$ as function of slag composition. In Fig. 1.7, this ratio is plotted against FeO content of slag for various *V* ratios. This figure shows that for each '*V*' ratio, there is an optimum 'FeO' content which is around 16 wt%, for obtaining maximum dephosphorisation. For predicting phosphorus distribution in case of complex slags, one makes use of ionic theory. Dephosphorisation reaction, according to this theory can be written as

$$2\,[P] + 5\,[O] + 3\,(O^{2-}) = 2\,(PO_4^{3-}) \qquad ...(1.61)$$

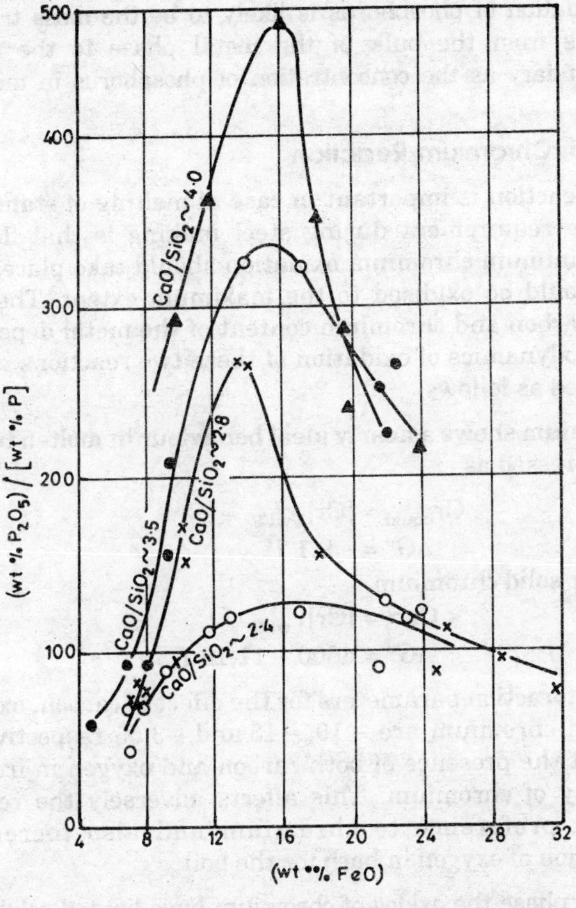

Fig. 1.7. Dependence of the distribution ratio, (wt%P_2O_5)/[wt%P], on the iron oxide content of slags with constant (wt%CaO)/(wt%SiO_2) ratios in laboratory experiments at a corrected temperature of 1685°C.

Equilibrium constant, K_P, for a complex slag is given by the expression :

$$\log K_P = 21\, N^*_{Ca^{2+}} + 18 N^*_{Mg^{2+}} + 13 N^*_{Mn^{2+}} + 12\, N^*_{Fe^{2+}} \ldots\ldots \qquad \ldots(1.61a)$$

Where, $N^*_{Ca^{2+}}$, $N^*_{Mg^{2+}}$, $N^*_{Mn^{2+}}$ and $N^*_{Fe^{2+}}$ are the electrically equivalent fractions of cationic species Ca^{2+}, Mg^{2+}, Mn^{2+} and Fe^{2+} respectively.

Relatively little work has been done on the study of kinetics of phosphorus transfer. It has been suggested that the rate controlling step in the oxidation of phosphorus is likely to be the mass transport of phosphorus from the bulk of the metal phase to the slag-metal phase boundary as the concentration of phosphorus in metal phase is low.

1.5.1.5. Chromium Reaction

This reaction is important in case of melting of stainless steel scrap. The requirement during steel making is that during the process minimum chromium oxidation should take place, whereas carbon should be oxidised to the maximum extent. The reaction between carbon and chromium content of the metal depends upon the thermodynamics of oxidation of these two reactions which will be discussed as follows.

Chromium shows a nearly ideal behaviour in molten iron which can be expressed as

$$Cr_{liquid} = [Cr]_{1\,wt\%} \quad ...(1.62)$$

With

$$\Delta G^\circ = -9.1\,T \quad ...(1.62a)$$

But for solid chromium

$$<Cr> = [Cr]_{1\,wt\%} \quad ...(1.63)$$

$$\Delta G^\circ = 4600 - 11.18\,T \quad ...(1.63a)$$

The interaction parameters for the effect of carbon, oxygen and sulphur on chromium are – 10, – 13 and – 3.55 respectively. This shows that the presence of both carbon and oxygen in iron lowers the activity of chromium. This affects adversely the removal of carbon in preference to chromium and also increases the concentration of oxygen in bath for the boil.

In slag phase the oxides of chromium have limited solubility. The form in which chromium is present in slag depends upon the chromium content of the bath. It has been found that for chromium contents below 9%, it is present as $Fe\,Cr_2O_4$ whereas for chromium contents above 9%, it is present as Cr_3O_4. Chromium distribution between slag and metal is controlled by the following slag metal reactions :

(*i*) For chromium contents below 9% :

$$(Fe\,Cr_2O_4) + 4\,[C] = [Fe] + 2\,[Cr] + 4\,\{CO\} \quad ...(1.64)$$

$$\Delta G^\circ = 198200 - 129.2\,T \quad ...(1.64a)$$

(*ii*) For chromium contents above 9% :

$$(Cr_3O_4) + 4\,[C] = 3\,[Cr] + 4\,\{CO\} \quad ...(1.65)$$

$$\Delta G^\circ = 210760 - 139.7\,T \quad ...(1.65a)$$

In Fig. 1.8 the relationship between carbon and chromium contents of the bath is shown for different temperatures. This figure shows that to produce high chromium low carbon steels, the temperature of the bath should be kept as high as possible. For making low carbon chromium steels, one in general makes use of low chromium content scrap which is oxidised at high temperatures to oxidise carbon. Finally the bath is enriched in chromium by means of low carbon ferrochrome.

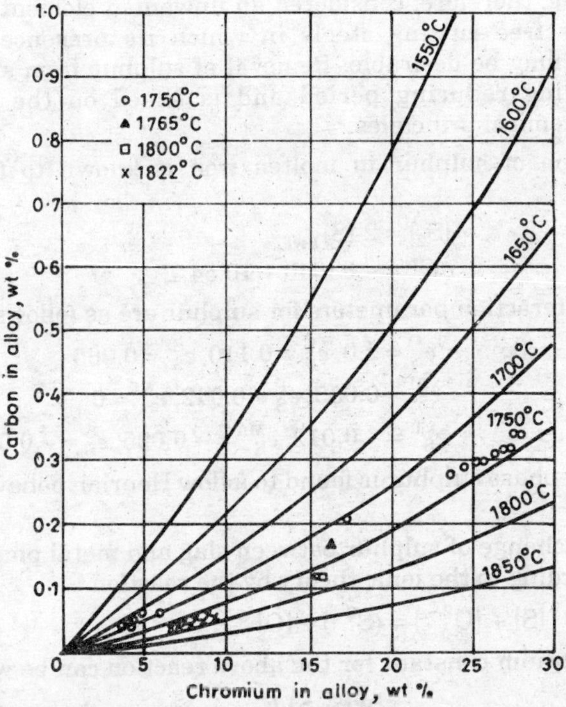

Fig. 1.8. Effect of temperature on the relation between carbon and chromium in solution in steel, at equilibrium with pure Cr_2O_3 and 1 atmosphere pressure of CO. The full curves represent equilibrium laboratory determinations while the experimental points are for small-scale oxygen lancing trials by Hilty.

Only a few studies on kinetics of chromium transfer have been made. These have shown that the rate controlling step is chromium transfer in the metal phase. Further it has been found that in practice chromium distribution approaches equilibrium value at the end of the heat.

After the unwanted metalloids are brought down by oxidation to a desired level, if required, a reducing period follows for the removal of sulphur.

1.5.2. Reducing Period

During this period the main reaction concerns the removal of sulphur. Apart from this reaction hydrogen pick by the metal bath also takes place during this period, but this reaction shall be discussed at a later stage under the subsection dealing with gases in steel.

1.5.2.1. Sulphur Reaction

Sulphur has deleterious effect on the mechanical properties of steel and is, therefore, considered an unwanted element in steels, except for free cutting steels in which its presence between 0.1-0.33% may be desirable. Removal of sulphur from steel takes place during reducing period and is based on the following physico-chemical principles.

Solution of sulphur in molten iron is known to follow the equations

$$\{S_2\} = 2\,\{S\}_{1\,wt\%} \qquad ...(1.66)$$

$$\Delta G^\circ = -\,63040 + 10.54\,T \qquad ...(1.66a)$$

The interaction parameters for sulphur are as follows

$$e_S^O = 2.0,\ e_S^C = 0.110,\ e_S^{Si} = 0.065$$

$$e_S^{Al} = 0.055,\ e_S^P = 0.042,\ e_S^{Ni} = 0$$

$$e_S^{Cu} = -\,0.012,\ e_S^{Mn} = -\,0.025,\ e_S^S = -\,0.03$$

In slag phase sulphur is found to follow Henrian behaviour upto 10 wt%.

The exchange of sulphur between slag and metal phases takes place according to the ionic theory by the reaction :

$$[S] + [O^{2-}] = (S^{2-}) + [O] \qquad ...(1.67)$$

The equilibrium constant for the above reaction can be written as

$$K_S = \frac{(\text{wt\% S})\,h_o}{h_s \cdot a_{O^{2-}}} \qquad ...(1.68)$$

For a complex slag the value of K_S at 1600°C can be calculated according to the relationship

$$\log K_S = -\,1.19\,N^*_{Ca} - 1.60\,N^*_{Fe} - 1.88\,N^*_{Mn} - 3.55\,N^*_{Mg} + 1.63\,N^*_{Na} \qquad ...(1.69)$$

For the sake of graphical representation Eq. (1.69) can be rewritten as

$$\frac{(\text{wt\% S})}{h_S} = K.\,\frac{a_{O^{2-}}}{ho} \qquad ...(1.70)$$

From slag theory it is clear that $a_{O^{2-}}$ is a function of slag basicity, whereas *ho* is a function of 'FeO' content of the slag. Therefore, in Fig. 1.9 (wt% S)/h_s is plotted against 'FeO' content of slag for various CaO/SiO_2 ratios. This figure shows that with increase in '*V*' ratio the slag distribution becomes favourable. It also shows that as 'FeO' content increase upto 3%, sulphur distribution coéfficient falls steeply. It remains approximately, constant for 'FeO' contents between 3 to about 20% and thereafter increases. The later effect is due to the dominating effect of FeO on basicity than an oxidising power. Thus it is seen that lower is the 'FeO' content in slag better is the desulphurisation. It is this fact which calls for the maintenance of reducing conditions for desulphurisation.

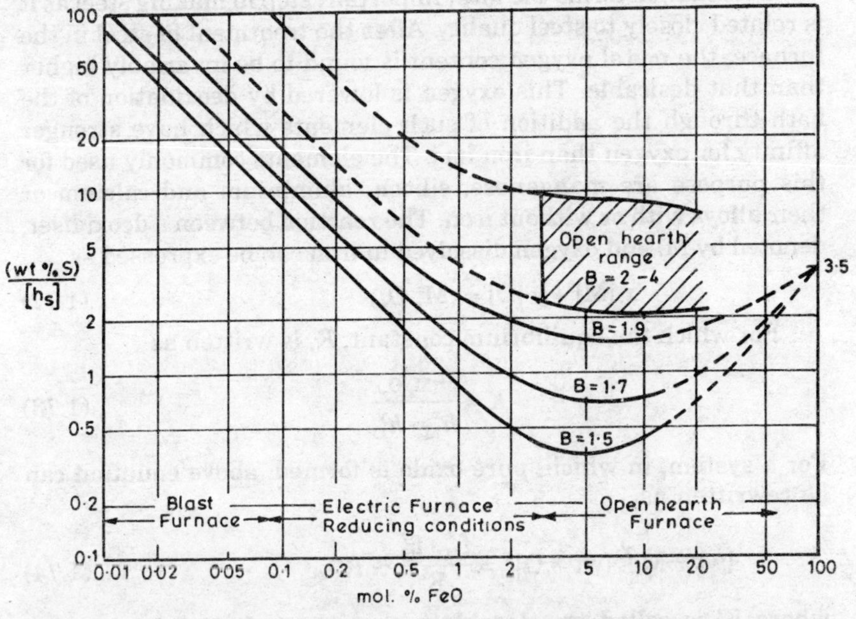

Fig. 1.9. Diagram showing the influence of basicity and iron oxide content of the slag on the distribution of sulphur between slag and metal at 1600°C. The lower curves refer to acid slags with values of the molar acid : base ratio, B = $(CaO + MgO)/SiO_2 + Al_2O_3)$, of approximately 1.5, 1.7 and 1.9. The range of basic open-hearth slags with B between 2 and 4 is shown cross-hatched.

Though the above treatment explains the behaviour of sulphur during refining period, some sulphur is also removed during laddle treatments. This aspect shall be discussed at a later stage.

The rate controlling step in kinetics of sulphur transfer from metal to slag is the transport of sulphur in the metal phase. On this basis, one can arrive at the following relationship for the variation of sulphur content in metal phase

$$\log \frac{[S]_t - [S]_e}{[S]_i - [S]_e} = -K\,t \qquad ...(1.71)$$

where suffices 'i', 't' and 'e' refer to initial concentration, concentration at any time 't' and equilibrium concentration respectively. K is the overall mass transfer coefficient and is a function of hydrodynamic conditions prevailing in the bath.

1.6. DEOXIDATION OF STEEL

Deoxidation forms the most important step in making steel as it is related closely to steel quality. After the treatment of steel in the furnace, the metal oxygen content is found to be invariably higher than that desirable. This oxygen is lowered by deoxidation of the bath through the addition of such elements which have stronger affinity for oxygen than iron has. The elements commonly used for this purpose are manganese, silicon, aluminium and calcium or their alloys with or without iron. The reaction between a deoxidiser, denoted by M, and oxygen dissolved in iron can be expressed as

$$x\,[M] + y\,[O] = (M_x\,O_y) \qquad ...(1.72)$$

For which the equilibrium constant, K, is written as

$$K = \frac{a_{(M_x\,O_y)}}{h_M^x \cdot h_O^y} \qquad ...(1.73)$$

For a system, in which, pure oxide is formed, above equation can be rewritten as

$$[\text{wt\% M}]^X\,[\text{wt\% O}]^Y = \frac{f_M^x \cdot f_O^y}{K} \simeq K' \qquad ...(1.74)$$

where K′ is called the deoxidation constant. It is this equation which helps in calculating the thermodynamic value of dissolved oxygen when a deoxidiser is added to a system. The value of deoxidation constants for Silicon, Aluminium and Boron are 2.8×10^{-5}, 1.02×10^{-14} and 4×10^{-9} respectively. As a ready reference one makes use of Fig. 1.10 to compare the deoxidation power of various deoxidisers and also to find out the amount of dissolved oxygen for a known concentration of a deoxidiser in iron.

Sometimes it is helpful to make use of two or more deoxidisers simultaneously. In such cases the reaction product may be a complex oxide leading to much stronger deoxidation than either of

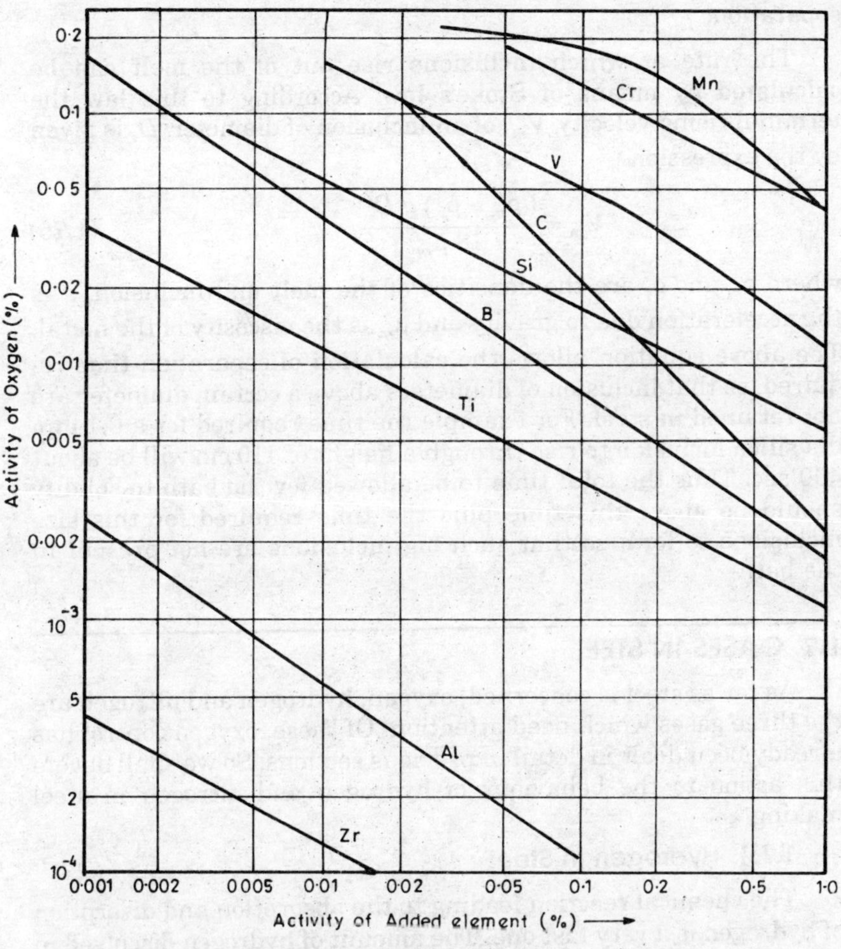

Fig. 1.10. Comparison of deoxidizing powers at 1600°C.

the oxides alone. Such deoxidisers are called complex deoxidisers. Two commonly employed complex deoxidisers are ferro-silico-manganese (Fe-Si-Mn) and calcium silicide (Ca-Si).

Though the dissolved oxygen content of the metal will depend upon the equilibrium with the product of deoxidation, the total amount of oxygen present in the bath will depend upon the rate of separation of the product of deoxidation. The amount of this product retained in the bath is termed as indigenous inclusions. A detailed treatment of inclusions is deferred to a later stage, but for the time being let us make some comments on their size and rate of separation.

The rate at which inclusions rise out of the melt can be calculated by means of Stoke's law. According to this law the terminal rising velocity, V_m, of an inclusion of diameter, D, is given by the expression

$$V_m = \frac{(\rho_m - \rho_i)\, g\, D^2}{18\, \mu_m} \qquad \text{...(1.75)}$$

where ρ_m and ρ_i are the densities of the melt and inclusion, g is the acceleration due to gravity and μ_m is the viscosity of the metal. The above equation allows the calculation of separation time required, so that inclusion of diameters above a certain diameter are not retained in steel. For example the time required for a 0.1 mm dia silica inclusion to rise through a height of 110 cm will be about 400 sec. Thus the total time to be allowed for the bath to solidify should be above this time plus the time required for this size inclusions to form so that such big inclusions are not present in the bath.

1.7. GASES IN STEEL

As far as steel is concerned, oxygen, hydrogen and nitrogen are the three gases which need attention. Of these, oxygen control has already been dealt in detail in previous sections. So we shall devote this action to the behaviour of hydrogen and nitrogen in steel making.

1.7.1. Hydrogen in Steel

The chemical reaction leading to the absorption and disorption of hydrogen is a very fast one. The amount of hydrogen dissolved in steel is controlled by the reaction

$$\tfrac{1}{2}\, H_2 = [H]_{1\,wt\%} \qquad \text{...(1.76)}$$

$$\Delta G^\circ = 7680 + 7.33\, T \qquad \text{...(1.77)}$$

However during steel making Hydrogen pick up takes place from water vapours either through the slag phase or directly. The direct pick up of hydrogen from atmospheric moisture is governed by the reaction

$$H_2O = 2\,[\,H\,]_{1PPM} + [\,O\,]_{1\,wt\%} \qquad \text{...(1.78)}$$

$$\Delta G^\circ = 47480 - 35.87\, T \qquad \text{...(1.79)}$$

The above equations show that the dissolved hydrogen will increase with increase in H_2O vapour pressure and temperature, but decrease with increase in oxygen content of the metal. The pick up of hydrogen by metal through the slag phase takes place by the reactions.

$$\{ H_2O \} + (O^{2-}) = 2\,(OH^-) \quad ...(1.80)$$

$$2\,(OH^-) = (O^{2-}) + 2\,[H] + [O] \quad ...(1.81)$$

The net result of these two reactions is the same as Eq. (1.78). But these two equations show that a basic slag will facilitate the transfer of hydrogen to metal. Further as dissolved oxygen will be a function of 'FeO' activity in slag, so one can infer that during oxidising period there will be little pick up of hydrogen whereas it will be appreciable during the reducing period.

The rate of hydrogen absorption or desorption is controlled by the mass transfer of hydrogen in metal phase. The removal of hydrogen from the bath is done by bringing the metal in contact with a phase having lower hydrogen partial pressure than that corresponding to equilibrium value for the dissolved amount of gas. This is done by either vacuum treatment or by purging with an inert gas or by autogeneous purging during carbon boil. Various degassification techniques shall be discussed in the next section, but let us consider the principle of degassification in this section.

Consider a bubble of volume dV of a gas rising through molten metal. Let us assume that hydrogen partial pressure in this bubble is p_{H_2} and is in equilibrium with the hydrogen content of the metal. Let this hydrogen partial pressure be maintained due to desorption from the metal phase. The material balance of hydrogen yields

$$\frac{2\,p_{H_2}^{dV}}{RT} = W \cdot d\,[\%H] \quad ...(1.82)$$

where W is the weight of metal phase.

As p_{H_2} is in equilibrium with dissolved hydrogen in metal, therefore one may write

$$p_{H_2} = \frac{[\%H]^2}{K} \quad ...(1.83)$$

Thus

$$dV = \frac{RT\,K\,W}{2} \cdot \frac{d\,[\%H]}{[\%H]^2} \quad ...(1.84)$$

If the removal of hydrogen takes place by purging argon through the bath, for a given volume, V, of purged gas, the change in hydrogen concentration can be calculated by the integration of the above equation which yields.

$$V = \frac{RTKW}{2}\left[\frac{1}{[\%H]_f} - \frac{1}{[\%H]_i}\right] \quad ...(1.85)$$

where suffixes '*f*' and '*i*' refer to final and initial hydrogen contents of the bath.

In case purging is autogeneous and is due to the oxidation of carbon, assuming p_{CO} inside the bubble to be equal to one atmosphere, one can write

$$dv = \frac{RT\,W\,d\,[\%C]}{12} \quad ...(1.86)$$

Substitution of this in Eq. (1.82) one gets

$$\frac{d\,[\%\,C]}{12} = \frac{K}{2}\,\frac{d\,[\%\,H]}{[\%H]^2} \quad ...(1.87)$$

or

$$\frac{d\,[\%H]}{d\,[\%C]} = \frac{[\,\%H\,]^2}{6\,K} \quad ...(1.88)$$

Above equation correlates the rate of carbon oxidation to that of removal of hydrogen. As Eqs. (1.87) and (1.88) are based on thermodynamic equilibrium for hydrogen, these lead to a lower hydrogen content for a given value of purged volume than that realised in practice.

In case of vacuum treatment hydrogen is first nucleated at active sites and then bubbles are formed, which get detached after attaining required size. This process is similar to the evolution of CO during carbon boil, hence the governing equation for hydrogen removal is of the same form, namely,

$$\frac{d\,[\%H]}{dt} = -K_1\,[\,[\%H] - [\%H]_e\,]^a - K_2\,[\,[\%H] - [\%H]_e\,] \quad ...(1.89)$$

The term containing K_2 refers to the evolution of hydrogen from the exposed surface of the metal. Integration of above equation yields the relationship between residual hydrogen and time.

1.7.2. Nitrogen in Steel

The raw materials and the atmosphere are the two main sources of nitrogen in most steel making processes. Nitrogen absorption or desorption by molten steel is governed by the reaction

$$\tfrac{1}{2}N_2 = [\,N\,]_{1\,wt\%} \quad ...(1.90)$$

$$\Delta G^\circ = 860 + 5.71\,T \quad ...(1.91)$$

The solubility of nitrogen is affected by the presence of alloying elements for which the interaction parameters are as follows :

$e_N^C = 0.25$ $e_N^P = 0.051$, $e_N^O = 0.05$ $e_N^{Si} = 0.047$

$e_N^S = 0.013$ $e_N^{Ni} = 0.01$ $e_N^{Al} = 0.0025$ $e_N^{Mn} = -0.02$

$e_N^{CR} = -0.045$ $e_N^{Nb} = -0.067$ $e_N^{Ti} = -0.64$

$e_N^V = -0.01$ $e_N^{Mo} = -0.011$ $e_N^W = -0.002$

The pick up of nitrogen during steel-making is also because of the consumption of active adsorption sites by oxygen.

The rate equation for removal of nitrogen from steel during various processes are similar to those for hydrogen.

TWO

Raw Material

2.1. INTRODUCTION

In India till late sixties there were only few ministeel plants, most of these plants were engaged either in the production of steel castings or special and alloy steels. However, only a few were melting steel scrap to manufacture small sized pencil ingots for rolling into bars, rods and sections. The integrated steel plants having an installed capacity of 8.9 M.T. produced only around 6.0 M.T. As a result of low production by integrated steel plants and spur in the construction purposes a great shortage of mild steel for construction purpose developed. The only alternative left to provide quick relief to the consumers of steel was to set up ministeel plants. Most of the new ministeel plants setup during late sixties were designed for manufacture of mild steel bars, rods and other light structurals. Mini steel plants in India can be broadly divided into the categories.

(*i*) Ministeel plants engaged in the production of alloy and special steels and steel castings.

(*ii*) Ministeel plants engaged in the production of mild steel bars, rods and light structurals for construction purposes.

Table I shows the installed capacity and production of integrated and mini-steel sector along with the yearwise demand of steel.

Table I. Steel production in Integrated & Secondary Steel Sectors

(in million tonnes)

Year	*Assessed*	*Production*		*Total Production*
		Integrated Steel Plants	*Secondary Steel Plants*	
1991-92	17.43	8.28	6.44	14.72
1994-95	22.00	13.40	9.52	22.92
1999-2000	31.00	15.82	12.38	28.20

Table II and III show the licensed capacity of various ministeel plants as per the category of steels and the regional distribution of ministeel plants in India.

Table II. Category-wise Mini-steel Plant Licensed Capacity

Licensed to produce	*No. of units*	*No. of furnace*	*Liquid metal capacity (in tonnes)*
Alloy Steel	10	21	472,000
Steel Castings	48	83	704,000
Ingots and Castings	18	52	667,000
Plain Carbon	117	159	2618,000
Special	6	11	54,000
Total	199	326	4,515,000

Table III. Regional Distribution of Electric Arc Furnaces in India

Region	*No. of units*	*No. of furnaces with capacity (tonnes)*				*Liquid metal capacity (tonnes)*	
		Below 5	*5-10*	*10-15*	*15 & above*	*Total No. furnaces*	*Total of capacity*
North	62	12	17	26	43	98	1,377,000
East	44	9	14	19	45	87	1,252,000
South	34	14	4	17	16	51	734,000
West	59	17	15	20	29	81	1,152,000
Total	99	52	50	82	133	317	4,515,000

The demand for steel by the year 1994-95 is expected to grow upto 23.00 million tonnes from about 12.00 million tonnes in 1983-84. The demand for different categories of steel in 1994-95 will be exceeding the production of steel in integrated steel plants by about 2.85 million tonnes. The increased demand can be met by :

(*i*) Obtaining increased output from the existing integrated steel plants and by putting up new integrated steel plants.

or (*ii*) Obtaining either full or part of the demand fulfilled from ministeel plants.

However, there were definite advantages of setting up new ministeel plants against new integrated steel plants which have been listed as given below :

(i) The capital cost in terms of per ton of steel produced in an integrated steel plant of 3.0 M.T. capacity is about 30 to 40% higher than a ministeel plant of 0.1 M.T. capacity routed through sponge iron unit, electric arc furnaces, concast unit and rolling mill.

(ii) Integrated steel plants heavily rely on large scale import of coking coal because of depleting reserves in India.

(iii) Ministeel plants have lower gestation period.

(iv) Regional demands are met by putting up plants at the consuming centres and thereby eliminating the problems and cost involved in the transportation of material.

(v) Ability to give short deliveries and change in the product mix if required.

(vi) Ability to expand melting facility if the market requirements so demand.

(vii) Utilization of local scrap.

As against the above advantages, following are the disadvantages which compared with advantages are not so considerable.

(i) Higher cost of production of steel in comparison with integrated steel plants.

(ii) Requirement of larger quantities of power when the country is already facing acute power shortage.

Production of steel in electric arc furnaces may be carried out either using acid refractory lining or basic refractory lining *i.e.*, in acid electric arc furnace or basic electric arc furnace. In acid steel making no refining takes place with respect to sulphur and phosphorus. This limitation requires very careful selection of raw materials to manufacture steel of desired specifications. Whereas, in basic electric furnaces refining of sulphur and phosphorus may be carried out along with controlled removal *e.g.* C, Si and Mn. The economic and technical obstacles to the use of selected scrap have led to large scale adoption of basic lined electric arc furnaces. This provides an option of using a vide variety of steel scrap and yet produce steel of specified chemical and physical characteristics.

The basic inputs of the ministeel plant are :

(i) Scrap
(ii) Sponge Iron (DRI and HBI)
(iii) Ferro-alloys
(iv) Oxidising agents
(v) Refractories
(vi) Fluxes
(vii) Electrodes
(viii) Power

All the above raw materials are equally important for satisfactory and economical production of steel both in quality and quantity. A detailed discussion on these materials is given below.

2.2. SCRAP

Remelting of scrap for steel making is very old practice. Scrap industry was established in India in late fifties. Scrap is nothing but the fragments of iron and steel which are discarded as waste at different stages. It is the single most important raw material used in EAF steel making and constitutes 65% to 75% of the raw materials cost in the steel making process. It has become indispensable and accounts for about 30% of all steels produced worldwise. Since the quality of the finished steel and economy of steel production depends upon the type of the scrap used, it is very important to select the scrap judiciously. To do this it becomes necessary to classify scrap in certain broad categories. Scrap may be classified on the basis of the following :

(*i*) Source of scrap
(*ii*) Physical form of scrap
(*iii*) Chemical nature of scrap.

2.2.1. Source of Scrap

Knowledge of the source of scrap helps identifying the characteristics of the scrap. The scrap may arise, in general, from three main sources.

(*a*) Internally circulating steel works scrap
(*b*) Process scrap
(*c*) Merchant or capital scrap.

(*a*) **Internally circulating steel works scrap**

The generation of internally circulating steel works scrap is associated with conversion of liquid steel into semifinished or finished products. Croppings to eliminate the ingot shrinkage or pipe defects and those arising from the mechanical deformation of slabs, billets, bars, together with trimming to length, can result under exceptional circumstances upto 50% of the original liquid steel product, although 30% is a more prevalent figure. In addition, the in-plant quality control systems which detect off-quality material and surface defects add to the circulating load. It is obvious that more stringent are the quality control norms higher will be the proportion of internally rejected material which goes into circulation.

It is atmost necessary that the composition of the internal scrap be closely monitored and the different grades segregated. This sort of structure is applied equally to plain carbon, low alloy and high alloy steels.

In addition iron can also be recovered from slag and casting pit residues. For example, steel containing 70-80% iron can be reclaimed from slag and wrecked ladles etc. by recovery processes employed in west.

A further source of iron is broken ingot moulds which are an inveritable consequence of the continuous operation of a steel plant which produces ingots.

Scrap recycling practice has an added advantage in that the plant house-keeping is remarkably good and safety hazards are entirely eliminated.

On the other hand, the internally circulating scrap is going to decrease gradually with increasing deployment of continuous casting in place of ingot mould casting in the ministeel plants.

(*b*) **Process scrap**

Process scrap is the second important category of scrap and essentially arises in the steel consumer industries during the manufacture of marketable products. It has been estimated that between 16 to 20% of the finished steel production becomes process scrap during manufacturing in the heavy engineering industries and in autobody pressing manufactured from flat rolled products. This scrap of rejected material is measurable in the engineering manufacturing process and is returned in different forms to the steel plants for re-melting. A high proportion of this scrap is of low density and is in the form of cuttings, stampings and off cuts which are not really suitable for fast, efficient consumption by the steel works. So the scrap dealers purchase this scrap and process it into more manageable shapes by compression in baling presses which render it more dense and uniform in size. Other residues such as punchings are often sold directly to steel works. The process industry will endeavour to recirculate scrap as fast as possible since the area available for stocking such materials is usually limited.

(*c*) **Merchant or Capital scrap**

The third of the three major categories of scrap is the long term material known as merchant or capital scrap which arises when steel products and engineering structures are discarded at the end of their useful life. These redundant items may vary from tin plate products, such as cars to household machines, motor vehicles and structural steel works. The major source of this type of scrap is undoubtedly the motor car or automobiles in European and other western, countries. This scrap is shredded and are used as a source of high grade clean scrap. Most of the scrap imported in India is shredded scrap to compensate the shortage of scrap.

2.2.2. Physical Form of Scrap

The most important factor which determines the scrap quality is its density. Density of the scrap is nothing but its mass per unit volume. Tap to tap interval of an arc furnace and thus the productivity depends upon the scrap density. Low density scrap requires more numbers of charges and therefore consumes more time.

A poor scrap mix directly increases the tap to tap time because of more number of recharges and larger refining time. The use of inferior quality scrap is not merely due to non availability of good scrap but also due to its price factor.

2.2.3. Chemical Composition of the Scrap

The composition of scrap affects the analyses of both slag and metal. The elements present, with the exception of carbon, can be divided into four groups as follows :

Group 1 : Elements almost completely taken up by slag *i.e.*, those elements which are oxidised under steel making conditions.

Silicon, Aluminium, Tatanium, Zirconium, Boron & Vanadium.

Group 2 : Elements which distribute themselves between slag and metal; manganese, phosphorus, sulphur and chromium.

Group 3 : Elements remaining almost completely in solution in the steel; Ni, Cu.

Group 4 : Elements eliminated from slag and metal zinc, cadmium, and lead.

On the basis of chemical composition the scrap is further classified into two groups :

(*i*) Plain carbon steel scrap

(*ii*) Alloy steel scrap.

Plain carbon steel scrap is cheapest of all scrap and is used in bulk.

As mentioned above the alloying elements in the scrap behave differently during the process of steel making and therefore the scrap should be segregated depending upon the alloying elements present. It is generally segregated as Ni, Cr, Ni-Cr, Ni-Mo, Cr-Mo, Ni-Cr-Mo etc. type of scraps.

Typical elements falling in group three are termed as residuals as they remain in solution in steel and are not removed during the process of steel making of that grade of steel which does not have these elements in its specification. Indeed, in steel, an element like nickel can be classed as a residual, but in another, it is an alloying element, the relevant criterion being its appearance or otherwise in the specification. Another way of describing this mixture of accidental and unwanted constituents is to call them tramp elements.

Control of those elements that can not be removed by chemical reaction in the steel making process ultimately raise the amount of these tramp elements because of successive recycling of the scrap and the desired properties of the steel are not obtained. To overcome this difficulty these residuals are diluted by using uncontaminated forms of iron as DRI pellets etc. The greater cost of such materials as compared to scrap creates the demand for processed and other forms of scrap segregated into appropriate quality grades. In alloy steel making, scrap containing Cr, Ni and Mo can be used effectively, but in making carbon steels, these elements are usually harmful.

Accumulation of residuals can only be attributed to contamination from merchant scrap and must be a function of inadequate segregation, because the large number of steel specifications either makes sorting too difficult or too expensive. Serious surface defects that first appear after billet casting or primary rolling can be attributed to the presence of residuals. These defects, which manifest themselves as surface tears and cracks, are usually transverse to the principal axis. Significant yield loss, and in extreme cases the scrapping of complete casts of affected steel, can occur due to the inability of the steel to deform uniformly during subsequent processing.

The non-metallic contaminants associated with scrap are many and varied and are related to the source of the material. They can indirectly affect the residual levels in EAF's (oxides, slag etc.). Metal turnings from the machinery operations can be contaminated with oil, thus increasing the sulphur input to the furnace. In addition, all grades of scrap will possess rust to some extent and may also be contaminated by dirt during transportation and handling. Moisture is another vital ingredient of purchased scrap which gives poor properties to steel.

Finally, when non-metallics are acidic oxides, additional time must be added to produce required slag composition, marginally increasing the energy demand and the iron losses to the slag.

The emphasis on scrap cleanliness and residual levels in the steel is important, since, apart from the physical condition of the steel, these are the two most important properties affecting steel quality and operational efficiency. It may finally be added that for heavy steel scrap, size is all important, whereas in case of light scrap, bulk density and residual elements like copper and tin must conform to certain limits.

2.3. SPONGE IRON (DRI AND HBI)

The secondary steel sector has been severely suffering due to scarcity of scrap. At the same time the number of ministeel plants mushroomed due to delicensing policy of the government to shorten the gap between demand and production of steel. This growth, in turn, increased the demand of scrap. The availability of indigenous scrap has been very much lower than the demand of secondary steel sector. The Indian steel scenario is in for a drastic change in the next decade. The conventional casting methods are going to be successively replaced by the continuous casting machines in the existing plants both in integrated steel sector and ministeel sector. At the same time with modernisation of steel plants the open-hearth furnaces are being replaced by BOF and casting is going to be through continuous casting route. Such trends are further going to affect adversely the indigenous generation of scrap and even integrated steel plants are going to face a short fall in this area.

The projected steel production in the secondary steel sector for the year 1993-94 is 5.6 million tonnes. It is estimated that the requirement of scrap for the entire steel industry will be 9.7 mill. tonnes in 93-94 and 10.83 M.T. in 1994-95. Against this, indigenous availability will be 4.85 M.T. and 5.04 M.T. respectively. To meet the balance requirement of 4.85 M.T. and 5.79 M.T. in two years, scrap import is must. At the same time the prices of scrap in the international market are going up.

Nearly 175 ministeel plants in the country are facing an acute shortage of steel scrap and many have cut down steel production. This includes some of the leading mini steel units like Mukand Special Steels Ltd., Zenith Steels Ltd. and Usha Alloys which have cut down production by more than 20%. According to Steel Furnace Association of India. During 1990-91, ministeel units produced only 3.6 million tonnes of steel as against the estimated 4 million tonnes. Metal Scrap Trading Corporation [MSTC], the canalising agency for the import of scrap, imported only 1.43 million tonnes of scrap during 1990-91 as against 2.84 million tonnes in 1986-87. The necessity to produce more steel in secondary steel sector to augment the steel requirement has resulted in increased import of scrap since the availability of indigenous scrap is limited. The imported scrap directly lays burden on the country's foreign exchange reserves which increases exponentially both due to increased demand of scrap and increased cost of scrap in the international market. The cost of steel produced in turn increases because of the devaluation of rupee and/or partial or full convertibility of rupee.

Besides the above constraint with respect to scrap, the other drawback of scrap circulation is the build up of tramp elements in the finished steel. This is primarily due to the indiscriminate use of unsorted scrap which results in steady rise in tramp elements. For example, it is reported that copper content in scrap which stood at 0.041% in 1920, rose to 0.15 in 1975. As the sophisticated high performance steels need raw materials free of such impurities which can not be removed during the process of steel making, the steel poducers have to be more selective in choosing the right type of scrap in particular. The sorted scrap is costlier than the normal scrap. Nevertheless, the persistent circulation of the scrap does increase the amount of tramp elements even when sorted scrap is employed as chief metallic charge. If due care is not taken to counter this effect, in the present decade, the problem of tramp elements would become very serious. If larger quantities of sophisticated high performance alloy steels are to be produced, the steel manufacturers have to go for a highly concentrated source of iron like sponge iron (DRI/HBI). On the other hand the secondary steel sector can not be allowed to suffer on account of nonavailability of required quantity of scrap as it is contributing above 46% of the total saleable steel produced in India.

In the light of above, it had been felt necessary to explore some alternative source of metallic raw material for ministeel plants to produce required quantity of steel to meet the indigenous demand of various grades. The use of sponge iron was explored a decade ago in India. However, now it has become an essential raw material due to better technical feasibility and inadequate availability, unstable pricing and nonuniform composition of ferrous scrap.

Sponge iron is the produce of direct reduction of iron ore carried out in solid state using different methods employing abundantly available solid fuels (non-coking coals) or gaseous fuels. As the reduction takes place at lower temperatures, the reduced iron absorbs neither carbon nor any other undesirable impurities like sulphur and phosphorus. The removal of oxygen leaves pores in the iron giving a sponge like appearance and thus the name sponge iron. The sponge iron produced by using gaseous reductant is generally compressed while at high temperature in the form of briquettes to make it more dense and is called Hot Briquetted Iron

(HBI). For sponge iron produced using coal normally higher temperatures are employed (1000°C) and is used as such. This is called Direct Reduced Iron (DRI). The temperature employed in this case is atleast 200°C higher than used in case of gas-based direct reduction processes. HBI is, however, a better product as compared to DRI, because it does not contain impurities like coal or ash. Nevertheless, under Indian conditions the feasibility of sponge iron project based on non-coking coal is much better as compared to gas-based projects. The reason being the abundant availability of non-coking coals in India that too at much cheaper cost. Whereas the availability of natural gas is limited and the cost of gas is very high. However, the carbon content of DRI is less than that of HBI which makes HBI more useful for steel making than DRI. Sponge iron has predictable price, is much cleaner than scrap, has known chemical constituents and is useful in producing high value-added steels. On the other hand scrap contains unknown contaminants and its chemistry can not be predicted. This shortcoming of scrap results into various operational problems in steel making.

As the sponge iron contains some unreduced iron alongwith the gangue, suitable adjustments are required to be made to reduce the iron oxide in the electric arc furnace and to keep the slag within the desired chemical and physical form. However, from an economic point of view, the reduction of oxide in the furnace and the addition of fluxes to counteract the influence of gangue on slag composition increases the energy consumption and hence the cost of production. However, these drawbacks are effectively counteracted by selecting proper quality sponge iron. The sponge iron quality depends upon its percent metallization. The residual oxygen content in sponge iron if in the form of FeO is not that harmful as when it is present as Fe_2O_3 or Fe_3O_4. The good quality sponge iron should have 92-95% metallization. The potential advantages of sponge iron are :

(*i*) It has known chemical composition and free from impurities.

(*ii*) Helps attaining stringent quality requirements.

(*iii*) Being of uniform size it is subject to continuous charging.

(*iv*) It is best metallic charge for UHP furnaces.

(*v*) Under optimum conditions it may increase the productivity of furnace.

Because of the above advantages of sponge iron as metallic charge for EAF units and rising scarcity of indigenously available scrap there is going to be an increase in the sponge iron proportion in the overall metallic charge of the electric steel making units in the times to come. The raw material consumption pattern in the indian ministeel industries in the present decade is shown in Fig. 2.1.

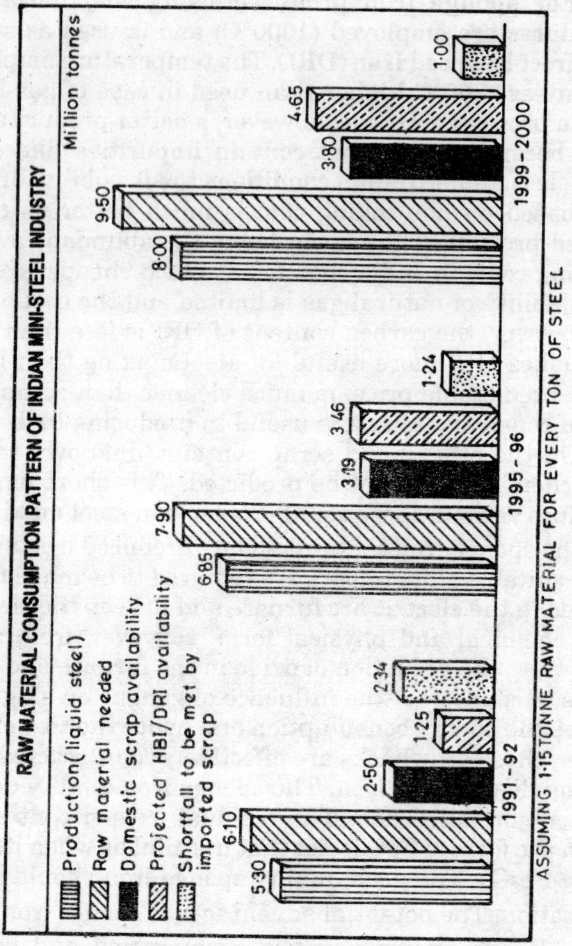

Fig. 2.1.

This pattern of raw material utilization in secondary steel sector is a necessity if the steel production targets are to be achieved to bridge the gap between production and demand of both plain carbon steels as well as alloy steels in the country.

The demand of sponge iron (DRI/HBI) is expected to rise to 8-9 millions tonnes for the secondary steel sector by the end of the decade. This provides an ample scope for the growth of sponge iron industry in India. The sponge iron industry in India started in the year 1980 having 30,000 T/year capacity at Sponge Iron Limited. The total installed capacity in 1991 rose to 1.5 million tonnes with six sponge iron plants operating as shown in Table IV.

Table IV. List of companies, plants and their modules operating at present (1991)

Sl. No.	*Name of Company*	*Location of plant*	*Process*	*Reactor*	*Fuel*	*Year of start - up*	*Capacity (000 lpy)*	*Technical Collaborators*	*Remarks*	*Source of Information*
1	Sponge iron India Ltd. (SIIL)	Paloncha Kothanguden (A.P.)	SL/RN	Rotary kiln	Coal	1980	1 × 30	Lurgi West Germany	First demonstration plant in India	Company report
2	Orissa Sponge Iron (OSIL)	Keonjar (Orissa)	ACCAR	Rotary Kiln	Coal	1984	1 × 120	Allis Chakners USA	Through the plant was to start on Nov. '82 it started regular production since April '84 only	do
3	IPITATA Sponge Iron Ltd. (IPITATA Sponge)	Joda (Orissa)	TDR	Rotary Kiln	Coal	1985	1 × 90		Indigenously developed technology using rotary kiln	do
4	Bihar Sponge Iron Ltd. (BSIL)	Chandil (Bihar)	SL/RN	Rotary Kiln	Coal	1989	1 × 150	Lugi West Germany		do
5	Sunflag Iron & Steel Co. Ltd.	Bhandara Nagpur	KRUPPCO DIR	Rotary kiln	Coal	1989	1 × 150	Krupp Codin Germany		do
6	Essar Gujarat Ltd.	Hazira (Gujarat)	Midrex	Shaft	Natural Gas	1991	2 × 440	Midex Corp USA		do
Total Installed Capacity							14,50, 000 TPY			

Table V. Expansion plans of existing sponge iron companies

Sl. No.	*Name of Company*	*Location of plant*	*Process*	*Reactor*	*Fuel*	*Year of start-up*	*Capacity (000 tpy)*	*Technical collaborators*	*Remarks*	*Source of Information*
1	Essar Gujarat Ltd.	Hazira Gujarat	Midrex	Shall	Natural gas	1992-94 (expected)	2 × 440	Indigenous	—	Media report
2	Bihar Sponge Iron Ltd.	Chandil Bihar	SL/RN	Rotary kiln	Coal	Not known	1 × 150	Lurgi	—	- do -
3	Orissa Sponge Iron Ltd.	Keonjar (Orissa)	—	- do -	Coal	- do -	1 × 100	Indigenous	—	- do -
Additional capacity after expansion in the existing plants							11,30,000 tpy			

Table VI. New sponge iron companies and their plants likely to come up by 1995

Sl. No.	*Name of company*	*Location of plant*	*Process*	*Reactor*	*Fuel*	*Year of start-up*	*Capacity (000 tpy)*	*Technical Collaborators*	*Source of Information*
1	Grasin Industries	Alibagh (Maharashtra)	HYL-III with HBI	Shaft	Natural Gas	1993 (expected)	1 × 728	HYL Mexico	Company Report
2	Tamilnadu Sponge Iron Ltd.	Salem (T.N.)	SL/RN	Rotary kiln	Coal	1992 (expected)	1 × 30	Sponge Iron India Ltd (SIIL)	- do -
3	Goldstar Steel & Alloy Ltd.	Vijaynagaram (A.P.)	Krupp. Codir	Rotary kiln	Coal	1992 (expected)	1 × 150	Krupp-Codir Germany	- do -
4	Nippon Denro Ispat Ltd.	Raigarh (Maharashtra)	Midrex	Shaft	Gas	1993	1 × 100	Midrex Corp. USA	- do -

Cont. table

Sl. No.	Name of company	Location of plant	Process	Reactor	Fuel	Year of start-up	Capacity (000 tpy)	Technical Collaborators	Source of Information
5	Usha Rectifier Corpn.	Amethi (U.P.)	Midrex (hot & cold DR)	Shall	Gas	1993 (expected)	1 × 880	Midrex Corp. USA	- do -
6	Hindustan Electrographite Ltd.	Borai Durg (M.P.)	SL/RN	Rotary kiln	Coal	1992	2 × 30	SIIL, India	- do -
7	Prakash Industries Ltd.	Bilaspur (M.P.)	SL/RN	Rotary kiln	Coal	1993	1 × 150	Lurgi GabH, Germany	- do -
8	Tamilnadu Steel	Tamilnadu	SL/RN	Rotary kiln	Coal	n.a.	1 × 30	SIIL India	Media Report
9	Kumar Metallurgical Industries Ltd.	Nalgonda (A.P.)	SL/RN	Rotary kiln	Coal	1992	2 × 30	SIIL	Company Report
10	Bellary Steel & Alloy Ltd.	Hospet Karnataka	SL/RN	Rotary kiln	Coal	1992	1 × 30	SIIL	Media Report
11	Raipur Alloys Steel Ltd.	Raiput (M.P.)	SL/RN	Rotary kiln	Coal	n.a.	1 × 30	SIIL	- do -
12	Singh Alloys & Steel	Durgapur (W.B.)	SL/RN	Rotary kiln	Coal	n.a.	2 × 30	SIIL	- do -
13	Jindal Iron & Steel	Raipur	Indigenous	Rotary	Coal	1992	1 × 100	Indigenous	- do -
Total capacity likely to be installed							32,80,000 tpy		

There are expansion plans for increasing the existing sponge iron producing capacity of some existing plants shown in Table V.

On the other hand new sponge iron manufacturing plants are going to come up by the end of 1995 to keep abreast the supply with demand. Table VI shows the list of new plants to be installed along with their capacity and the process to be used for production of sponge iron.

2.4. FERROALLOYS AND DEOXIDISERS

A third minor source of iron is provided by the deoxidisers and alloy additions, generally in the form of ferroalloys, that are used to control the casting process or to improve the metallurgical and physical properties of the steel produced. Although the iron arising from ferroalloys and deoxidisers is a fundamental feature of steel making, it represents only 1-2% by weight of the total iron input for most steels. An exception to this is the manufacture of highly alloyed steels such as Hadfield manganese steel, for which significant additions of ferromanganese are required to produce the requisite composition.

The desired chemical elements may be added to the steel at various stages during the steel making operation, for example, in the charge, bath, ladle or moulds. In selection of ferroalloys mainly two factors are considered :

1. ease of addition to consistently achieve the desired steel composition.
2. cost of making such additions.

The ferroalloys are usually available in different grades depending upon the purity and selection is done considering the cost and satisfactory performance. It is of prime importance that the ferroalloys selected should not contain objectionable amounts of undesirable elements, especially carbon, phosphorus and sulphur. The melting point and heat of solution of ferroalloys are also important considerations in their selection, particularly when large quantities are involved. The melting point of ferroalloy rises as its purity increases. The method and time of addition depends largely upon the affinity of the element for oxygen and on the oxygen content of the bath at the time when addition is made. Alloys which are slow to dissolve and have lesser affinity for oxygen than iron may be added in the initial charge or just after melt down. However, if an alloy is readily oxidizable, to avoid excessive loss and uncertain recovery, the addition may be made during the finishing or reducing period of heat, or wholly or partly in the ladle during tapping or in the mould during teeming. The size and size mix and density of ferroalloys play an important role on the recovery of alloying elements. Furnace additions on usually made in the form of large lumps whereas additions made prior to tapping or in ladle or in ladle or mould are in general of fine size to achieve fast dissolution in the molten bath.

Ferroalloys such as ferromanganese, ferrosilicon and silicomanganese are universally used as deoxidisers alongwith aluminium. They are also used as alloying additions where required. Different types and qualities of ferroalloys additions are made in manufacture of various grades of steel.

2.5. FURNACE REFRACTORIES

The lining of the electric arc furnace as well as runners and ladle are composed of heat resistant materials called refractories. The refractories used should not react chemically with either metal or slag, withstand mechanical abuse during charging and temperature fluctuations. The refractories in electric arc furnace are subjected to decidedly more severe conditions as compared to other steel making furnaces. The type of refractory used in the furnace is dictated by the type of steel making practice employed in the plant. At the same time the refractory material should be available in plenty and reasonably inexpensive. Refractories are manufactured by using common minerals such as magnesite, dolomite, chromite, fireclay and quartzite. Magnesite and dolomite are the principal basic raw materials while fireclay and quartzite or silica are principal acid raw materials.

The subject of refractories for basic arc furnace has been given considerable attention in recent years. Refractories in many of the newly constructed plants are being subjected to much more severe conditions than hitherto; this, together with the need for speedier and less costly repair schedules, demands an entirely new approach.

Increase in production in the recent past has been achieved by installing high capacity furnaces ranging between 80 to 150 tons with high power inputs (HP and UHP furnaces) along with the use of oxygen for speeding up the refining. From this it will be realised that conditions are very much different from those in smaller and older units.

Now-a-days one knows exactly that the refractory lining has not only to withstand. The attack in the different areas of the furnace but has also to meet the actual metallurgical requirements. Dolomite has been found to be most widely used refractory material in various steel making units using basic slag practice.

Furthermore, in modern refractories technology it is possible to adjust desired CaO/MgO ratio in manufacture of basic bricks by blending dolomite and magnesia. The refractory linings in different areas of the vessel may be selected using correct CaO/MgO ratio keeping in view technical and economical aspects. The kind of refractories used depend upon availability. For example in Austria magnesite or high lime and dicalcium ferrite bonded "Magnet" is used for lack of availability of dolomite whereas in other European countries dolomite is the preferred refractory material. In U.S.A., Japan and Australia magnesite is the most commonly used lining for vessels. In Europe there was a change from enriched dolomite and upto 100% magnesite linings where excessive wear requires this. But in due course this trend is again reverted to use of dolomite by changing the techniques.

Assuming these prerequisites one might observe two different trends to meet a zoned wear adjusted lining.

1. One way is to use pitch-bonded magnesite bricks having high MgO contents upto 99%.

2. The renewed application of dolomite in those places where it was severely attacked and was replaced by magnesite.

The preheating procedure when using dolomite bricks is less problematic and independent of the size of the vessel. Normally it needs not more than 2-3 hrs. With regard to preheating the hot strength is quite important.

However the most important criterion is the efficiency of lining. One tries to get a most uniform wear and a low amount of reclaimed bricks. This is achieved by zonal lining using different types of refractories.

The tuyere area is more prone to wear. In this area the lining thickness is increased (450 to 500 mm). The tuyere area in the lower cone is now lined with bricks made out of the fused magnesia-dolomite clinker, pitch-bonded magnesite enriched dolomite bricks instead of pitch-bonded dolomite bricks. Such lining increased the number of heats from 50 to 70 to more than 100-106. The tuyere bricks are usually made of pitch-impregnated burnt magnesite enriched dolomite bricks. One will take advantage of the slag refusing property of the pitch impregnated bricks.

Use of direct bonded high fired bricks having low porosity, high heat resistance, good dimensional accuracy and low shrinkage during use is responsible for satisfying lining performance.

It is possible to counteract the wear with an appropriate slag formation. Different slags influence the lining in different manner and their composition and basicity is relevant to the wear mechanism.

2.6. LADLES

Following the rapid development of secondary steel making techniques, the ladles were transformed from simple transporting vessel to a reaction unit. The refractories had to meet these new requirements. The basic lined ladle is the prerequisite to get maximum from the ladle with regard to desulphurisation, homogenisation, alloy recovery and low oxygen content.

With the perfection of vaccum treatment, more and more important metallurgical processes which hitherto took place only in the furnace were transferred to ladle. These are deoxidation and alloying, desulphurisation, degassification, purging, and oxygen refining. Dolomite bricks are preferred for their high thermal capacity which allow tapping temperatures to be reduced once the ladle is preheated prior to taking the melt.

2.7. USE OF WATER-COOLED PANELS

Refractory wear, particularly in the regions of sides and roof, is primarily due to radiation because of which the refractory lining is heated to very high temperatures. At such high temperature the attack of fumes, oxides and slag becomes quite significant resulting in excessive wear of refractory lining. Such refractory wear can be minimised by cooling the refractory lining externally. The modern electric arc furnaces have been provided with water-cooled panels both around the side walls as well as roof of the furnace. These furnaces are high power and ultra high power furnaces. The provision of water cooled panels results in the operator being able to drive the furnace harder giving higher outputs while at the same time reducing the refractory wear in the roof and side walls. Water cooled panels may be in the form of steel tubes or curved panels of cast iron, welded steel plates and copper plates. Panel plate thickness is important criterion in the system; thin plates may be likely to get damaged by the arc or oxygen lancing and thick plates may get cracks during operation. Refractory coating is given on the surface of WCP facing the arc heat of the furnace. Apart from hearth and the side wall adjacent to the slag line where refractory bricks shall continue to be used, it is sufficient to 'gun' the water cooled panels by a refractory coating (dolomite), or allow them to become coated with slag to provide the necessary campaign life. It is not uncommon to find the furnaces with WCP running for six weeks and giving around 600 casts for each compaign before attention is required for refractories. WCP have life from 2500 to 5000 casts depending upon water quality. Careful monitoring of water circulation is required for water temperature. Special shape studs welded on the inner panel surface help accumulate and retain an insulating layer of the slag which effectively protects the panel and reduces the heat losses of the furnace. Over slag line and hot spots copper panels with a layer of bricks between copper and gunned coating are now being employed. These are highly reliable to withstand severe thermal shocks. The roof may also be provided with water cooled centre portion or complete water cooled roof. It consists of a water-cooled outer ring and inner water-cooled ring, which is divided into three parts. Each of these parts are insulated from each other. In side, the roof is protected by high alumina ramming mass with binder. The roof dome may be constructed flatter which reduced refractory consumption and electrode oxidation. Water used for cooling should have *ph* between 7 and 8 its temperature should not exceed above 50°C.

2.8. FLUXES

The function of fluxes is to form a slag, of desired physical and chemical characteristics, by combining with a product of oxidation of varying amounts of the silicon, manganese, iron and phosphorus in the charge during melting and refining.

2.8.1. Lime and Lime Stone

These raw materials are the source of CaO which is needed as a flux to form slag. Lime stone may be added along with the charge while burnt lime is used for later additions, particularly for making second slag. A good electric arc furnace grade lime stone should contain 98% $CaCO_3$ with sulphur less than 0.10%. The size and porosity of lime stone is also important. The size should invariably be less than about 12 cm and porosity of lime stone varies from about 1 to 5%.

Burnt lime is used extensively in electric arc furnace. Silica, Alumina and Sulphur content of burnt lime impair its properties and should be as low as possible. Burnt lime being hygroscopic in nature should be kept in air tight containers and should be heated before adding to the furnace to achieve low hydrogen contents in the steel produced.

2.8.2. Fluorspar

Fluorspar in steel making is used mainly as slag conditioner. When added to basic slags it increases the fluidity and helps dissolution of lumpy lime which gets heated with dicalcium silicate. Fluorspar usually contains about 5% silica more than 90.0% CaF_2 and sulphur is about 0.01%.

2.8.3. Sand

In reducing slag practice burnt lime, sand and coke are added after removing the oxidising slag to form reducing finishing slag. Sand is also used in acid arc furnaces for slag additions. The quantity of sand used is not large but care is taken to use a good grade of sand. It is a common practice to heat the sand to remove moisture before adding to the furnaces. It is desirable for proper slag control to use low iron oxide content sand for finishing slag.

2.9. COKE

Carbon in the form of coke and in many other forms is used in electric arc furnace for two purposes :

(*a*) To form a carbide type slag

(*b*) To carburize the steel in furnace or ladle.

The other forms of carbon used are coke breeze, petroleum coke, graphite, crushed electrodes and fine coal. Coke breeze is most commonly used for slag making. The size, sulphur content and ash are important considerations while selecting. Size should preferably be less than 6 mm but with proportion of fine as low as possible. Sulphur content should be below 0.1%. Silicon carbide has also been used along with coke to aid in shaping the carbidic slag.

Fine coal, petroleum coke, crushed electrodes and graphite are used generally for carbon addition in furnace and ladle. Petroleum coke is generally available in uniform size but differs in its sulphur content from batch to batch. Coal powder is normally not used in arc furnaces owing to its varying ash and sulphur content, graphite and crushed electrodes are more pure form of carbon and are used more frequently.

2.10. ELECTRODES

The necessity of producing quality steel using high technology and economically has lead to an upsurge in the technological upgradation and modernization of electric arc furnaces. This has brought high power and ultra high power electric arc furnaces in operation. Electrodes carry current from the transformer into that part of the furnace where electrical energy is converted into heat for melting the charge material. Earlier self-baking type of electrodes were used but now all electric arc furnaces employ carbon or graphite electrodes.

Carbon electrodes are amorphous in nature. They are manufactured by extrusion of a mixture of calcined anthracite or petroleum coke of definite size proportion and pitch or tar, the binder, and subsequently baked at a temperature of 1200°C under non-oxidising conditions.

Amorphous carbon when subjected to temperatures of the order of 2000° under non-oxidising conditions renders it crystalline. This is usually carried out by passing current through the amorphous carbon electrodes, packed and sealed in a special furnace.

All modern furnace are designed for graphite electrodes and there are only few using amorphous electrodes these days. Given below in Table VII are some characteristics of carbon and graphite electrodes.

Table VII. Comparison of Carbon (Amorphous) and Graphite Electrodes

Properties	*Graphite Electrodes*	*Large Carbon Electrodes*
Specific resistance (ohm-inches)	0.0004	0.0017
Specific resistance (ohm-cm)	0.0010	0.0043
Weight (lb. per cu. ft.)	96	99
Apparent density (grams per c.c.)	1.52	1.57
Tensile strength (kg/mm^2)		
Lengthwise	40	16
Crosswise	30	12
"Practical" oxidation point	470°C	370°C
Ash	0.50%	7.0%

The electrical resistivity of carbon electrodes decreases with increasing temperature, whereas of graphite electrodes it decreases at temperature upto 600–800°C but increases at higher temperatures and reaches its initial values at 1000–1700°C.

Electrodes get consumed during operation and the new electrode is attached to the working electrode with the help of a nipple having cylindrical or conical thread. The electrodes have threads provided in their ends for nipples to fasten the electrodes. The factors that determined the consumption of electrodes are as follows :

(*i*) Oxidation of surface of electrode in furnace by air sucked-in.

(*ii*) Losses owing to fracture of electrodes or nipples.

(*iii*) Atomisation by the arc.

(*iv*) Dissolution in molten baths.

The consumption of electrodes depends on the furnace capacity. The main rules of proper maintenance of electrodes during operation are as follows :

(*i*) Electrodes must be kept in dry place because cracks might develope in electrode because of moisture.

(*ii*) Leakage of gases from furnace through gaps in roof should be eliminated because it will lower the heating of the electrode and subsequent oxidation by atmospheric air.

(*iii*) Electrodes should be screwed properly to each other.

(*iv*) Electrodes should be positioned properly for their free movement through electrode holes in roof otherwise due to lateral pressure they may break.

(*v*) Electrode diameter should correspond to the current supplied as the high current density would lead to excessive heating of electrodes and greater electrode diameter would increase the power consumption.

The current density of the electrode decreases with increase in diameter, typical figures for 225 mm and 600 mm diameter electrodes are 22 amp/cm^2 and 12 amp/cm^2 respectively. For large electrodes there are two grades, namely a regular grade and premium grade. The premium grade is a high strength, high electrical conductivity electrode made from very best raw materials and usually is pitch-impregnated before graphitization.

Current density of electrodes :

Regular quality	18 A/cm^2
Pitch impregnated	24 A/cm^2

However, in light of changing scenario of EAF steel making technology requirements of electrode specifications have become more stringent. With the introduction of HP and UHP furnaces for achieving rapid melting rates, the power levels have approached 1000 KVA/t. The current densities necessary to be carried by the electrodes are much beyond the safe levels of operation for regular electrodes. The electrodes for HP and UHP furnaces should meet the following requirements :

(*i*) Operate at high current density levels with attendant lower resistivity.

(*ii*) Have low coefficient of thermal expansion.

(*iii*) Have high resistance to thermal spalling.

(*iv*) Should have high bulk density.

(*v*) Must have excellent mechanical strength.

(*vi*) Resistant to surface oxidation.

It has been observed that the electrodes with lower coefficient of thermal expansion (CTE) are in general associated with poor mechanical strength and the ones having high mechanical strength show higher CTE values. Indian calcined petroleum coke (CPC) yields graphite electrodes of regular grades with high CTE. These electrodes are not suitable for HP/UHP furnaces. For these furnaces needle coke (CPC) is required as raw material for manufacturing graphite electrodes, which is an imported raw material.

Graphite electrodes form one of the significant costs in EAF steel making. Typically, electrode cost may be about 6-8% of the total production cost. Electrodes consumption, however, depends upon number of factors and thus varies from plant to plant and from one steel grade to other.

The electrode consumption occurs due to oxidation, atomisation and breakage during the furnace campaign. Typical data on electrode consumption is given below :

Surface oxidation	40-50%
Loss through tip	35-45%
Breakage/stub end losses	10-15%

Furnace electrodes wear out mainly at the tip by the action of arc and by oxidation over the whole surface area within the furnace. The loss of electrode at tip is related to the current carried by the electrode. The loss due to oxidation is a function of furnace temperature and the oxidation nature of the furnace gases. The rate of wear at the tip of the electrodes is greatest during melting period of the charge when the highest power inputs and arc currents are used. At this stage the furnace temperature is fairly low (900–1500°) and so the oxidation loss at the surface of the electrodes is small. After clear melt, the power input to the furnace is reduced to about one third of the maximum and arc current is correspondingly decreased so that the rate of wear at the tips of the electrodes is considerably reduced. Conversely, furnace temperature increases during the refining period from about 1500 to 1700°C and the rate of electrode surface wear is increased.

It is observed that 70% of electrode loss is by surface oxidation during refining period, therefore a considerable reduction in operating costs could be achieved if electrodes wear due to oxidation could be reduced or eliminated. There are three possible ways of doing this.

(*i*) By coating the surface with a high melting material which is not easily reduced by carbon at high temperatures.

(*ii*) By impregnating the open pores of the electrodes with a refractory substance which can fuse on the graphite surface and inhibit oxidation by excluding the ingress of air or oxidising gases.

(*iii*) By incorporating suitable substances in the electrode mix during the manufacturing process.

Out of the three possible means by far the electrode coating has attracted the attention most. The coating material should meet the following requirements.

(*i*) It should have high melting point and should not be reduced by carbon at furnace temperatures.

(*ii*) It should be chemically stable and unaffected by iron oxide, slag and metal attacks.

(*iii*) It should be resistant to mechanical abrasion and form a strong rigid bond with electrode surface.

(*iv*) It should have coefficient of expansion compatible with that of electrode to prevent its flaking and spalling.

(*v*) It should be easy to apply and if nonconductor it should be able to be applied by spray insitu on the furnace.

Different types of materials have been used for coating the graphite electrodes but each have some shortcoming on the other.

1. Materials like silica, Alumina, Magnesia and titanium dioxide etc. give good protection but application is difficult and they have tendencies to flake and spall.
2. Metallic coatings like iron, Nickel, Stainless Steel etc. are good conductors but here again the high coefficient of expansion leads to spalling which is more than oxide coatings.
3. Refractory carbides and nitrides would make a good coating but there are difficulties in the method of application. Silicon carbide is coated by induction heating and gives extremely good protection against oxidation but cost of application of the coating is so great that it overweighs the savings in electrode consumption.

Graphite electrodes have porosity of the order of 20 to 30% and many of the pores are interconnected and open to the surfacc. The size of the pores varies from 0.5 to 100 microns and if filled by impregnation with solutions or colloidal suspensions the permeability of the graphite electrode can be reduced to very low values. Liquids such as ethyle silicate, sugar solution and furfuryl alcohol are generally used but they inhibit oxidation only upto 800°C.

Other possibility is the introduction of some oxidation resistant material into the electrode mix during manufacture. It is possible that carbides such as those of boron or titanium may form oxides during graphitization and provide protective action during use.

THREE

Production of Plain Carbon and Low Alloy Constructional Steels

3.1. GENERAL

The Electric Arc Furnace can readily tackle the complete range of plain carbon steels from low carbon (< 0.10%) to high carbon (upto 0.85%). The low carbon spectrum takes in reinforcing, mesh and galvanised fencing wire. From 0.1 to 0.2% C, a wide range of bars, flats, angels and channels used in basic construction are produced. But one of the limitations of the electric arc furnace/continuous casting route is its inability to produce rimming steels. The electric arc furnace/continuous casting route also readily handles low alloy and micro alloy steels. Examples of these are the metal inert gas welding compositions (1% Si, 1% Mn) and HYLA (high yield low alloy) compositions (1.2/1.5% Mn with Nb and V) to produce high yield reinforcing steel.

Generally two slag practice is followed for the manufacture of quality carbon and low alloy constructional steels. But single slag practice is also used for the manufacture of open-hearth grades of steels. The melting practice varies widely depending on the grade of steel, characteristics of charge and to some extent on the preference of various operators. In the present text, "double" and "single" slay practice commonly employed for the production of various grades is discussed in brief.

Modern innovations like the computer control of Electric Arc melting operations, Oxy-fuel burners, scrap preheating, continuous feeding, use of Directly reduced iron etc. have made the process more efficient and versatile.

3.2. DOUBLE-SLAG PRACTICE

3.2.1. Selection and Preparation of Charge : Considerations

Ideally, the melted metal should be close to the desired analysis. Hence, segregated scrap of known composition must be employed to obtain consistent melting practice and to meet the specifications during melting and refining. The size, cleanliness and carbon content of scrap must be regulated and for most portion of the charge, heavy scrap should be preferred, if available. It is because, if dirty rusty and light scrap is employed during melting, both the slag volume and also the amount of recarburisers needed to achieve a fixed opening carbon at the melt out stage would increase. However, 20% of metallics as "light scrap" is always needed to protect the furnace roof refractories. All the metallics charged to the furnace must be accurately weighed.

3.2.2. Charging

The following charge mix is employed :

(*i*) Heavy scrap - 35% — in the form of blooms, butts, crops and billets.

(*ii*) Medium scrap - 40% — chopped rejected plates, *I* & *U* sections etc.

(*iii*) Light scrap - 25% — flashings, bundles, turnings, borings etc.

Following factors must be taken into consideration.

3.2.3. Melt Down

Maximum voltage and current is employed in this stage in order to achieve maximum possible rate of melting. For this reason, power and electrode consumption is maximum during the melt down period.

Timing of lime and ore additions vary considerably. No lime and ore is added in the initial charge. When the charge is 2/3rd melted, additions of lime + ore + balance of scrap are then made. This practice gives rapid melt down because the presence of lime and ore in the charge prior to the formation of pool, decreases the conductivity of arc. With recent trend of using "gaseous oxygen" ore is added only to cool off the bath. Lime, however, is added early enough to prevent acidic products of melting from attacking the refractory lining and its quantity amounts to 2 to 5% of the weight of metallics. Alloys such as Ni & Mo, which are not oxidised during melting, are included in the initial charge itself.

3.2.4. Oxidation Period

Oxidation period begins as soon as the scrap is nearly all melted and impurities such as P, Si, Mn and C are oxidised. Any oxidisable elements present in the scrap are also oxidised and lost to the slag initially, the temperature of the bath is low and therefore, dephosphorisation is possible even by a low basicity slag.

The first sample is drawn for estimating C, P, Mn, S, Cr, Ni, Mo etc. Then additional lime is added for maintaining proper *V* ratio. Gaseous oxygen is then blown through consumable seamless steam pipes to achieve desired rate of carbon removal. As a result, bath temperature increases and vigorous carbon-boil starts. Oxidation continues till carbon below the specified limit is obtained. Besides other advantages, vigorous carbon boil effectively brings down hydrogen to < 2 ppm level.

If sulphur is to be < 0.01 max, a Fe-Mn or spiegel addition 1/2 hr before slag off eliminates upto 50% Sulphur from the bath and facilitates further de-S during reducing period.

3.2.5. Slag-Off

Deoxidation prior to slag-off by Fe-Mn addition brings the bath oxygen level in equilibrium with the carbon content of the bath and the carbon boil effectively subsides. But care has to be taken to ensure that slag is not deoxidised, otherwise slag/metal equilibrium may be disturbed and rephosphorisation of bath may occur. At this stage, the electrodes are lifted and slag is raked off the furnace as completely as possible using steel or wooden rabbles. Practically, no slag should stay in the furnace. Analysis of bath is known, through the sample that was drawn prior to the commencement of "slag off". Carbon is adjusted by making additions to bare bath followed by ferro-alloy additions based on lower limits of specified composition. The deoxidation of bath is completed by the help of Fe-Si, Ca-Si and/or Al additions.

3.2.6. Reducing Period

Fluxing materials are added in the following ratio to form reducing slag :

Calcined Pebble lime 5 to 8 parts	+	Silica Sand $\frac{1}{2}$ to $1\frac{1}{2}$ parts	+	Flourspar $\frac{1}{2}$ to 2 parts.

Calcined pebble lime is added first, followed by silica sand and flourspar, and it takes about 30 minutes to shape the slag properly. Total weight of reducing slag amounts to nearly 2% of metal bath and its thickness may be about 1 inch. Pulverised coke is added throughout the reducing period to engage oxides and this coke forms CaC_2 under the action of arc. CaC_2 disintegrates to a powdery form upon cooling and gives a typical acetylene smell when brought in contact with water. In low carbon steels, powdered Fe- Si may be used instead of pulverised coal to avoid the danger of carbon pick-up. Typical analysis of a carbidic slag is as follows :

CaO - 60/75	FeO - 0.5 max
SiO_2 - 18/22	S - 0.3
MgO - 7/9	MnO - 0.1
CaC_2 - 1-2.5	Al_2O_5 - 0.3

The bath is quiet during the reducing period and therefore the rate of slag/metal reactions is decreased, and also it results in stratification with respect to both composition and temperature. Therefore, after 20 to 40 minutes, the bath is agitated with the help of induction stirrers.

The reducing period may extend upto 2 hours depending upon the final sulphur specification level needed. Extra care is needed to affect desulphurisation if carbon level is below 0.3%. This aspect is discussed in subsequent sections. Final test analysis is obtained and the requisite alloy additions are made. Bath temperature is checked and final deoxidation with Fe-Si and Al is carried out 15 minutes before tapping. The heat is then tapped.

3.2.7. Tapping and Ladle Deoxidation

Preheated and perfectly dry ladles must be used. Loose refractory is cleaned from furnace tap hole and paper plug is placed. Heat is tapped with minimum possible turbulence in the stream to avoid secondary oxidation of liquid metal. Also the stream must fall in the centre of ladle to avoid refractory erosion.

The ladle deoxidation practice mainly depends upon the bath carbon content, the grain size required in the solidified ingot & other additions needed. Small additions of Zr, B, Ti, V etc. may be required to be made to the ladle. For coarse grained steel, no Al is added. For fine grained, medium carbon steel nearly 0.3 kg of Al per tonne is added and for fine grained low carbon steel, 0.5 kg of Al per tonne is added. For alloy heats, as much as 2-2.5 kg of Ca-Si alloy per tone is added in the ladle uniformly throughout the tapping to obtain most effective deoxidation of the bath.

3.2.8. Temperature Control

Close temperature control is necessary throughout the processing of heat. Temperature control after slag off is hampered by the quiescent bath. During the reducing period it becomes increasingly difficult to maintain a uniform bath temperature. Usually, rabbling is carried out or induction stirrers are applied to overcome this problem. These days Pt-Pt/Rd type quick immersion parameters are commonly employed to measure the bath temperature accurately. Some operators prefer to employ sufficiently high temperature before slag off to permit the addition of ferroalloys to the bath and yet tap the heat within desired temperature range. For different grades of steels, specific temperature levels are considered ideal. A few example are given below.

	EN-24	*A1S1 8620*	*A1S1 52100*	*EN 18D*
At slag off	1640°C	1625°C	1605°C	1620°C
At finals	—	1605°C	1580°C	1595°C
At tap	1610°C	1595°C	1555°C	1595°C

3.2.9. Control of Phosphorus (Fig. 3.1)

A minimum basicity of 2.2 (CaO/SiO_2) and preferably greater is required to keep the P in the metal under 0.020%. The ratio of P in the slag to the P is the metal, as jointly affected by lime-silica ratio and iron-oxide in the slag is shown in the figure (Fig. 3.1) below :

$$2\,[P] + 5\,(FeO)\,(slag) \rightarrow P_2O_5 \quad \Delta H_P = -\,290 \text{ K.cals/mole } P_2O_5$$

$$P_2O_5 + 4\,CaO \rightarrow 4\,CaO.\,P_2O_5$$

presence of free or excess lime in the slag is essential and also FeO in slag is necessary for any extensive removal of P.

P may be removed by a slag relatively low in basicity if the FeO level is high, conversely, if the FeO level is low, (high C steels), the basicity must be high.

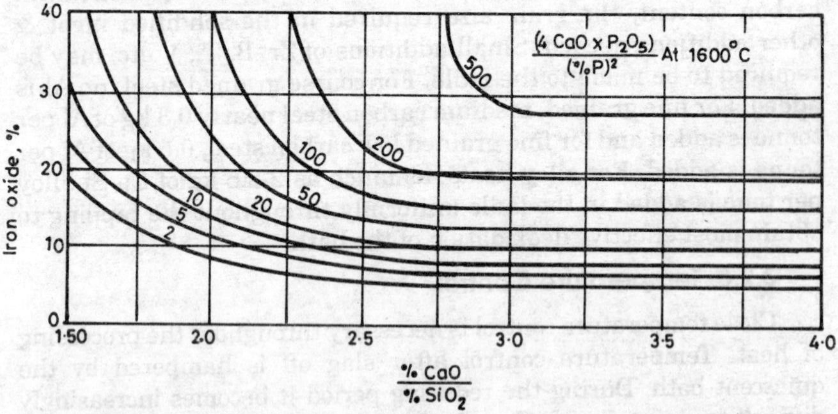

Fig. 3.1. Distribution of phosphorus between slag and metal at 2910 F (1600°C) showing the joint influence of the iron oxide content and the basicity of the slag.

A certain minimum combination of FeO and CaO is necessary to hold P in the slag after it has been removed from the steel. Also, low temperatures help effective dephosphorisation.

P removed from Elect. Fce. is fairly rapid under a slag that is neutral (or semi-basic) with high FeO content (first melted heat) and when, temperatures are low. As oxidising period progresses, bath temp. increases, and since the supply of oxygen is limited (unlike in OH), the FeO content of slag may decrease. Therefore, sufficient lime must be added to compensate for lower FeO content. Silica content of slag must be kept low, specially if a large amount of P must be removed. Low Si charges and a minimum of SiO_2 from other sources result in a high basicity of slag for a given slag volume.

When extremely high Si charges are encountered, use an extra slag-off early in the heat, and then make-up a new slag, which is fairly basic.

3.2.10. Control of Sulphur

Optimum desulphurisation is achieved when the following conditions are met :

(*i*) high basicity

(*ii*) low FeO content of slag (consistently < 0.5%)

(*iii*) intimate mixing of slag and metal

But it must be mentioned that extra care is needed to affect desulphurisation if the carbon content of bath is below 0.3%.

Certain important factors which control/improve desulphurisation at carbon contents less than 0.3% are discussed below.

3.2.10.1. Effect of Carbon Content

Data above shows (Fig. 3.2 & 3.3) equilibrium conditions existing only when slag/metal are intimately mixed *e.g.* Perrin Process (S reduced to < 0.005% within seconds). In actual practice— slag/metal equilibrium is not achieved during refining, hence FeO content of slag is not a reliable measure of *O* content of bath. Oxygen level of bath is more dependent on *C* content of bath. Data for a number of Elec. Fce. heats show that the *C* & *O* values approach the equilibrium curve. (Fig. 3.2)

De-sulphurisation in normally greater on high — *C* heats since the state of oxidation is a function of the C level of the bath.

Relationship between ladle S and ladle C shows that average S remains constant at ≈ 0.013% on heats having C contents of 0.3% or higher, but S content increases rapidly on low C heats. It is with low C grades, that greatest difficulty is experienced in meeting low S requirements. (Fig. 3.3)

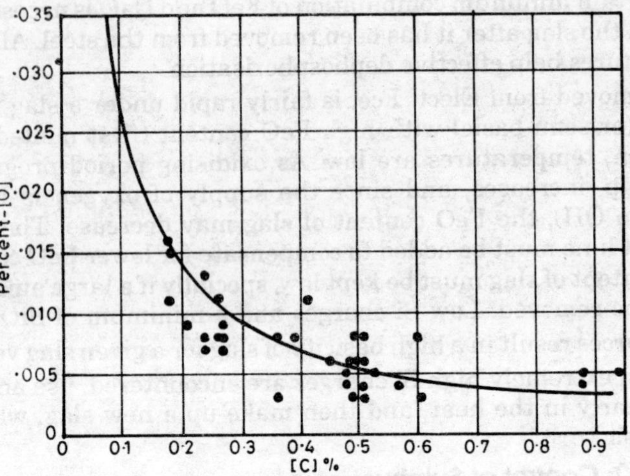

Fig. 3.2. Oxygen-carbon data for the electric furnace finishing period.

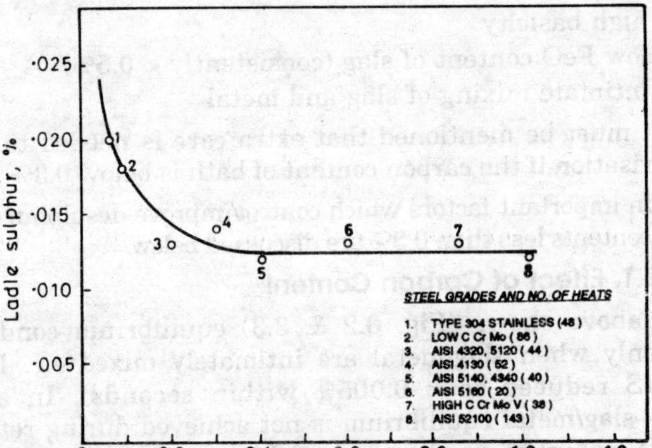

Fig. 3.3. Relationship between ladle sulphur and ladle carbon.

Fig. 3.4 shows the amount of S removed in making an AISI-4130 grade as compared with 0.13% max. C steel of similar Cr and Mo analysis. In higher C heats, 50% of melt S is removed, whereas in low C grades, only 35% S is removed. (Fig. 3.4)

3.2.10.2. Effect of Si before Tap

De-S of low C grades can be improved most readily by addition of a deoxidiser like Fe – Si during refining period to reduce the *O*

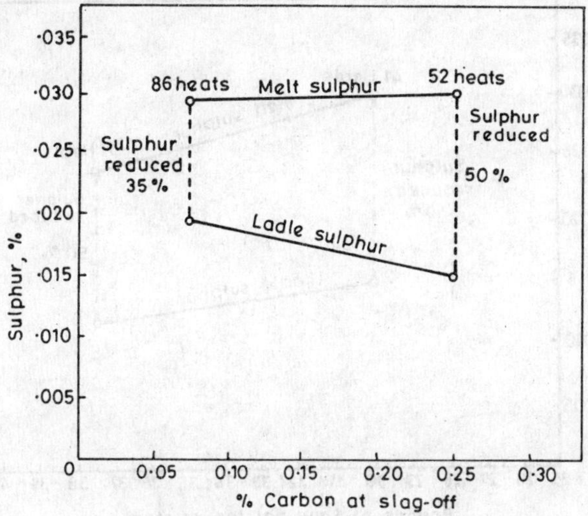

Fig. 3.4. Effect of carbon content at slag-off on melt and ladle sulphur in Cr-Mo steels.

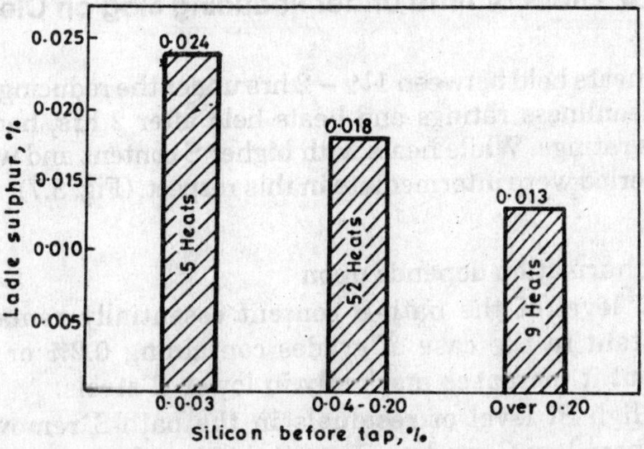

Fig. 3.5. Relationship between silicon and ladle sulphur contents in a low-carbon Cr-Mo grade.

level of steel bath. Higher Si residuals at tap result in much lower S levels. (Fig. 3.5)

3.2.10.3. Effect of Lime Addition

Amount of additional lime at charge or additions before slag off, affects both slag volume and the basicity of oxidising slag. The combined effect results in lower S at slag-off (Fig. 3.6).

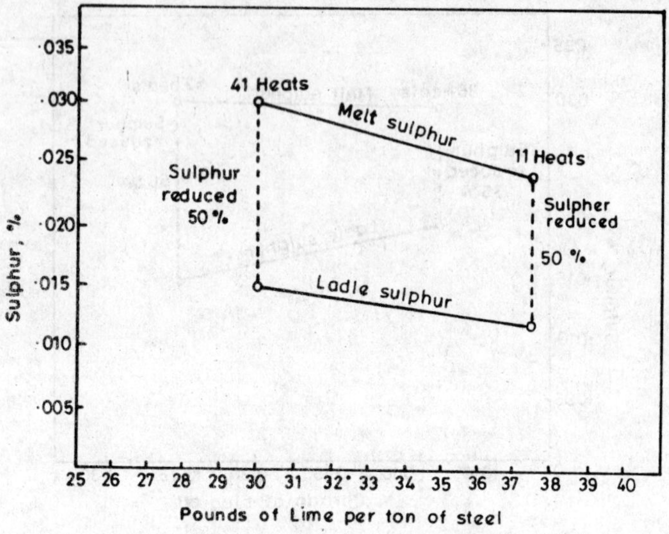

Fig. 3.6. Effect of lime added before slag-off on melt and ladle sulphur in AISI 4130 grade steel.

3.2.10.4. Effect of Time Under Reducing Slag on Cleanliness (Fig. 3.7)

Those heats held between 1½ – 2 hrs under the reducing slag had the best cleanliness ratings and heats held after 2 hrs, had poorest cleanliness ratings. While heats with higher S content and with short reducing period were intermediate in this respect. (Fig. 3.7)

Summary

Desulphurisation depends upon

(*i*) C level of the bath-S content essentially remains constant in the case of grades containing 0.2% or greater, but it increases markedly in lower C steels.

(*ii*) High Si level of residuals in the bath-S removed from these low-C grade = that attained at the higher C levels.

(*iii*) Higher slag volumes and higher basicity of the oxidising slag aid S removal.

(*iv*) For high S melts or for heats being melted to very low S levels, use a second reducing slag.

3.2.11. Control of Oxygen During the Reducing Period

After slag-off, the steel is in a condition represented by an excess of oxygen over the equilibrium value as controlled by the C-content. In addition to the oxygen in metal, Iron oxide, absorbed by

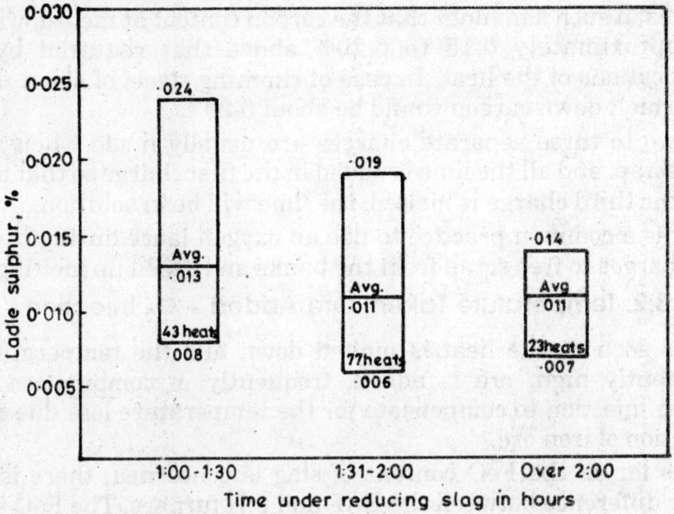

Fig. 3.7. Per cent sulphur in ladle vs time under reducing slag for AISI 52100 grade steels.

the furnace banks may be redissolved under reducing conditions requiring additional deoxidisers.

3.3. SINGLE SLAG PRACTICE FOR RIMMING AND SEMI-KILLED STEELS

These steels are commonly employed for the production of sheets, strips, wires etc. The rules which apply to the manufacture of rimming steels in the open hearth furnaces apply equally well in EAFs. Attaining a high productivity in EAFs is the only problem.

3.3.1. Practice

After the previous heat has been tapped and the banks repaired, the furnace is charged as rapidly as possible usually with a miscellaneous assortment of scrap including bundles and light scrap. A typical charge is as follows :

Mould scrap and pig iron	10.6%
No. 1 steel	11.1%
Bar crops	11.5%
Thin bar	15.8%
No. 1 bundles	51.0%
Ore (% of metallic charge)	3.6%
Burnt lime (% of metallic charge)	2.4%

Pig bar, broken moulds, coke or broken electrodes are usually charged in such a manner that the carbon content at melt-down will be approximately 0.15 to 0.20% above that required by the specifications of the heat. In case of rimming steels of about 0.07% C, the melt down carbon would be about 0.20%.

Two to three separate charges are usually made when using light scrap, and all the lime is added in the first charge so that by the time the third charge is melted, the lime will be in solution.

It is a common practice to use an oxygen lance during the last two charges to free scrap from the banks and speed up melting.

3.3.2. Temperature Taken, Ore Added + O_2 Injection

As soon as the heat is melted down and the temperature is sufficiently high, ore is added, frequently in combination with oxygen injection to compensate for the temperature loss due to the reduction of iron ore.

As far as the FeO content of slag is concerned, there is only minor difference between the EAF and 0.H furnace. The FeO in the slag tends to be slightly higher on low-carbon heats in the EAF compared with the 0.H furnace probably because a higher CaO/SiO_2 ratio can be attained in the EAF. (Fig. 3.8)

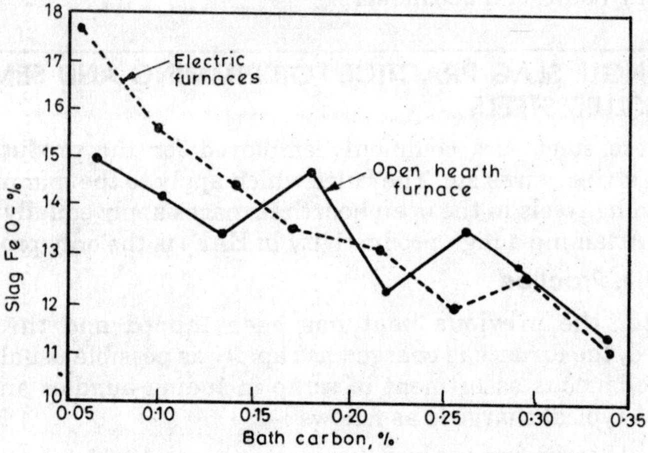

Fig. 3.8. The FeO content versus bath carbon in the electric and open-hearth furnaces.

Because of the differences in oxidation during melting and differences in charge make-up, the manganese residual at a given carbon content is considerably higher in the EAF compared with the 0.H practice for a similar grade. (Fig. 3.9)

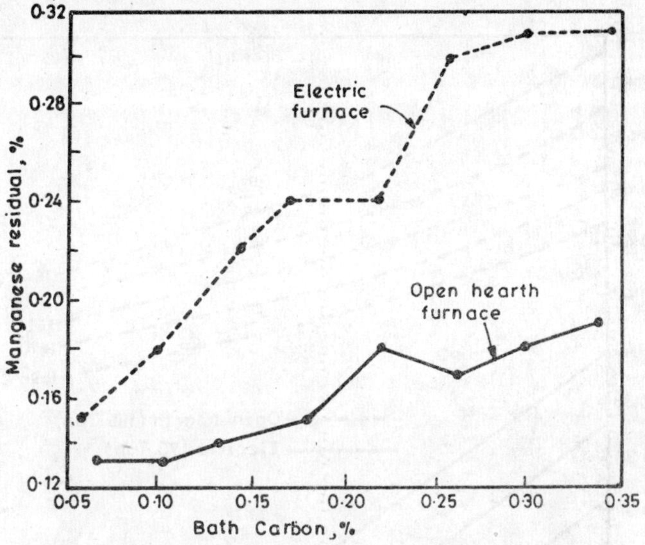

Fig. 3.9. Manganese residual in electric and open-hearth furnaces of different carbon levels.

3.3.3. Tapping the Heat

After carbon and Manganese preliminaries are obtained, the heat is tapped into a ladle and poured in the usual manner. Tapping temperatures are on the average 15°C higher than 0.H practice for the same grade.

3.3.4. Deoxidation

The amount of Al to be added to the ladle to produce semi-killed steels may be somewhat lower in EAF than that used in the 0.H furnace as shown in Fig. 3.10. For rimmed steels, the relationship may be reversed — the average Al in EAF being 0.095 lbs/t (0.043 kgs/t) for 0.065% C steel and 0.073 lbs/t (0.033 kgs/t) in an 0.H fce for the same carbon content.

Deoxidation of rimmed steel varies from shop to shop. No. Al is used on 0.2% carbon steel, whereas as much as 0.4 lb/t (0.186 kgs/t) is required on low carbon heats. Typical Al addition for 0.08 to 0.13 percent carbon steels are given below. One pound per ton (0.45 kgs/t) Ferrocarbon Titanium is sometimes used with the Aluminium.

Typical Al Additions

% FeO in Slag	*Al, lb/t (kgs/t)*
upto 20	0.3 (0.136)
20 - 23	0.4 (0.1816)
24 - 27	0.5 (0.227)
28 and above	0.6 (0.272)

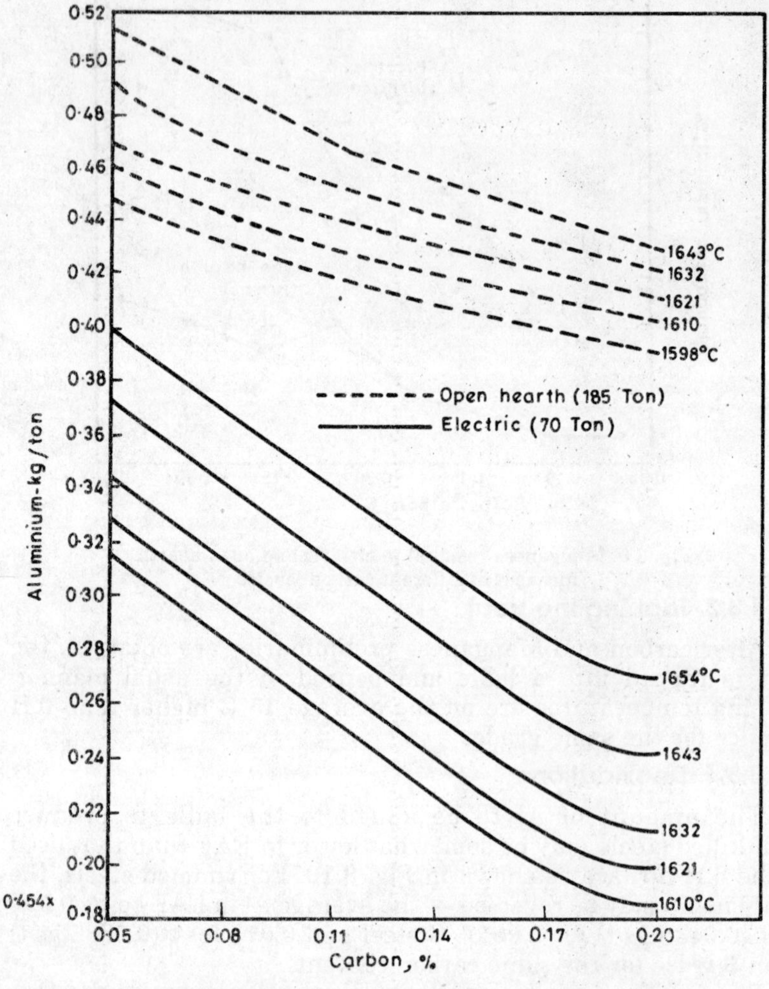

Fig. 3.10. Amount of aluminium required to produce semi-killed steel in open-hearth and electric furnace.

3.3.5. Rimming Action

Rimming action in the mould is affected by several variables *e.g.* temperature (optimum being 1593°C), FeO content (optimum being Fe as FeO = 35 to 40%) and the carbon content (lesser the carbon better the mould action). Mould action is classified into five different grades, namely, light, light to moderate, moderate, moderate to vigorous and vigorous.

FOUR

Manufacture of Special Alloy Steels

4.1. INTRODUCTION

The steels other than stainless steels and tools alloy steels having substantial quantity of alloying elements are used for specific purposes. These steels have been classified as

(*i*) The standard constructional alloy steels
(*ii*) The carburizing steels
(*iii*) The high carbon-bearing steels
(*iv*) The high strength heat-resisting steels
(*v*) Special alloys steels

The melting practices of above steels have many similarities, however, each type must be treated as a typical melting problem because of varying composition, processing technique or applicable quality requirements.

4.2. GENERAL MELTING PRACTICE

4.2.1. Charge and Melting

In earlier practice charge composed of carbon and/or alloy steel scrap was fed to the furnace and melted down. As soon as the heat started melting, a chemically reducing slag was prepared and the heat finished under it. Unquestionably economics tended to dictate this type of practice. As the tonnage of alloy melting increased, the analyses become more stringent. The typical electric furnace operator could no longer depend on select scrap but had to consider the effect of bath contaminants, such as sulphur and phosphorus and the various alloying elements contained in the remelt scrap, and an unpredictable melt-in carbon level in the heat. Thus, the melter turned to the use of iron ore, scale, and finally gaseous oxygen to aid in controlling and adjusting the composition of the melt-in bath. This led to present day practice in which both process and purchased scrap are segregated into categories according to alloy content in order to get the full economic advantage of alloy recovery in the melt.

The charge consists of (1) a reasonable percentage of revert alloy and carbon steel scrap, (2) as much additional purchased alloy scrap as the metallurgical quality control of the analysis will permit, and (3) carbon steel scrap to complete the balance of the charge weight. In addition to these, fluxes are frequently added either in the initial charge or during back-charging as described in an earlier chapter. Carbon in the form of scrap electrodes, coal, petroleum coke, or graphite is added with the charge when higher carbon steels are to be made to permit the heat to melt-in within the range of the aim carbon of the analysis. Nickel, copper, ferromolybdenum, molybdic oxide, and ferro-chromium can be added to the charge in addition to the scrap, but the usual practice is to add these after the melt-in bath chemistry is known.

Physical shape of the charge must necessarily conform to the type of scrap available in the individual plant, keeping in mind both the economic aspects of the scrap's alloy content and best arrangement of heavy melting material and light scrap for fast, efficient meltdown.

4.2.2. The Refining Period

After melt down, the first consideration is that of phosphorus control when this element is known to run high in the scrap. This is done by addition of sufficient iron ore or scale or scale and lime in the charge to guarantee a melt-in slag of atleast a 2.2 : 1 V ratio with enough FeO to hold the phosphorus in the slag. This slag must then be flushed before the bath temperature rises too high to prevent reversion of the phosphorus back into the bath. Once the bath becomes too hot the only means of attack is to deslag, cool the bath, and start over with more lime and ore. If phosphorus control is no problem, this step can be by-passed.

After the slag flush-off, additional lime or limestone and oxygen in the form of iron ore, clean scale, or gaseous oxygen are added and sufficient power is applied to elevate the bath temperature to 1590 C or higher, depending upon the grade of steel being melted. During this period a steady boil should be in evidence as the bath carbon reacts with the FeO present, forms CO, and moves to the furnace atmosphere. At the same time removal of chromium must be considered when this element is a problem as indicated by the sampled bath chemistry.

As the bath increases in the temperature and slag begins to shape up, the carbon-oxygen reaction becomes the primary one in the furnace. Many operators favour placing sufficient carbon in the charge to guarantee the bath melting in at least within range of the aim analysis. A number preferred to be 20 to 40 points high. The heat is worked using ore, scale, or gaseous oxygen until the carbon content has been dropped 0.10 to 0.15 percent below the specification. This permits some recarburization on the bare bath, and the reaction, being endothermic, helps to cool the bath from its peak temperature prior to slag-off to the lower temperature level desired in the finishing period. The over-oxidized heat usually ends up so low in carbon that it requires an abnormal addition of recarburizer to bring into range. Similarly, heats which melt in at specification and are finished off without adequate boil frequently show up poorly in later quality rating.

When the bath chemistry is in proper shape and the temperature has been adjusted to the desired level the heat is ready for deoxidation. This is initiated before slag-off by the addition of some ferromanganese, silicomanganese, or spiegeleisen where the manganese range of the final chemistry permits. After this reboil, the slag is skimmed and the formal deoxidation cycle commences. In some shops this is initiated by dipping the electrodes until the boil subsides. Recarburizers are then added in conjunction with small additions of ferrosilicon or calcium-silicon. Where dipping the electrodes is not standard practice, an extra amount of recarburizer must be added above that theoretically needed for adjusting composition because some carbon is consumed in deoxidizing the bath. This has a desirable effect in cooling the bath slightly and forms a gaseous reaction product rather than non-metallic inclusions. Studies have shown that this deoxidation closely follows the theoretical C-O curve. Naturally, if careful control of the order of the additions is not maintained, the silicon will deoxidize the bath preferentially and the subsequent carbon addition will overshoot the specification due to the greater efficiency of recovery. Aluminium can be poled at this point for total bath deoxidation but many operators prefer to wait until later in the bath to add this element when melting quality steels.

This bare-bath deoxidation is properly followed by the addition of the finishing slag. Either carbide slags or lime-silica slags are characteristically used in finishing off the alloy melt, choice of these depending on the carbon level of the grade being melted. In some cases a lime-alumina slag is used on very low carbon chromium-bearing heats similar to the slags used on some stainless grades. Bath chemistry samples are taken as soon as the slag is fluxed followed by thermocouple check of bath temperature. Aluminium and additional silicon frequently are added at this point followed by the ferroalloys required to adjust the composition to specification. After the analysis results have returned, final alloy adjustments are made, the bath and slag are shaped up, and final bath temperature checks are made with an immersion thermocouple preparatory to tapping the heat. Ladle additions are usually restricted to final additions of special deoxidizers such as calcium-silicon, and aluminium when used for grain size control. Though a number of exothermic alloys are available which can adequately be dissolved in the ladle, it is felt that this practice is not always satisfactory for quality specifications. As an example melting practice of a typical steel is discussed below.

4.3. MANUFACTURE OF THE HIGH-STRENGTH, HEAT RESISTING STEEL

The steels included in the group are principally low-carbon chromium-molybdenum steels varying in chromium content from 0.50 to 9.00 percent and molybdenum from 0.50 to 1.00 percent as the intermediate alloy grades. These steels are used for a number of applications in the steam and power generating and chemical industries, primarily in super-heaters, heat exchangers, and furnace construction, and for high-temperature piping.

The charge is carefully computed to take advantage of recovery of chromium and molybdenum available in alloy scrap up to the limit of the chemical specification of the heat to be melted. Since substantial alloy additions in the form of reducing agents, etc., are to be added in later stages of the heat, the stoichimometric balance must be considered when determining the total weight of carbon scrap to be added to the charge. Frequently, additions are made of revert straight chromium stainless scrap or high-carbon ferro chromium to attain the lowest possible alloy cost balance. Care must be taken as with the stainless steels, to ascertain that all the charge materials are low in phosphorous content because of the subsequent reduction technique to be followed. As the charge is fluidizing, some lime stone or lime is added to prevent the melt down slag from becoming siliceous yet taking particular care to keep slag volume as low as is feasible, then oxygen in the form of iron ore, scale, or gaseous oxygen is added for carbon reduction. Since the steels in this group must normally meet a 0.15 percent maximum carbon specification, the carbon reduction is carried on much more stringently. Tests are taken at frequent intervals as required until the carbon has reached the 0.15 - 0.10 percent level at which point decarburization is stopped by the addition of a reducing mix composed of burnt lime, ferrosilicon, and/or ferrochrome silicon. Degree of deoxidation depends entirely on the aim chemistry of the melt in progress. Furthermore, heats can occasionally melt in with low residual chromium contents due to mix-up in scrap segregation. Heats of this nature need not be reduced but may be slagged off in the normal manner because of lack of availability of chromium in the slag.

When the reducing mix has been dissolved and bath indicates the reduction has been properly carried to completion, the bath is slagged off completely. Necessary molybdenum is added in the form of molybdic oxide briquettes or oxide concentrate and the reducing slag ingredients to make either a lime-alumina or a lime silica mix are added. If the residual silicon content of the bath is determined to be quite low as determined by observation of a spooned sample, at least 0.10-0.15 percent silicon should be added to the bare bath prior to the slag mix. This will aid slag control during the finishing period, greatly enhance the desulphurizing potential of the slag, and speed the fluxing action of the slag.

When the tests are taken, 2–2.5 kg of aluminium per ton is poled to the bath, the temperature is checked with an immersion thermocouple, and any chromium still needed is added to adjust the composition to the low side of the range. Here care should be used in choice of alloys to blend or intermix the additives in such a manner as to attain the lowest alloy cost. It is for this purpose that the extra effort is made in decarburizing earlier in the heat cycle. Upon receipt of chemical analysis results, the final composition is adjusted, the balance of the silicon is added, and the heat is prepared for tap. With increasing chromium content, the alloys in this series are increasingly susceptible to nitrogen contamination. It is necessary, therefore, to exert extreme care in the use of raw materials during the finishing cycles of the heat and also to spoon pour a test ingot of some substantial weight prior to tap to ascertain the tendency of the melt to bleed during solidification. If the test ingot shows signs of being gassy several techniques can be used for treatment including purging with "dry air", or with argon, or addition of titanium or zirconium-bearing alloys which tend to combine with the free nitrogen present, thus lowering the partial pressure of nitrogen in the system.

Steels in this group are occasionally alloyed with amounts of columbium or titanium. The method of addition used is the same as the technique applied in the stabilized grades. When alloying with the titanium it is desirable to decant the reducing slag and add the lime-alumina "neutral" slag before making the ferrotitanium addition. Granulated aluminium is then used in place of ferrosilicon to keep the slag from oxidizing the titanium before it can dissolve in the bath. On all grades in this series it is advantageous to deoxidize with aluminium in the ladle at tap to eliminate the possibility of abnormality of grain size and guarantee a consistency of mechanical properties in the finished products.

These steels at the low-chromium end of the seried tend to behave like most other low-carbon steels, being prone to all the ingot defects that characterize the type. Care must be used in choice of nozzle size, temperature control, etc., to ensure the ingot product will roll without excessive product loss on the mill. The higher chromium and those containing columbium, titanium, higher than normal silicon, etc., will be identical to the ferritic stainless steels. Care must be taken in thermal handling following solidification to prevent the ingots from cracking due to stress build-up. Careful cooling or short track time is most important.

4.4. MANUFACTURE OF BORON STEELS

There has been an increasing interest in the steels containing boron between 0.003 to 0.005%. The presence of boron in such small amounts considerably increases the hardenability of steel. As compared to manganese the effect of boron is about 200 times.

The complexity of production of boron steels lies in the uniform distribution of boron with maximum yield. Boron addition has to be done in such a manner so that the loss of boron is minimum and assimilation in steel is at the guaranteed level. This is carried out by use of silico-calcium injection along with the boron addition. Injection alongwith SiCa is a must to achieve the desired level of boron in steel with uniform distribution. SiCa provides necessary reducing atmosphere to eliminate the loss of boron by oxidation. Besides SiCa, addition of boron is always done with such elements like aluminium and titanium. Silicon and aluminium protect the boron from oxygen and titanium protects it from nitrogen since it is strong nitride former. The amount of titanium added is such that the residual Ti in steel is below 0.04% to achieve good hardenability.

Melting Practice

Good quality sorted scrap is selected with respect to its chemical composition and size so that the opening analysis of melt is of desired level. The meltdown carbon is always kept about 20 points higher than the final carbon at the finishing stage. Carbon boil after meltdown homogenizes the temperature and composition of the melt along with lowering the nitrogen and non-metallic inclusion at desired level in the steel. After the carbon boil subsides, the slag is flushed to remove all oxidising slag. The heat is then blocked by addition of ferrosilicon and tapped at silicon level of about 0.20%. The blocking of heat is done above the tapping temperature to ensure desired uniform temperature throughout the bath.

The heat is tapped in a preheated ladle to minimise the drop in the heat temperature. The ladle is equipped with slide gate mechanism. The heat in the ladle is purged with argon to attain temperature uniformity. The oxygen content of the heat is determined and depending upon the oxygen level necessary amount of aluminium is added.

After the heat is deoxidised to the required level CaSi injection is started. After having ascertained low oxygen levels titanium and boron additions are made along with CaSi injection. The recovery of B and Ti is found to be around 50%. The heat in the ladle is always kept covered with insulating and reducing slag.

The heat is poured in the ingots under the nitrogen gas shroud to avoid oxidation of boron by atmosphere.

FIVE

Melting of Stainless Steels

The category of steel which is termed as stainless steel contains 12 to 30% chromium and not more than 0.12% carbon. A variety of stainless steels contain about 0.07% carbon while a large tonnage of chromium-nickel steels contain 0.03% maximum carbon. Owing to the fact that many a types of stainless steels contain high chromium and low carbon, the melting practice adopted is unique to stainless steels.

First commercial production of stainless steel called 'dilution process' made use of low-carbon ferro-chrome for chromium addition to melt, obtained primarily from the plain carbon steel scrap charge. However, with growing demand of stainless steel and huge quantity of stainless steel scrap accumulating, attempts were made to utilize stainless steel scrap and recover the valuable alloying elements present in scrap to the maximum extent. There are, therefore, basically two types of melting processes employed for melting of stainless steel.

(*i*) Dilution process
(*ii*) Oxidation-reduction process

5.1. DILUTION PROCESS

The process consists of dilution of low-carbon ferrochromium with iron and, when used, with nickel. Neither carbon removal by intentional oxidation in the presence of chromium nor remelting of stainless steel scrap characterizes the process. The dilution process was characterized by a meltdown in the absence of stainless steel scrap. Consequently, stainless steel scrap accumulated in the plants of the producers of low-carbon stainless, and the question of its eventual utilization began to be a very serious problem.

The inevitable pickup of carbon by low-carbon stainless steel scrap, or by low-carbon ferrochromium in the furnace charge, led to two modifications in the production of low-carbon metal. First, stainless scrap was not used in the charge. Second, lump low-carbon ferrochromium was added to the molten bath under conditions carefully controlled to minimize carbon pickup. For the production of stainless steel to 0.20 percent carbon or more, it was always possible to use some stainless scrap in the charge. When necessary even recarburization was employed.

5.1.1. The Meltdown and Slag-off Period

In the production of stainless steel by the dilution process, plain carbon steel scrap is first charged into the furnace along with iron ore and burnt lime (CaO), the exact amount depending on the type of scrap available. After the scrap is substantially melted, the carbon is eliminated by use of the ore and/or oxygen until the carbon content is 0.01 to 0.02 percent. The electrodes are then raised, and the slag is removed as completely as practicable. Care must be taken to have the bath hot enough to prevent freezing during slag-off.

5.1.2. Finishing Period

About 0.5 percent silicon (as ferrosilicon) is added to the bath and a second slag is built up by charging about 40 kg burnt lime and 8 to 10 kg fluorspar per ton of metal. Power is applied to fuse the slag-making constituents and to raise the temperature of the bath to a point short of refractory damage, but high enough to avoid freezing the bath when the ferrochromium is added. Judicious small additions of crushed ferrosilicon or calcium-silicon are also made from time to time to reduce undesirable amounts of iron oxide in the slag, and to form a "white" slag of the disintegrating calcium silicate type. This indicates that the chromium oxide plus iron oxide content of the slag is less than 5 percent. The calculated weight of lump low-carbon ferrochromium (or ferrochromium pellets) is then charged in two or three separate batches. A relatively heavy addition of this ferroalloy is required. Care must be exercised to avoid too much chilling of the metal bath on adding ferrochromium, which at once sinks to the furnace bottom. Melting any large quantity of metal frozen on the bottom is a slow process and demands extended heat time. While chromium is being incorporated in the bath, the slag is kept in shape and very fluid by addition of lime and, if necessary, fluorspar and ferrosilicon. The slag should be maintained in a sufficiently fluid condition to cover the metal directly beneath the electrodes. If the slag becomes stiff enough to be repelled by the force of the arc, thereby exposing bare metal, carbon pickup by the metal bath can take place.

Because of the heavy additions of low-carbon ferro-chromium, the bath must be thoroughly mixed by extensive rabbling, paddling, or induction stirring before samples are taken for preliminary analysis. After the preliminary analysis is known, the final additions are made; the heat is brought to the desired tapping temperature and is tapped in the usual manner.

Chromium in the percentages which characterize stainless steel is powerful deoxidizer, so that neither ferrosilicon nor aluminium is needed to "kill" the metal although when used with manganese, both are effective in altering the shape and distribution of nonmetallic inclusions. In other method ferrochromium is charged onto the slag close to the banks so as to cause the slag to freeze and support the ferrochromium. Three locations are preferred, namely, those portions along the banks which are on the same radii as the electrodes. At these points heat radiation is maximum. The ferro-chromium is heated rather effectively before it sinks below the slag, thereby exerting a less pronounced chilling effect on the metal bath.

5.2. OXIDATION-REDUCTION PROCESS

This process was developed to utilize the accumulating huge quantity of stainless steel scrap. Melting of stainless steel using stainless steel scrap could be accomplished due to the fact that oxidation of carbon is possible at elevated temperatures before complete oxidation of chromium may take place. It has been established that as the temperature increases the carbon content of the bath decreases and the loss of chromium is minimised as shown in Fig. 5.1. Such high temperatures required to keep carbon level low and higher recovery of chromium were not used earlier.

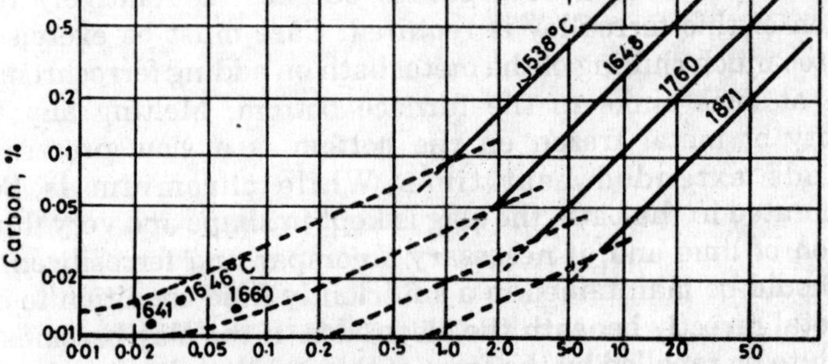

Fig. 5.1. Approximate Cr-C temperature relations for all chromium contents of interest in the melting of chromium steels. (From Hilty, Healy, and Crafts.)

5.2.1. Oxidation of Carbon from Molten Stainless Steel

The oxidation of carbon from a high-chromium steel bath was accomplished by recourse to oxidizing conditions at temperatures considerably higher than has hitherto been employed in electric arc furnace practice. At these higher temperatures carbon could be removed by oxidation from the metal bath down to very low percentages, without complete loss of chromium. As oxidizing agent a very large quantity of iron oxide in the form of roll scale or iron ore was employed as a charge constituent together with stainless steel scrap and high-carbon ferrochromium (4 to 6 percent carbon). To withstand the action of the superheated iron oxide slag and metal, the furnace bottom and side walls to a height above the slag line were constructed of chrome brick. So long as very cheap stainless scrap was available, 70 percent of the total chromium was obtained from such scrap and the remaining 30 percent from high-carbon ferrochromium. The process possesses flexibility with respect to proportions of stainless scrap and high-carbon ferrochromium. This process has come to be commonly known as the "Rustless Process".

Chrome ore was used to repair the bottom and banks of the chrome brick lining used in the Rustless process. It was found from experience that the quantity of chrome ore initially charged in this manner could be increased to advantage to the point where it constituted as major instead of a minor source of chromium. At the same time, bottom life was greatly prolonged. In the Rustless process for melting low-carbon stainless steel, iron oxide also is used as the oxidizing agent. About 30 to 50 percent stainless steel scrap is charged along with high-carbon ferrochromium and/or chrome ore. The chrome ore is first charged evenly onto the banks and against the side walls, using a so-called "dolomite" slinger. The ore becomes thoroughly dried by contact with the highly heated refractories before the charge is melted.

While the carbon is removed from the system as gaseous carbon monoxide and reduces only a minor fraction of the iron oxide of the slag to metallic iron, Chromium on oxidation not only reduces a large quantity of iron oxide to metallic iron but replaces the iron oxide in the slag stoichiometrically with chromic oxide (Cr_2O_3). This chromic oxide increases the melting point of the slag and, if in sufficient quantity, converts the slag to a thick pasty or crusty condition. Unless iron oxide is present in sufficient amount to keep the slag in a reactive condition despite its chromic oxide content, carbon removal by oxidation either ceases or proceeds toward the desired endpoint at a rate too slow to be practical.

In actual practice by the Rustless process the desired low-carbon content of the metal bath is reached by the time the charge has been melted down. This objective is attained by the use of a high power input during the meltdown period and an adequate quantity of iron oxide in the initial charge. The high temperature necessary for the elimination of carbon in the presence of chromium is attained both by the high power input and the oxidation of chromium by the iron oxide charged.

The temperature of the metal bath at the end of the oxidizing period of the Rustless process is too high to permit the use of the usual platinum thermocouple technique. Because of high temperature to use a magnesite or chrome-magnesite furnace bottom have been completely unsuccessful. Such a bottom fails to survive a single heat, whereas the chromite bottom exhibits an unusually long life.

Gaseous oxygen

As oxidising agent, instead of employing large quantities of iron ore or roll scale in the initial charge, pure gaseous oxygen under pressure is injected directly into the metal bath through an iron pipe. By this procedure heat is generated rapidly by oxidation of the metal bath, thereby raising the metal temperature, and, when the necessary high temperature had been reached rapid oxidation of carbon occurs.

The oxygen-lance process is a method of removing carbon by oxidation at high temperature without regard to subsequent reduction of the oxidized chromium from the slag. In other words, the chromium-containing oxidizing slag is removed, and a new calcium silicate slag was formed on the low-carbon bath. The requisite of low-carbon ferrochromium is then added to bring the bath to the desired analysis. At a later date, however, the process was modified to include a reduction period during which oxidized chromium was recovered from the slag by silicon reduction, after which the slag was removed.

The usual practice in the oxidation of carbon with gaseous oxygen is to raise the electrodes and start the oxygen blow somewhat before the heat is completely melted. Oxygen can be introduced either by means of a water-cooled lance, or by the use of black iron pipes. The pipes are sometime coated with a refractory slurry or small sleeve brick to slow down the rate of melting or oxidation of the pipe. Oxygen is introduced through 25-40 mm pipe and is blown at or just below the slag-metal interface. Blowing is continued usually for one hour or until the carbon content has fallen to some what below the value desired in the final product. At this time the temperature, chromium, and carbon content of the bath are related as shown by Fig. 5.2.

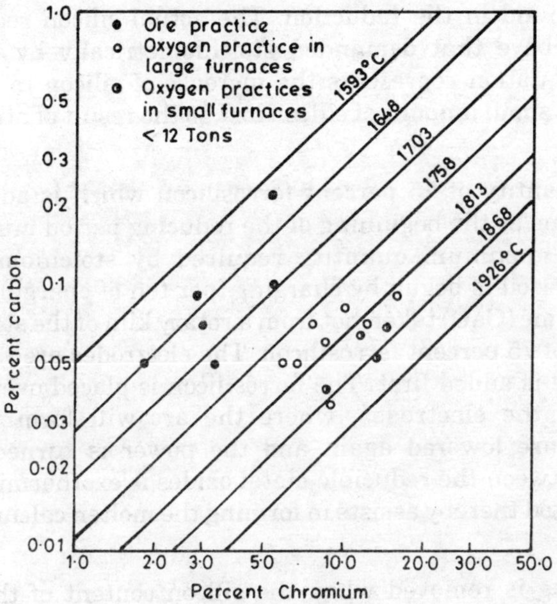

Fig. 5.2. Indicated temperatures at end of oxidizing period in commercial heats made by ore and by oxygen practices.

Reduction of the Oxides of Chromium and Iron

Whether iron oxide or gaseous oxygen is used for the removal of carbon during the oxidizing period, the slag contains relatively large quantities of chromic oxide (Cr_2O_3) at the end of the oxidizing period along with iron oxide. To discard this slag with its potential value as a source of chromium is uneconomical. Therefore, the next step in the melting procedure is to cause these reducible oxides to react with ferrosilicon thereby reducing the chromium and iron from the slag into the bath. The chemical reaction is as follows :

$$FeO \cdot Cr_2O_3 + 2Si = 2\,Cr + Fe + 2\,SiO_2$$

In furnaces lined with basic refractories the silica formed by this reaction remains in the slag and is fluxed with lime and fluorspar to form a dicalcium silicate slag with a ratio of CaO to SiO_2 between 1.5 and 2.0. Crushed ferro-silicon, usually containing 75 percent Si, is added to the oxidizing slag. Some of this ferrosilicon dissolves into the bath. This silicon then reacts at the slag-metal interface to aid in the reduction. The actual silicon requirement over and above that demanded stoichiometrically by the above chemical equation represents the increase of silicon in the metal bath, and a small amount of silicon lost as the result of atmospheric oxidation.

The quantity of 75 percent ferrosilicon which is added along with the lime at the beginning of the reducing period must be held below the maximum quantity required by stoichiometry. The reducing period is begun by charging (per ton of charge) 350 kg of dry burnt lime (CaO) taken hot from a rotary kiln of the storage type and 80 kg of 75 percent ferrosilicon. The electrodes are raised, and all the lime is added first. The ferrosilicon is placed over the lime and under the electrodes, where the arc will form when the electrodes are lowered again and the power is turned on. The reaction between the reducible metal oxides is exothermic, and the heat liberated thereby assists in forming the molten calcium silicate slag.

The slag is removed when the silicon content of the bath is judged to have dropped to about 1 percent and the slag tests show that the chromium oxide in the slag is at least as low as 5 to 6 percent. The furnace is tilted, and the slag is skimmed off until the bath is clean. Immediately after slag-off the finishing slag is built up by charging lime and fluorspar on to the metal bath with whatever small, judicious quantities of crushed ferrosilicon may be called for to maintain optimum slag conditions. The slag is "shaped up" and test samples for preliminary analysis are then taken. When the analyses of these samples are reported by the chemical laboratory, furnace additions are made to bring the metal bath to the desired final composition, the temperature of the metal bath is measured by immersion thermocouple technique and the heat is tapped into the preheated ladle.

Phosphorus is retained almost completely in the system. It becomes an integral part of the ingots, since none is eliminated by the slag. Sulfur is eliminated quite effectively as is shown in Fig. 5.3, and

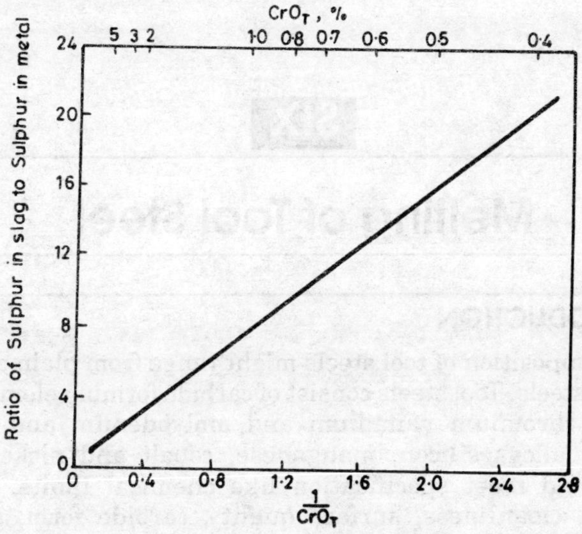

Fig. 5.3. Effect of chromium oxide content on desulfurlzing power of neutral slags in electric furnace.

is accomplished at an adequate CaO/SiO_2 ratio, almost exclusively by the amount of reducible oxide present.

SIX

Melting of Tool Steel

6.1. INTRODUCTION

The composition of tool steels might range from plain carbon to high alloy steels. Tool steels consist of carbide forming elements like tungsten, chromium vanadium and molybdenum and solution hardening alloys silicon, manganese, cobalt and nickel. These steels should meet specification like chemical limits, porosity standards, cleanliness, surface quality, carbide form, size and distribution and hardenability. Raw materials, melting and refining, tapping temperatures, casting temperature, mould and mould steel design, hot tops all have effect of the quality of steel.

6.2. RAW MATERIAL

Scrap used should be low in phosphorus with carbon content low or high. Manganese content should be below 0.4%. Residuals should be as low as possible. The scrap should have a proper size mix to facilitate charging and melting. The scrap should be properly segregated to prevent contamination with undesirable alloy content and for maximum recovery of those with are required in the final steel composition. It is always desirable to use the scrap of similar composition as that of the steel to be made to get most from the furnace and minimum melting and refining time by suitably modifying the practice.

6.3. TYPE OF MELTING PRACTICE

Depending upon the composition of charge and the type of steel to be made, any of the following two methods of the manufacture may be used for tool steels.

1. Single Slag Process : Utilizes an oxidising slag during melt down that is converted to a reducing slag during the finishing phase of the heat. This procedure is used when the charge contains valuable alloys that would be lost in a two-slag process.

2. **The two-slag process** : Utilizes first an oxidizing slag that is raked off after the desired oxidizing reactions are completed and is followed by a reducing slag.

6.3.1. Single-Slag Practice

When quantities of alloy-bearing tool steel scrap and ferroalloys are used in the charge, the single-slag practice makes possible almost complete (97 percent or higher) recovery of such elements as tungsten, molybdenum, chromium, vanadium, cobalt, etc.

The slag-making materials, pebble lime or limestone and spar, can be put in as soon as a stable arc is obtained or with the charge if desired. Enough lime and spar is added to give a working slag of 60/65 percent CaO that will be approximately 8 cm in depth. On melting down with single-slag practice, a moderate oxidizing condition can be established through the medium of carbon steel mill scale or other metallic oxides added with the charge which will cause or be sufficient to remove residual silicon and minimize the danger of hydrogen pickup. An oxygen lance can be used in place of solid oxides to give the desired reaction.

After clear melting, stirring, and sampling, the oxides in the slag can be quickly eliminated by treating the slag with 6 mm and down coke, calcium carbide or 10 mm × 12 mesh 75 per cent ferrosilicon, either single or in combination at 10 min intervals in heats ranging from 3 to 10 tons. These reducing agents are spread lightly over the slag in quantities of 5 to 15 kg at a time, depending upon the size of the heat. As the heat size increases, the size of these additions increase. At all times proper basicity (CaO/SiO_2 ratio of approximately 3/1) should be maintained. A slag too high in SiO_2 will be thin, flat, and will cut the banks, the arc reflection will cause overheating of the roof and side walls, and the metal will pick up both carbon and silicon. On the other hand, too high a CaO content may result in a "foaming" slag that may reach a depth of 30 cm to 50 cm, particularly if the slag is high in WO_3 or a pasty slag that will not react with the metal.

The slag colour is indicative of the degree of reduction of metallic oxide. A pure white slag is not always free of oxides whereas a gray/blue coloured slag will be free of oxides to have high affinity for sulphur. For example, a black slag in a heat of high-speed steel contains WO_3 up to 15.00 percent, Cr_2O_3 up to 4.00 per cent, and V_2O_5 up to 2.00 per cent immediately after melting down. A light brown slag on the same type of heat will be free of WO_3 but will contain Cr_2O_3 and V_2O_5. A gray-coloured high-lime carbide slag will be entirely free of these oxides and is usually the ideal slag condition. High-lime slags, light gray in color, are highly reducing, free of undesirable metallic oxides, and have a high affinity for sulfur. However, dark gray/blue slags have excessive carbon and may cause carbon pickup, particularly in steels under 0.75 per cent carbon.

6.3.2. Double-Slag Practice

When the charge is free of valuable alloying elements a two-slag procedure is frequently used. The first slag is an oxidizing slag. The oxidizing slag after clear melting will contain approximately 60 per cent CaO and 15 per cent Fe_2O_3. This will produce a suitable boil and the reaction will remove phosphorus, manganese, and residual silicon along with some carbon, chromium, and molybdenum. It will also lower the hydrogen and nitrogen content of the metal. A slag depth of approximately 8 cm is desirable. The degree of oxidation necessary is usually obtained within 15 to 30 min after clear melting. After the desired reactions have been obtained, the oxidizing slag is raked off and a second slag of lime and spar is added. The reducing agents used are usually fine coke 6 mm and down, calcium carbide of 10 mm and 12 mesh 75 per cent ferrosilicon. These may be added single or in combination. From this point on furnace manipulation is the same as that used with single slag practice. A two-slag heat is also used when large quantities of scheelite ore are used.

6.3.3. Composition Control

By raising the metal temperatures to tapping temperature level as quickly as possible, chemical reactions and solutions of subsequent alloy additions will be accelerated. In all instances vigorous stirring of the slag and metal of approximately 10 min intervals promotes rapid completion of slag-metal reactions.

The speed of reduction of the metallic oxides is in the following order : WO_3, MoO_3, Cr_2O_3 and V_2O_5. The speed of these reaction rates in a tungsten bearing hot work steel is illustrated in Fig. 6.1. Bath agitation is equally important in expediting the solution of the ferroalloys. This is particularly true with ferro-tungsten as its solution rate is low due to its high melting point. Vigorous stirring also insures the obtaining of representative samples for chemical analysis taken early in the heat. This is particularly true of carbon preliminaries. Because of the narrow carbon range in most tool steels the charge is calculated to melt down about 10 points below the desired carbon range.

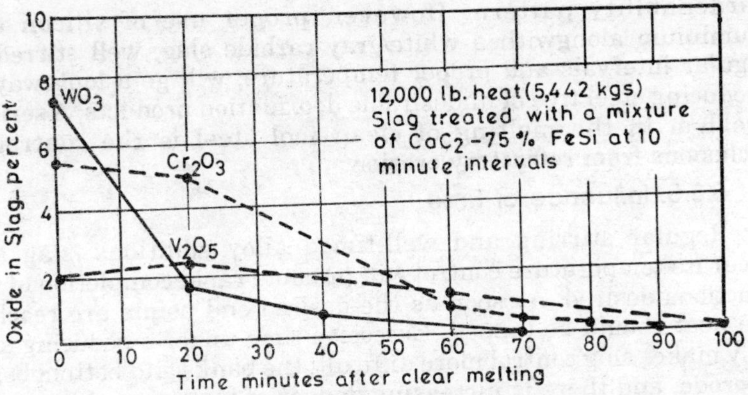

Fig. 6.1. Rate of metallic oxides reduction

The test sample can be taken with a spoon. The metal may be cast into the form of a test block. Usually all ferroalloys, including those containing silicon and manganese, are added in the furnace. Aluminium and the special deoxidizers may be added either in the furnace or in the ladle.

After the carbon preliminaries have been received, carbon is added by the use of low-phosphorus/low-silicon pig iron or carbon blocks. Both of these have their shortcomings in tool steel practice. If a pig iron addition is made, compensating alloys must also be added to offset the alloy dilution. The problem with carbon blocks is their slow rate of solution. In steels containing chromium the carbon addition can be quickly and accurately made by utilizing the carbon contained in the high-carbon ferro-chromium. Additions of carbon by injection may also be used.

Carbon control of tool steel is critical, particularly in those grades where a range of 0.06 per cent is all that is allowed. For this reason quick solution of the carbon additive is desired. Carbon additions to the slag must be carefully controlled since a slag too high in carbon can cause carbon pickup in the metal. The electrode controls must be in good adjustment to prevent "hunting" since this could cause dipping of the electrodes, particularly on low voltage, resulting in carbon pickup.

6.3.4. Deoxidation

In tool steel practice all steels must be completely killed and be as free as possible from centre porosity, large dendrite patterns, centre segregation and undesirable inclusions. Many tool steels have a silicon range of 0.15 to 0.35. Silicon is supplemented with aluminium in quantities up to 1 kg per ton to promote through deoxidation and for grain size control. Supplemented additions of aluminium could change the inclusion pattern as well as the hardenability pattern. However, proper use of silicon and aluminium alongwith a white/gray carbide slag, well stirred at regular intervals and proper temperature, will go a long way in producing steel free of undesirable deoxidation products. A serious problem in the melting of clean tool steel is the control of inclusions from refractory erosion.

6.3.5. Influence of Time

Regular stirring and well-timed alloy additions, slag and accurate temperature control will promote rapid completion of the reactions desired. As soon as the desired end points are reached the heat should be tapped. Excessive time under a reducing slag only makes slag control more difficult; the banks and bottom begin to erode, and there is increasing danger of hydrogen pickup.

6.3.6. Temperature Control

Each type of tool steel has its "best" pouring temperature and is usually limited to 10°C range. The tapping temperature of tool steel is closely related to carbon content. For example, tool steels ranging from 0.20/0.60 per cent carbon, a tapping temperature of 1590/1620°C may be used ; for 0.60/1.20 per cent carbon steels, 1566/1590°C ; and for 1.20/2.25 per cent carbon steels, 1480/1540°C. There is usually a temperature drop of 40/80°C between tap and the first ladle opening depending on heat size and ladle preheat temperature. The metal temperature to the mould for 1 per cent carbon steels and high-speed steels is usually in the range of 1480/1510°C. Low-carbon hot work steels are usually poured in the range of 1510/1540°C.

6.4. LADLE AND NOZZLE PRACTICE

To maintain uniform temperature throughout the casting of the ingots, the ladle should be thoroughly preheated to 650/750°C. This degree of preheating is also helpful in minimizing the danger of hydrogen absorption from moisture in the ladles. Each type of tool steel has its own characteristic viscosity and best pouring temperature. Careful selection of nozzle size for a given composition and set of ingots sizes is an effective means of controlling the smoothness of the ingot surfaces and helps to keep pitting, splash, sub-surface inclusions, and sub-surface blowholes to a minimum.

Tool steel ingots range in size from 150 to 10,000 kg. Nozzles usually range in size from 25-50 mm in diameter for ingots up to 5000 kg. Larger ingots require larger nozzles depending on the pouring characteristics of the metal. On a 1 per cent carbon steel heat a group of 25 cm square ingots can be poured with a 25-30 mm diam nozzle. On the more fluid high-carbon compositions the nozzle size can be reduced somewhat, while on low carbon heats, particularly those below 0.50 per cent carbon, the nozzle diameter must be increased.

The pouring temperature also influences the depth of the columnar grains, centre soundness, segregation, the susceptibility to hanger cracks and inverse bottom ruptures. Forgeability of all tool steel is closely related to ingot surface conditions and the internal ingot structure which, in turn, is related to tapping temperature, pouring temperature, and speed of filling the molds.

6.5. MOULDS

Tool steels are poured in big end-up moulds using either plug bottom, or open bottom, set on cast iron or copper stools. Mould wall thickness, taper and length of ingot in relation to its cross section influence centre soundness, freedom from shrinkage cavities, and centre segregation. A general rule is that as length of the ingot increases, the danger of centre weakness and centre segregation also increases due to the possibility of "bridging" during solidification. The use of generous mould taper is an effective means of promoting sound ingot centres.

6.6. HOT TOPPING

All tool steel ingots are fitted with a hot top to promote centre soundness and freedom from centre segregation. Hot tops vary widely in design; cold tops, preheated tops, electric and gas heated tops, tops made of insulating material, and those utilizing exothermic mixture or linings are used. Anti-piping compounds may also be used. There is no general rule for hot-top practice. A given hot-top practice is based on individual experience with many compositions, pouring temperature, and mould design.

6.6.1. Mould Wash

In tool steel ingot practice a number of mould coatings are employed which include resin smoke; powdered pitch; mixtures of aluminium, graphite, lampblack, and kerosene silicones; also special manufactured preparations. In all instances every effort is made to obtain a smooth ingot surface free of tears, pits, sub-surface blowholes, scabs, rolls, and splash. The moulds should be preheated to 90/150°C before using.

6.6.2. Stripping Practice

Tool steel ingots should be stripped as soon as solidification is complete. A holding time in the mold is approximately 3 min per 50 kg is necessary. After stripping, the forgeability of high-alloy ingots can be improved by very slow cooling through the 700/250°C range.

SEVEN

Casting Pit Practice and Ingot Defects

7.1. GENERAL

There are two major kinds of casting pits employed in practice, namely (*a*) Sunken pit design and (*b*) Moving pit in the form of "open bogies". Modern steel making shops need to cast as much as 200 to 300 t of liquid steel/hr. and for this purpose, only "Bogic Casting System" is suitable. Sanken pits are used only in small shops having EAFs these days. The entire teeming equipment includes : a spout (launder), teaming ladle, Tundish, ingot moulds and continuous casting machines.

7.2. TEEMING LADLE

The refractory lining may be bricked, rammed or sprayed ; bricking however being most common. Generally, high grade fire bricks are commonly used to line the ladle. Lining is carried out in two parts (bottom and side walls). The bottom is made sloping towards the nozzle. Several brick shapes are used *e.g.* normal, rectangular, tapered or curved with interlocking arrangement. Good quality cement must be used as a mortar while laying the bricks. The joints between the bottom and the side wall happens to be most vulnerable spot for failure of lining. Since more than one layer is used in lining the ladle, the joints must be staggered. Also the joint must be tight fit to prevent metal penetration. The ladle wall is lined in three courses. A thin layer of ganister next to the steel shell is followed by a layer of "safety lining" and then the "working lining". The working lining at the normal metal surface level is made of high Al_2O_3 bricks rather than usual firebricks to resist the attack of Oxidising Slag, a little of which always comes into the ladle from furnace. Recently, spraying of refractory material to build up the ladle lining has been developed. Similarly; ramming silicious lining has been used and found satisfactory. The ordinary fire clay brick lined ladle still offers the best choice, since ladles give a life of nearly 10 to 15 heats and the refractory consumption amounts to 3 to 5 Kgs/t depending upon the quality of refractories and the process of steel treatment in the ladle.

A freshly lined ladle is thoroughly dried before use to achieve good life and to eliminate the possibility of hydrogen pick up by the metal. The ladle is dried and preheated by the help of oil burners.

The nozzle diameter decides the time required to teem a given size mould. The initial teeming temperature be such that adequate superheat is still left at the time of teeming the last mould. If teeming with single nozzle prolong the teeming period beyond nearly 45 minutes, two-nozzle ladles are used to teem two moulds at a time. The interrelationships of mould size, nozzle dia, number of nozzles, initial teeming temperature etc. are established to fix up the correct teeming practice in relation to the ladle size, which is generally equal to the furnace capacity.

The metal flow is principally controlled by the nozzle shape and size (diameter) rather than the stopper position. The nozzle should last without excessive erosion and appreciable change in diameter until all the metal in the ladle is teemed. It is essential to teem the ingot mould with a pipe like stream of metal flowing at a constant rate to minimise possible ingot

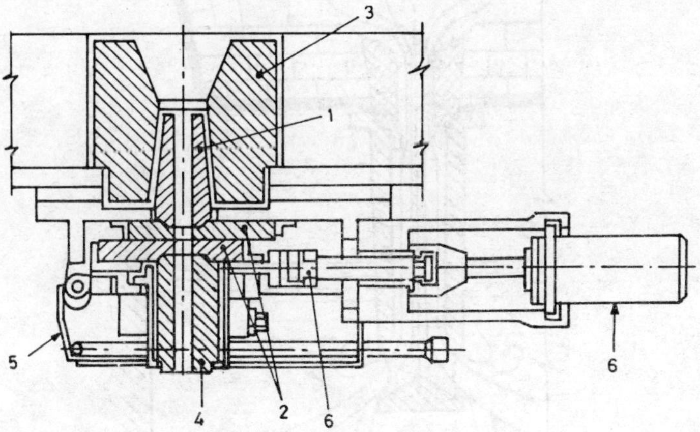

(*a*) gate type, 1 - teeming sleeve of ladle; 2 - plates; 3 - shaped brick; 4 - outer teeming sleeve; 5 - protective metal casing; 6 - plate-moving device

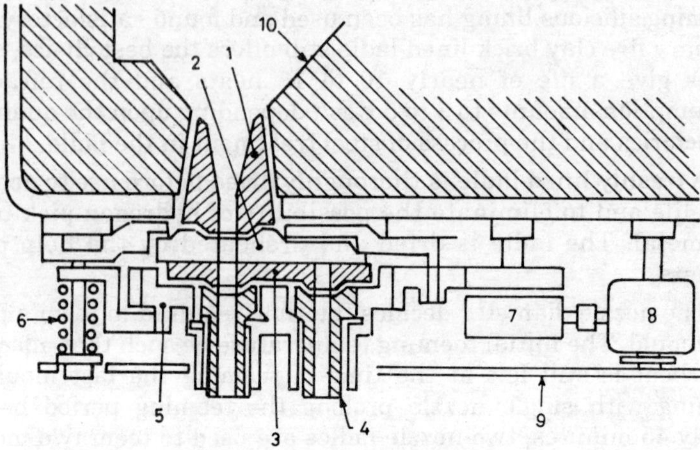

(*b*) rotary type, 1 - upper sleeve; 2 - fixed plate; 3 - rotating plate; 4 - collector; 5 - rotor; 6 - springs; 7 - reducer; 8 - electric motor; 9 - protective casting; 10 - intermediate ladle lining.

Fig. 7.1. Slide Gate Systems.

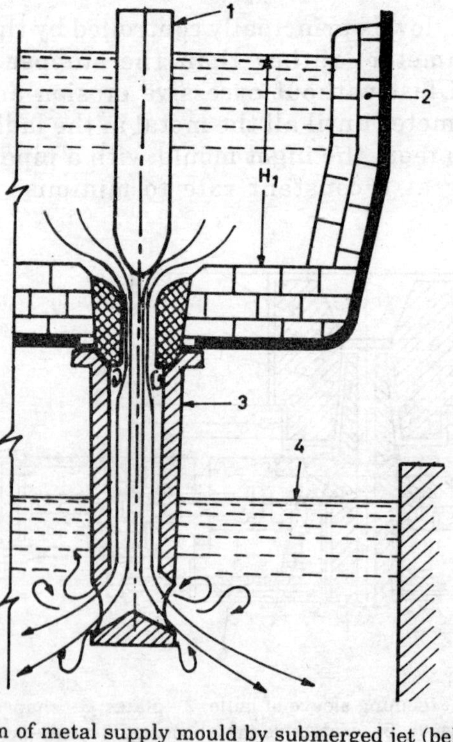

Fig. 7.2. Diagram of metal supply mould by submerged jet (below metal level)
1 - stopper; 2 - ladle; 3 - long teeming sleeve; 4 - molten metal level in mould.

defects. In the modern practice, smooth or flawless teeming is ensured by the use of "slide-gate" systems (Fig. 7.1). Slide gate systems commonly employed are of two types namely (*a*) gate type stoppers and (*b*) rotary disc stoppers. In the latter case, a number of holes of various diameters may be provided in the disc. Disc stoppers make it possible to maintain nearly a constant flow rate by using holes of increasing diameter as the ladle is being progressively emptied. Also, in the modern times, nozzles having an increased length have been used so that liquid metal passes through them without contacting the surrounding air. Thus Secondary oxidation of liquid metal stream is avoided. Such nozzles are also termed as "Sub-entry nozzles" or "submerged nozzles". (Fig. 7.2)

7.3. THE INGOT MOULD

The moulds are themselves massive castings of more uniform shapes with cross sections like square, rectangular, round, polygonal etc. Ingot with square cross-section is used for rolling into billets, rails, structural sections, girders, merchant products etc. Rectangular shape (slab) is used for producing flat products like sheets, strips and plates. Round ingots are not used much but can be used for tube making. Polygonal ingots are used for tyres, wheels, forgings etc. A 3-5 t ingot of mild steel was once commonly used for rolling but it has now been replaced by 5-20 t ingots. Some of the alloysteel ingots for rolling and forging may be only few hundred kgs. Casting of 300-400 t forging ingot is not an exception any more.

The "forging ingot" is different from a normal rolling ingot. It has two parts : the main portion which is actually forged is called "Chill" and the portion which is to be held in manipulator grip (for small forgings) or inporter bar (for large forgings) is known as "head".

The inner walls of the mould may be plane, cambered, corrugated or fluted as shown in Fig. 7.3. The plane walled mould has minimum of surface per unit weight of ingot it produces and hence the cooling rate of steel is least. In corrugated and fluted ingot mould surface, faster cooling is promoted and therefore, the initial solidified skin thickness increases. Corrugated moulds are more commonly used than fluted ones. The height of corrugation varies from 10-20 mm and the wavelength of corrugated shape varies between 75-150 mm.

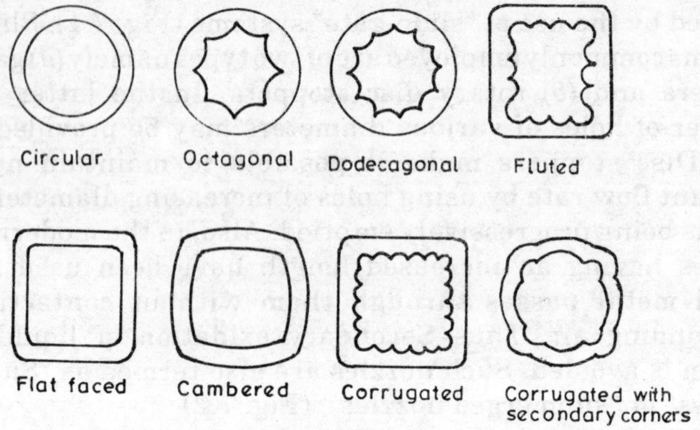

Fig. 7.3. Different designs of inner walls of the mould.

The mould in general should be massive enough to possess sufficient "heat capacity" so that the heat evolved during the solidification does not heat the mould. The massiveness is measured in terms of mould wall thickness. Small moulds may have walls as thin as 8 cms which increases upto 30 cms for big moulds. Since mould consumption is an important cost item in the production of ingots, the mould wall thickness should be minimum commensurate with adequate mould life. In general, the ratio of mould weight to ingot weight is 1.1 to 1.5 for open top moulds and the corresponding values are between 1.2 to 2.0 for hot-topped forging ingots. The height of mould in general varies between 1.5 to 2.5 meters. Rimming steels rise during solidification by about 10 to 25 cms and hence this much extra height of the mould is left after teeming is over.

The design of mould has a profound influence on the solidification characteristics of the ingot. Since the ingot shrinks on solidification, an extra allowance of 2% in the mould cross section is allowed in the design. Similarly, 1-2% taper is provided in the inner walls of the mould. Also the corner radius is important since there the cooling rate is much higher compared to the flat walls. Usually a corner radius varying between 1.5" (38.1 mm) to 2.0" (50.8 mm) is provided in practice.

7.3.1. Mould Material

Cast iron has good thermal shock resistance and is hard enough to resist grazing. Earlier moulds were made of Cupola metal but now these are increasingly being made of blast-furnace hot metal after careful control of its composition and temperature. These are cheaper and low in sulphur content than those of cupola metal.

Since phosphorus content of metal affects mould life, it is usually kept less than 0.05%. Higher P contents reduce the resistance of mould to high temperature conditions, and therefore promote premature failure. Typical analysis of mould material is given in the following table :

S. No.	*Total Carbon*	*Silicon*	*Sulphur*	*Phosphorus*	*Manganese*
1.	3.76	1.41	0.047	0.050	0.59
2.	3.60	1.75	0.050	0.048	0.48
3.	3.74	2.06	0.061	0.037	0.48
4.	3.77	1.95	0.049	0.046	0.76

The Russion practice mentions the composition of ingot mould as follows :

$$TC = 3.8\%,\quad Si = 1.8\%,\quad Mn = 0.9\%,\quad P = 0.2\%$$

and Sulphur content as low as possible. A certain quantity of chromium in pig iron increases the life of moulds.

Cast iron has a different coefficient of thermal expansion from that of steel and therefore the mould shrinks away from the ingot face on solidification separating the two readily. The sticking of ingot to the mould is further minimised by the application of mould dressings.

7.3.2. Mould Life and Causes of Failure

Mould consumption is an important item in the overall cost of ingot production. The total requirement of moulds is generally about 2% of the weight of ingot production. In means the mould can be used for about 50 pours, assuming that the weight of the mould is nearly equal to the weight of the ingot. In very efficient plants, the mould is used for as much as 100 pours with usual intermediate repairs.

Moulds fail in service primarily due to (*a*) Fire cracking or major cracking and (*b*) grazing or minor cracking of the mould chamber walls. Major cracks are usually initiated near the bottom and spread upwards till the entire surface may show a network of these cracks. Some of these cracks may be deep enough to obstruct smooth stripping of the mould. It then becomes necessary to discard the mould. Fire cracking also affects the surface quality of ingots and even though mould stripping may not be a problem, the mould has to be scrapped to maintain good surface quality of the product. This is more true in case of hot topped moulds.

Defective practices such as late stripping, too short an interval between use leading to inadequate cooling between two consecutive pours, teeming of ingots at high temperature, pouring of ingots in close proximity with each other, fast teeming etc. contribute to fire-cracking of moulds. Accidental bumping and dropping of the mould may aggravate the problem of cracking leading to early discarding of mould. Alloy steels may be teemed at relatively higher temperature compared with mild steels and this again will affect the mould life.

Grazing or fine cracking results in the working face of the mould becoming covered with a network of fine cracks, especially in the middle of each working face where the temperature remains high for a longer period of time. When grazing starts, it is cumulative and the network on the mould surface develops so much that the surface of ingot is impaired. At this stage, the mould is rejected.

Cast iron has maximum resistance to grazing, when it is mechanically hard *i.e.* pearlitic or cementitic base matrix, close grained and low in C and Si and hence contains a small amount of graphite, which should be in a fine state of division. Growth is greater when C and/or Si is high and when graphite is greater in amount and coarser. But iron has maximum resistance to thermal shock when it is mechanically soft and ferritic.

7.3.3. Mould Conditioning

Following methods are commonly used :

(*i*) Cooling of molds by water sprayers

(*ii*) Assembly

(*iii*) Mold washing, generally with tar or pitch (dressing)

After spray cooling, tar or pitch coating is applied by portable sprayers which are lowered in the mold from above. It is best coated when the mold surface is > 95°C and < 230°C for tar and pitch coating respectively.

As the molds get older, these temperature are increased to 150°C and 230°C, so that excess volatile matter does not get trapped in surface cracks. Carbonaceous coating prevents splashed steel from adhering the walls and due to its gasification, pushes the metal away from the walls as it rises in the mold. The thickness of the coating should be adequate to generate necessary gas pressure to push steel away from the mold-wall so that it produces cleaner and smoother skin.

Common mold dressing are : anhydrous tar (dehydrated tar)
Bituminous paint

Graphite wash

Montan Wax

Pitch

Lime wash

Al paint (Water and spirit based)

Volatile mold dressing *e.g.* dehydrated tar derivatives are effectively used to minimise

— lapping — ripping

— splashes and — double skins

Occurrence of lapping is prevented by the use of volatile dressing when deliberately lower pouring speeds and lower temperatures are used to form thick initial skin.

Still Non-active dressing *e.g.* soot forming, reeking compounds and various graphite coatings still find favour. These provide Non-Wet Coating which discourages the adherence of splashes and laps.

Aluminium coatings are often used where a possible pick-up from the coating in the ingot surfaces is considered undesirable.

Volatile dressings and their interaction at the metal/mold interface helps to reject oxide and refractory inclusions to the top surface of ingot. (These refractory and oxide inclusions are termed as teeming scum.).

An ideal dressing must have the following properties :

(*i*) Prevention of atmospheric oxidation of the steel and its alloys.

(*ii*) Should not contain V.M. which can give rise to blow holes by becoming trapped below the rising liquid steel-level.

(*iii*) Prevent "teeming scum" from becoming trapped in the ingot surface.

A material placed in intimate contact with the total surface of the rising steel has the best chance of preventing surface oxidation.

Recently free flowing powder materials based on fly-ash have been found to be effective. Its quantity employed is approx. @ 31b/t of steel. They partially fuse at metal/powder interface providing a perfect means of mopping up and preventing scum formation.

The insulating nature of such top-covers prevents heat losses and provides protection from re-oxidation. So, the absence of volatiles, completes the three conditions mentioned above and therefore it produces best ingots. But, there should not be any turbulence in slag cover, during uphill teeming, so teeming speeds are lower to ≈ 8″/min. The surface condition of bottom six inches of ingot is usually prone to a certain degree of slag entrapment. This is due to initial turbulence in bottom pouring.

Presently, synthetic formulations are being tried.

7.4. HOT TOPS OR FEEDER HEADS

In early days of "heading ingots" reservoir of liquid metal was maintained in a "head box", lined with refractory brick or clay, the contained metal being covered with powders like vermiculite, kieselguhr, charcoal or coke-dust which improved yields. Boxes used many times with minor repair. But

(*i*) Refractory lining extracts considerable heat to satisfy their own thermal capacities.

(*ii*) When hot, both refractory and inferior feeding compounds used are poor insulators and much useful heat is lost from the head by conduction.

(*iii*) Also hot tops are cumbersome to use and apply.

Recently - expendable exothermic linings for ingot heads have been used. Main requirements :

(*i*) It burns raising its own temperature to that of molten steel or above, just before the head is filled. So, the head lining does not extract heat from the head-metal and on the contrary contributes heat to the hot-top when high cal-value exothermics are employed.

But the exothermic residues after combustion have relatively poor insulating properties and so, the initial benefits are rapidly lost by conduction of heat to the head-box container or to the atmosphere.

(*ii*) Duplex form is more efficient, where head is made from a good refractory or insulating material faced with exothermic to prevent early chilling of head metal. Costs reduced by using cheap backing material *e.g.* sand instead of part of the exothermic, but sand is not a good insulator and such heads are unsatisfactory on large molds.

Ingots upto 250 mm sq. could be headed satisfactorily with simple exothermic shapes and larger ingots with exothermic faced insulating material. Economics is poor on low grade steels.

The main problem is with medium and large ingots, where head metal must be held liquid for over 3 hrs.

If the high cost of exothermic was to be avoided, the thermal capacity of head-lining material must be low and the lining themselves should be easily handelable and robust. Based on these ideas was born the "super-insulator" PROFAX. (Fig. 7.4)

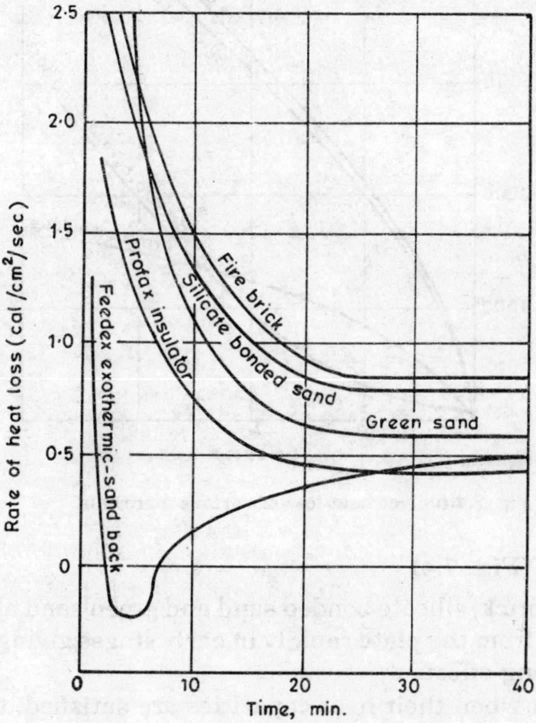

Fig. 7.4(a). Rate of heat loss of various materials.

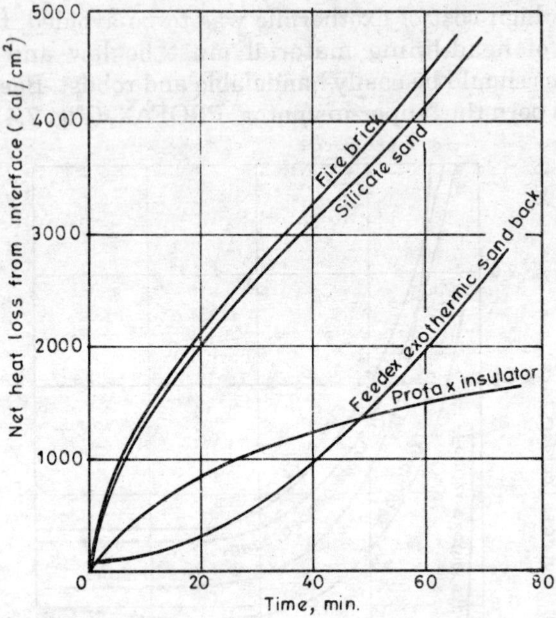

Fig. 7.4(*b*). Net heat loss of various materials.

Conclusions (Fig. 7.4)

1. Firebrick, silicate bonded sand and green sand all extract heat from the plate rapidly in early stages giving a initial chilling effect.
2. Even when their heat capacities are satisfied, they conduct heat to their environs rapidly.
   "Profex" super insulator with low thermal capacity gives some initial chilling effect, but once its temperature is raised to that of the metal, further heat loss is slow.
3. For ingots, which take > 50 mits to solidify, 'profex' can be expected to give better feed patterns and better yield than even the exothermics.
   Exothermic heads are extremely important in small ingots of special steels. And exothermics are frequently unsatisfactory on large ingots due to poor insulating props of their residues. So upto 50 mits (from the time of teeming), exothermics are better, and beyond 50 mits, the super insulator will be better. A 335 mm square (14" square) ingot will solidify in 50 mits.

For larger ingots :

(*a*) Duplex exothermic head can be used, provided the thickness of exothermic is increased or the insulating props of backing material is improved.
(*b*) But this kind of duplex system is more expensive than super insulating one.
(*c*) Duplex heads have been used upto 20t.
(*d*) For ingots > 401 mm square, better use super insulator. Upto 250 mm square ingots, use duplex system.

Solidification Time of Ingots

$$T = \left[\frac{D'}{2}\right]^2 \text{ or } = \left[\frac{D''}{5}\right]^2$$

T = Solidification time, minute

D' = Ingot section, inches

D'' = Ingot section,.cms.

Hot-top Volumes and feeder Ratios

$$\% \text{ Net feeder ratio} = \frac{\text{Ingot head volume}}{\text{Ingot body volume}} \times 100$$

$$\% \text{ Gross feeder ratio} = \frac{\text{Ingot head volume}}{\text{Ingot body vol.} + \text{Ingot head vol.}} \times 100$$

for small molds	———	gross feeder ratio
(high ht : breadth ratio)		4%

Typical gross feeder ratios with efficient

exothermic hot tops	4–10%
with refractory hot-tops	13–18% for mill ingots
With Profax (super insulator)	9–12%

Yield Consideration

If a refractory hot top using say 16% gross feeder ration can be replaced by profax hot-top of 11% ratio, a saving of 5% of metal is achieved provided both the ingots are of similar quality below the shoulder.

Since segregation pattern is better with profax than with refractory, the steel below the shoulder is of better quality. On smaller ingots, a 16% gross feed vol may be replaced by 8% exothermic yield resulting in a saving of 8% metal volume.

Antipiping Compounds (Fig. 7.5)

25 to 30% of total heat losses from a brick lined hot-top without feeding compound cover are by surface radiation.

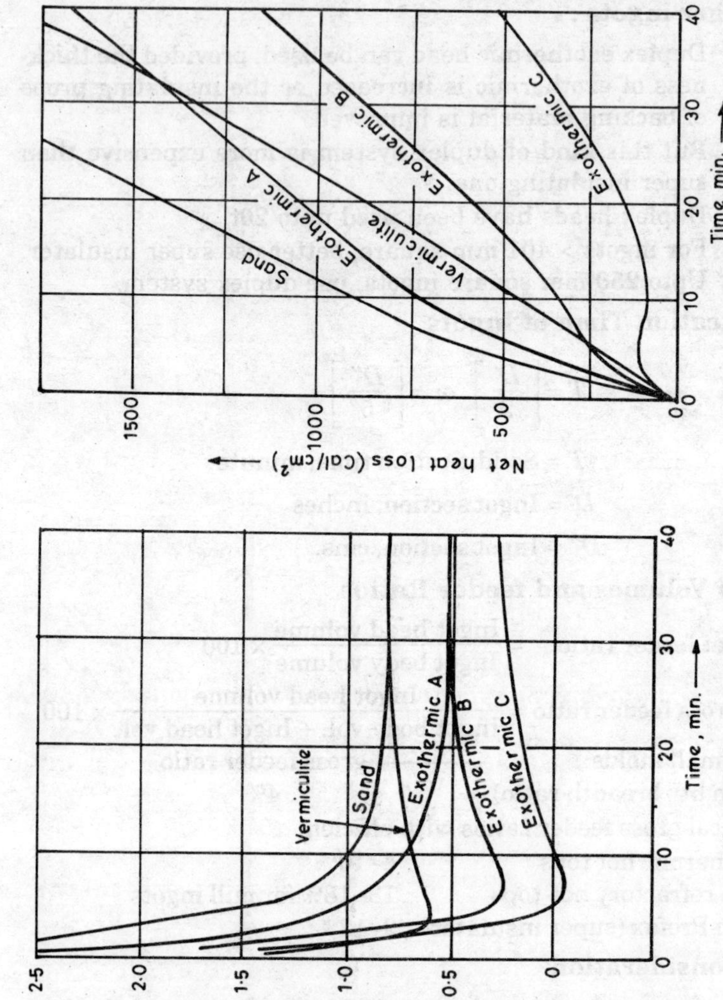

(a) Thermal characteristics of a variety of feeding compounds.

(b) Thermal characteristics of a variety of feeding compounds.

Fig. 7.5.

In case of exothermic or super insulator lined hot-top, the loss of heat is ≈ 40%. Before the use of exothermics, vermiculite (low heat capacity material) was used to cover the top surface of metal in brick and clay hot tops.

Conclusions (Fig. 7.5)

1. Since none of the compounds is preignited, each extracts some heat from the plate.

Sand and exothermic powder A chill the plate heavily until they have reached the temperature of liquid steel.

2. Exfoliated vermiculite, because of low density and conductivity, chills only slightly and prevents heat losses in the early stages. Then vermiculite starts to fuse and its conductivity increases. After 30 mits, the cover is no better than sand.
3. Exothermic powders *B* and *C*, on burning, rapidly lower the chill effect and in case of *C* the compound actually contributes heat. After burning, the powder residue is relatively insulating and heat is lost only slowly.

On smaller hot tops - the powder should be sensitive and give a high and rapid heat output to prevent initial chilling of head metal and to contribute heat to metal. With larger ingots - the exothermic compounds on burning should satisfy the heat capacity of the covering layer to prevent heat loss from the head metal. So, residues should be highly insulating.

7.5. INGOT DEFECTS AND REMEDIES

7.5.1. Introduction

An effort is always made to produce ingots which are free from common defects such as physical and chemical inhomogeneity, ingotism, segregation and non metallic inclusions, shrinkage cavities, blow holes, internal fissures, longitudinal and transverse cracks and numerous other defects and their possible remedies are discussed in the following paragraphs.

7.5.2. Chemical Inhomogenity in Ingots (Zonal Segregation)

Zonal Segregation

Pure metal separate out first from the liquid known as "Selective Solidification". When the cooling rate is fast (early stages of solidification), the effect of segregation is only weekly pronounced. When cooling rate decreases, segregation becomes more noticeable. At this stage "dendritic segregation" occurs. Segregates are lighter than pure iron and float to the top portion of ingot. This difference in chemical composition of metal in different zones of ingot is termed as "zonal segregation".

Table 1. Variations of Characteristics of an Alloy on Addition of Various Elements to Molten Iron

Element	Lowering of melting point on adding 1% of element, °C	Segregation coefficients		
		K	*1 – K*	
		δ-iron	*δ-iron*	*γ-iron*
Sulphur	30	0.02	0.98	0.98
Oxygen	65	0.02	0.98	0.98
Boron	90	0.05	0.95	0.96
Phosphorus	28	0.13	0.87	0.94
Carbon	73	0.13	0.87	0.64
Titanium	18	0.14	0.86	0.93
Nitrogen	—	0.28	0.72	0.46
Hydrogen	—	0.32	0.68	0.55
Copper	7.0	0.56	0.44	0.12
Silicon	12	0.66	0.34	0.50
Molybdenum	2	0.80	0.20	0.40
Nickel	3.5	0.80	0.20	0.05
Manganese	3	0.84	0.16	0.05
Cobalt	1.8	0.90	0.10	0.05
Vanadium	2	0.90	0.10	—
Aluminium	3	0.92	0.08	—
Tungsten	< 1	0.95	0.05	0.50
Chromium	1	0.95	0.05	0.15

Gas evolution takes place upon solidification because of decreased solubility of gases in metal with temperature. These gas bubbles interact with segregating impurities. In most cases, the interfacial tension on gas (gas bubble) - inclusion boundaries, σg-in is small than that on metal-inclusion boundaries, σ_{M-in}. So, σg-in < σ_{M-in}. So, inclusion is more easily wetted by gas than by metal. So, rising gas bubbles carry along various segregates and impurities. By this mechanism, bands of impurities are formed in the ingot ($\wedge$ - shaped segregation or 'whiskers'). So the shape of bands shown in Fig. 7.6(*a*) depend on the correlation between the process of solidification, gas evolution and metal motion in the mould.

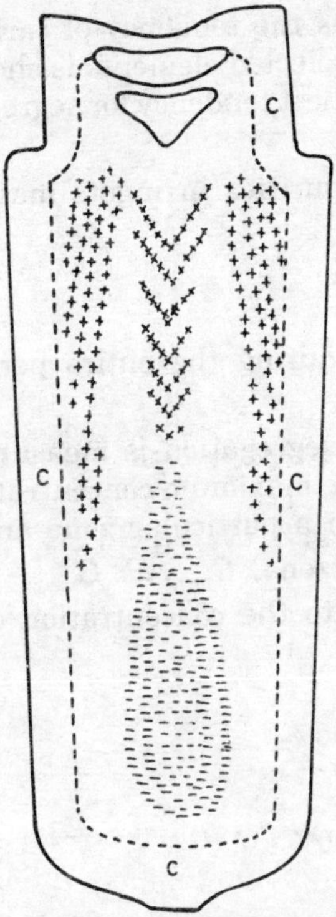

Fig. 7.6. (*a*) Zones of segregation.

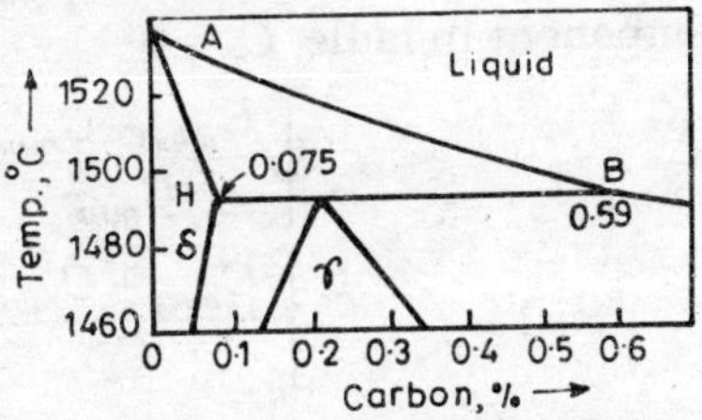

Fig. 7.6. (*b*) Top portion of Fe-C constitutional diagram.

The liability of a particular element to segregation is characterised by the segregation coefficient, *K*.

$$K = \frac{\text{Maximum solubility of an element in } \delta \text{ or } \gamma \text{ iron}}{\text{Concentration of that element in the last droplet of solidified metal}}$$

Segregation can be higher at a lower value of *K i.e.* at a higher difference (1–*K*). Segregation is possible only in cases when the concentration of impurities in the metal at the temperature of solidification is higher than their maximum solubility in the solid solution.

Considering the left hand corner of Fe-C phase diagram (Fig. 7.6(*b*)) in the region of δ iron for molten steel with 0.075% C (point H) the concentration of carbon in the last solidifying droplet is 0.59%. So, the segregation coefficient for carbon

$$K = 0.075/0.59 = 0.13 \text{ and so, } (1 - K) = 1 - 0.13 = 0.87.$$

The difference (1–*K*) characterises the tendency of carbon to segregate in iron. Values of (1–*K*) for selected elements is shown in Table-1, where it may be seen that highest tendency for segregation is exhibited by C, S and P.

The degree of segregation of impurities in ingots may vary within wide limits depending upon :

(*a*) the duration of solidification

(*b*) the intensity of cooling

(*c*) intensity of metal stirring during the entire period of solidification.

In actual practice, the degree of segregation is measured by determining the difference between the maximum concentration of a given component (impurity) C_{max} in a particular zone and the minimum concentration in another zone, C_{min}, *i.e.* $C_{max} - C_{min}$. This difference is related to C_{min} or to the concentration of the component in ladle, C_l.

$$\left[\frac{C_{max} - C_{min}}{C_{min}}\right] \times 100\%$$

or

$$\left[\frac{C_{max} - C_{min}}{C_l}\right] \times 100\%$$

The segregation may turn out to be positive or negative. Segregation is usually positive in the middle and upper portions of ingot *i.e.* these portions have a higher concentration of segregated impurities. In the bottom portion of ingot, the segregation is usually negative.

Negative segregation is attributed to the following causes :

(*a*) Accelerated cooling of the metal by the massive mould bottom plate.

(*b*) Solidified crystals of metal, which have a higher density than that of the still liquid solution, sink to the bottom. This forms "Settling Cone". Since the axes of growing crystals are purer than the remaining liquid, the metal in the settling zone is usually purer than elsewhere.

Segregation of C, S and P in a 2.5t big end-up ingot headed with a super insulator is shown in Fig. 7.7. The steel quality specified a carbon level between 0.95 and 1.05% and it can be seen that at positions 3 and 8, C segregation has been excessive. Sulphur and Phosphorus also occurs in great concentrations in the upper, central part of the head. When 140 mm square billets were rolled from identical ingots and analyzed for C content of surface and centre at various distances from the head, C and S variations existed. Carbon levels at corresponding positions in billets made from ingots with refractory, insulating and duplex exothermic heads are compared in Fig. 7.8. Since the maximum C level acceptable in this instance is 1.05%, the discard necessary with exothermic head is 7.8% and with super insulator it is 9.0%. The choice between exothermic and super insulator would depend on whether the 1.2% difference in yield is greater in value than the difference in the heading cost.

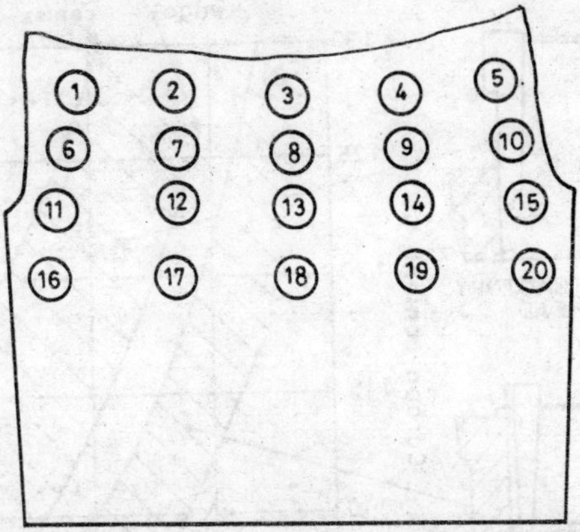

2.5 ton big-end-up ingots headed with super-insulator. Position of drilling holes.

Drill Analyses

Position	*1*	*2*	*3*	*4*	*5*	*6*	*7*	*8*	*9*	*10*
Carbon %	0.98 0.92	0.98 0.98	1.15 1.15	0.95 0.95	0.95 0.94	0.95 0.95	0.98 0.98	1.02 1.04	0.98 1.00	0.98 1.00
Sulphur %	0.008	0.008	0.016	0.008	0.007	0.005	0.007	0.008	0.007	0.008
Phosphorus %	0.010	0.010	0.012	0.010	0.010	0.010	0.010	0.010	0.010	0.010

Position	*11*	*12*	*13*	*14*	*15*	*16*	*17*	*18*	*19*	*20*
Carbon %	1.00 1.02	0.95 0.96	0.95 0.95	0.98 0.99	0.92 0.98	1.00 1.00	1.00 1.00	0.96 0.95	0.98 0.98	0.93 1.00
Sulphur %	0.009	0.007	0.008	0.007	0.007	0.008	0.008	0.007	0.007	0.008
Phosphorus %	0.010	0.010	0.010	0.010	0.010	0.010	0.010	0.010	0.010	0.010

Fig 7.7. Analysis results of 20 samples taken from 2.5 ton big-end-up ingots showing segregation of C.S. and P.

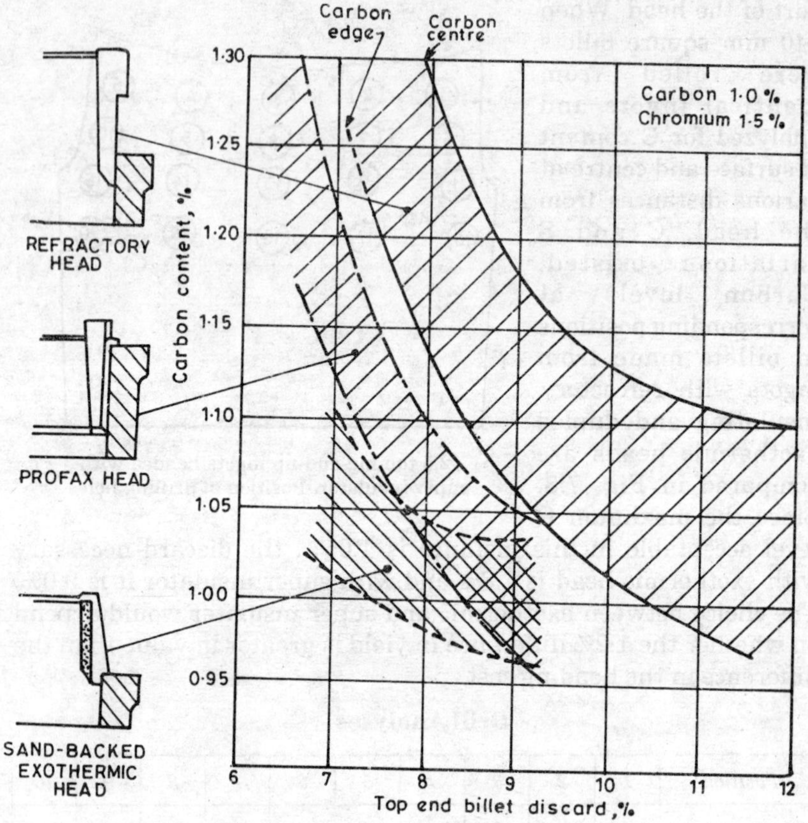

Fig. 7.8. Top end billet discard in %.

7.5.3. Surface Defects

7.5.3.1. Longitudinal casting cracks

These cracks form at the initial moment of crystallization. They usually appear at corners of faces of ingots (Fig. 7.9). It is because the outer crust of ingots which is uneven in thickness and has detached from the mould owing to shrinkage, fails to withstand the ferrostatic pressure of molten metal. It can be prevented by :

(*a*) rigid stream of metal strictly at the axis of ingot.
(*b*) larger perimeter of mould.
(*c*) slow rate of teeming.
(*d*) use of square moulds with concave or corrugated faces.
(*e*) increase the viscosity of liquid metal by changing the composition.

So longitudinal cracks are only seldom found in ingots of high speed, refractory and stainless steels.

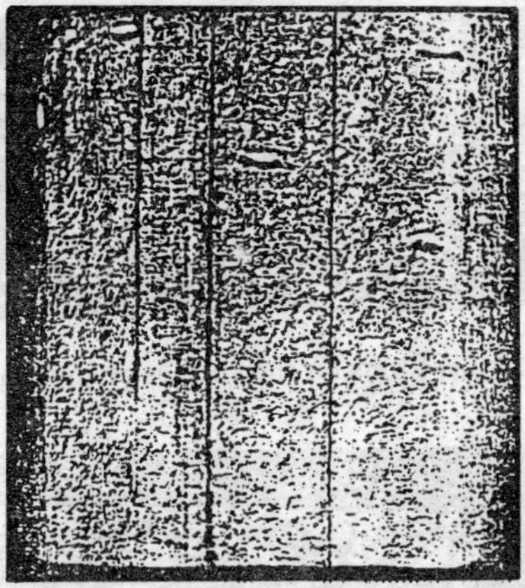

Fig. 7.9. Longitudinal cracks in an ingot.

7.5.3.2. Lateral casting cracks

Deep lateral ruptures are formed due to constricted shrinkage as a result of hanging of an ingot in mould. Usually found in widest portion of ingot just under the head portion. Common causes of hanging are entry of metal into the joints between mould and hot top, splashes of metal reaching the upper surface of hot-top, recesses in mould and rough places on the working surface of lining.

Accurate machining of the matching surfaces of mould and hot-top is necessary to prevent this defect.

7.5.3.3. Surface fire cracks (Longitudinal and Lateral)

Usually such cracks are of zig-zag shape.

They appear at 500-600°C due to thermal stresses if the ductility of the material is too low.

So transfer hot ingots to soaking pits of cogging mill to prevent the formation of such cracks.

7.5.3.4. Films (Fig. 7.10)

Films are formed on the surface of ingots due to splashes onto mould walls due to off centre stream of metal or due to metal boil at mould wall with excessively high rate of teeming. Splashed metal solidifies quickly and gets oxidised which does not dissolve in molten metal. This oxidised film does not weld together with the main metal.

Fig. 7.10. (*a*) Films on the surface of steel ingot.

Fig. 7.10. (*b*) Flaws in a rolled billet resulted from crust turns in an ingot.

7.5.3.5. Crust - Turns (Fig. 7.10)

These are rests of oxidised metal crust left in the ingot body; they produce flaws on rolling. To prevent this problem, the metal should be heated properly and the speed of mould filling should be controlled accurately. It is also possible to form a fume coat above the metal in the mould (due to burning of dressing or wooden frames), which protect the metal against oxidation.

7.5.3.6. Subcutaneous blow holes

Appear due to a thick layer of dressing in moulds or due to quick teeming when the dressing continues to burn in the metal. At the beginning of teeming, metal splashes adhere to moulds and are oxidised.

If the temperature of metal is high the frozen droplets of metal will be dissolved. In moderately heated metal, the frozen droplets can not be redissolved and remain as foreign inclusions in the body of ingot.

In both cases, the oxygen in droplet reacts with the elements of molten steel and form CO and NMI. So, the bottom portion of ingot will be contaminated with NMI and zone of columnar crystals will be porous. These result in fine hair cracks in rolled billets. (Fig. 7.11)

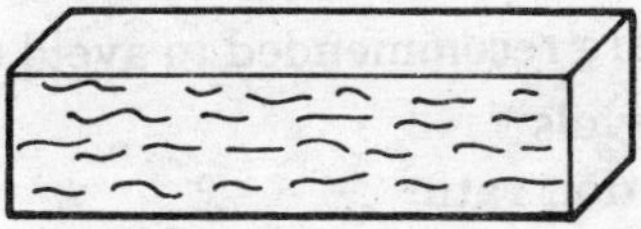

Fig. 7.11. Hair cracks in a rolled billet.

7.5.4. Internal Defects

7.5.4.1. Internal (intercrystalline) cracks

Cr - Ni steels with > 4% Ni and other complex alloyed steels are especially prone to form intercrystalline cracks. These cracks from due to :

(*a*) Internal stresses appearing in ingot due to different rates of cooling of outer and deeper layers.

(*b*) Inhomogeneity in metal composition resulting in that the metal in dendritic axes and interdendritic spaces pass through the critical points at different time.

(*c*) Sulphide and Aluminium Nitride which segregate at grain boundaries weaken the cleavage between grains.

Formation of cracks begins from the end of crystallization and may proceed for a rather long time during storage of ingots in cold state. These cracks at the axis of ingot can cause lamination in the fracture of alloy steel. Intercrystalline spider-shaped cracks at the ingot axis can be welded in by rolling with high reduction ratio. (Fig. 7.12)

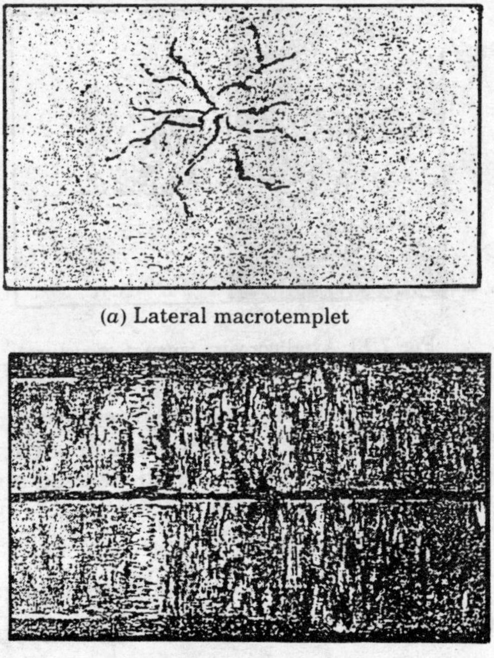

(*a*) Lateral macrotemplet

(*b*) Longitudinal fracture

Fig. 7.12. Spider-shaped cracks in a billet of chrome-nickel-tungsten steel.

Following steps are recommended to avoid the problem :

(*a*) Lowest S levels

(*b*) High reduction ratio

(*c*) Heat the top of moulds so as to minimise the rate of heat transfer and improve the supply of metal to the axial portion of ingot.

(*d*) Transfer hot ingots soon after complete solidification to soaking pits.

(*e*) Control properly the temperature and rate of teeming so as to ensure good filling of shrinkage voids.

7.5.4.2. Internal blow holes (Fig. 7.13)

Spongy structure in cross section. This defect being due to high gas content or improper deoxidation of metal before tapping, a moist launder, ladle or bottom plates.

If the gas content of metal is high, the ingots will start to rise in a few minutes after teeming.

With low Si in metal, last ingots tend to grow because insufficient amount of Si in the end is not able to prevent reaction between C and O forming CO. If the last ingot grows, the first ones will be spongy. Grown and porous ingots are rejected.

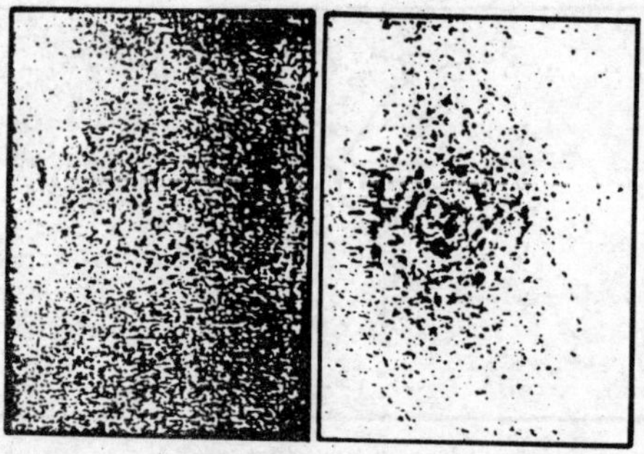

Fig. 7.13. Axial porosity (spongy structure)
(*a*) index 2; (*b*) index 5

Even high Si steels (≈ Si = 4%) also tend to grow. Some high grade Fe-Si tend to introduce much of gases in particular Hydrogen in metal. So better 45-65% Fe-Si should be used and it should be burned and added late to the metal.

7.5.4.3. Internal hair cracks

Detected by magnetic flaw detectors or step-turning. (Fig. 7.14)

Non metallic oxide inclusions can accumulate near these cracks.

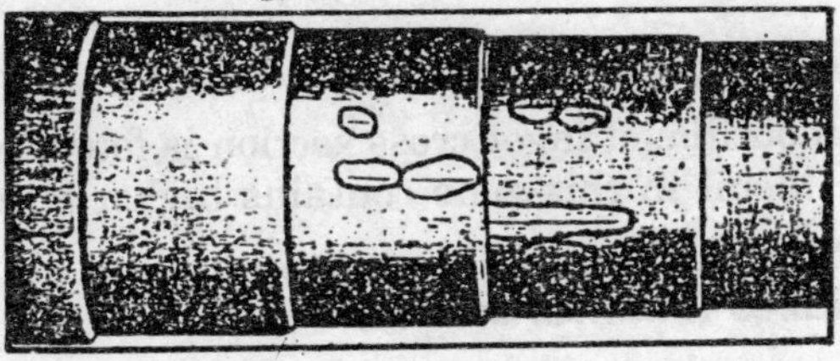

Fig. 7.14. Hair cracks in machined rod seen after step turning.

These cracks are directly linked with gases in steel *i.e.* they can form through rolling of gas blow-holes.

So, gases and oxide inclusions must be minimum in the steel. Formation of oxide films on the rising surface of steel in the ingot should be prevented.

7.5.4.4. Shrinkage porosity (Axial porosity) Fig. 7.13

Due to poor supply of liquid metal to the axial zone. Mainly observed in high C alloy steels and steels of high viscosity. Temperature and fluidity of metal diminish during the course of teeming.

Taper of ingots have some effect on axial porosity.

7.5.4.5. Spot segregation (Fig. 7.15(*a*))

Observed quite often in Al alloyed electric steel. Coarse spot segregation has adverse effect on mechanical properties and this may result in rejections.

High viscosity of molten metal due to presence of Al_2O_3 inclusions hampers gas evolution and removal of segregation. In this spot, there is increased content of C, S, and P and in Al alloyed steels of Al_2O_3.

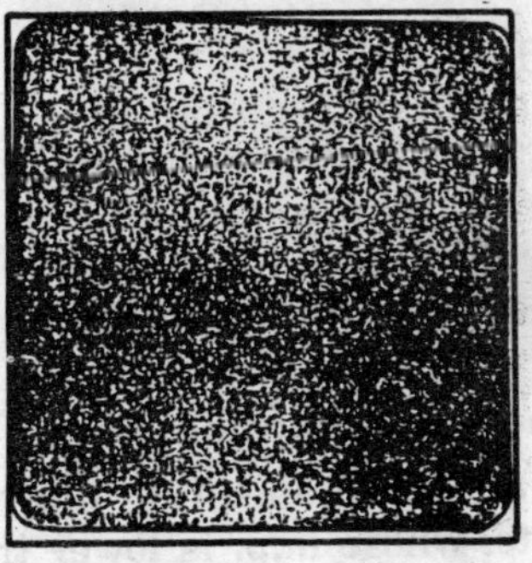

(*a*) Spot segregation in (180 × 180 mm billet).

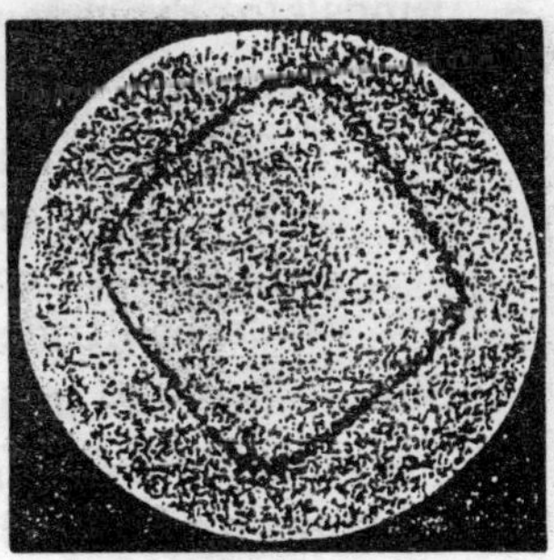

(*b*) Segregation square.

Fig. 7.15.

With low gas content and S, this defect is rarely seen.

7.5.4.6. Segregation square (Fig. 7.15(*b*))

Having an outline of ingot cross section is found in forged and rolled billets. Metal in this zone contains higher S & P and may have high NMI.

7.5.4.7. Flakes (Internal cracks)

Appear in billets or forging during cooling. As the volume of internal voids diminish upon thermo-mechanical treatment of ingots, the pressure of molecular hydrogen increases. And when, this exceeds the strength of steel, the metal fractures internally and forms flakes. (radial cracks)

These are also promoted by internal stresses caused by structural transformations in a chemically heterogeneous ingot or fast cooling of billet.

Flakes can cause sudden failures of parts operating under variable loads.

So, H_2 in the finished steel should be reduced. For this, bath should be boiled intensively and dry burned materials must be used. The ingots must be annealed at high temperature (1100–1150°C) for a long time.

7.5.4.8. Non-metallic inclusions

— Initiate or promote other defects such as hair cracks, internal cracks etc.

— Detrimental in themselves.

— In working surfaces (parts of roller and ball bearings, crackshafts, gun barrel etc.) are the starting point for corrosion and wear, stress raisers and may result in cracking upon heat treatment.

— Kinds - (*a*) Exogenous (*b*) Endogenous.

— Composition -

Oxides, sulphides,
Nitrides, Carbides
Complex Compounds such as oxy-sulphides,
Carbonitrides etc.

— Shape - depends mainly on their melting point.

(*a*) Ferrous oxide, iron and manganese silicates, alloys of ferrous oxides with MnO etc. whose m.p. is lower than that of iron, precipitate in liquid state from solidifying steel and therefore have the shape of spheres or globules in cast steel. (Fig. 7.16).

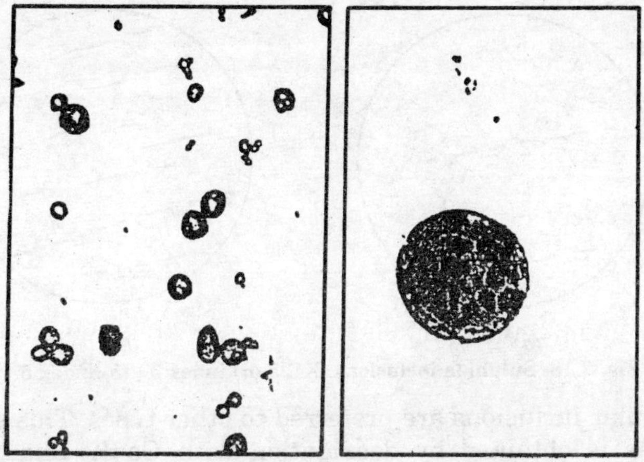

(a) (b)

Fig. 7.16. Globular silicate inclusions (a) X400 ; (b) X1000

(b) High m.p. substances, such as Alumina, high silicon silicates, manganese sulphide, nitrides precipitate from molten steel in solid state and form inclusions of irregular shape. (Fig. 7.17)

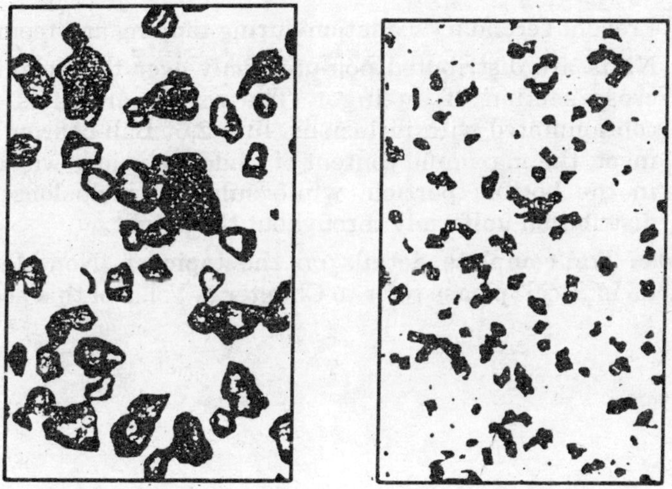

(a) Alumina inclusions, X600. (b) Titanium nitrides, X400.

Fig. 7.17.

(c) The inclusions which are liquid or plastic at that temperature become stretched as sulphides. (Fig. 7.18)

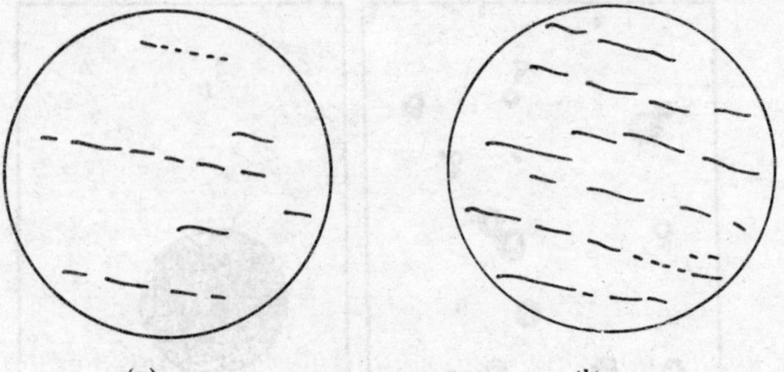

(a) (b)
Fig. 7.18. Sulphide inclusions, X125 (a) index 3 ; (b) index 5.

Globular inclusions are preferred to other types. This shape of inclusions is obtained by deoxidation with Ca-Si. Sharp edged inclusions as that of Al_2O_3 are most unfavourable. (Fig. 7.17)

Low S < 0.01 % ensures that ingots will be free from sulphide inclusions. Also oxide inclusions must be low. Addition of Si-Mn at the end of oxidising period favours fluxing of Al_2O_3 and spinels with liquid manganese - silicates and their removal to slag.

— or, hold the metal under reducing slag or treat the steel with synthetic slag.

— Prevent secondary oxidation during tapping and teeming.

— NMIs are distributed non-uniformly over the height and cross section of an ingot. The axial portion is more contaminated with inclusions. In a 2.6t Ball-bearing steel ingot, the maximum content of oxide inclusions was found in the bottom portion, while sulphide inclusions were distributed uniformly throughout the height.

Note. For complete details on the topic of "Non-Metallic Inclusions in Steel" please refer to Chapter 5, Vol. 2 of this book.

EIGHT

Steel Casting and Related Problems

8.1. GENERAL

While manufacturing steel castings, the following aspects have to be considered carefully :

(*i*) Correct selection of the quality of sand, mold additives, molding process, mold-washes or coatings etc.

(*ii*) Correct selection of the quality of core sands, core-additives, core-washes, core-blowing equipment etc.

(*iii*) Correct methoding systems to provide the following :

(*a*) necessary machining allowances on the pattern.

(*b*) design of proper feeding and gating systems in order to permit proper solidification patterns in the entire geometry of the casting such that the last areas to freeze are adequately fed by the feeders. In essence, the unidirectional solidification of casting must be achieved leading to the feeder irrespective of the variations in casting section thickness. Hence, proper planning of solidification is a must to achieve a perfectly sound casting.

(*iv*) Freedom of casting from numerous defects.

(*v*) A proper control of metal chemistry, degree of superheat and fluidity so that it does not present problems during subsequent operations of pouring the liquid metal in molds.

(*vi*) Cleaning and inspection and heat treatment of castings.

Since superheated liquid steel imposes very severe conditions on the molding materials and because of its typical freezing characteristics, a number of problems are faced in actual practice.

Only a few important ones of them are being discussed in the following sections.

8.2. MOLD AND CORE FACTORS

8.2.1. Metal Penetration

Because sand in contact with steel may be heated to an excessively high temperature, the molding sand must be highly refractory so that it does not fuse together or deteriorate. The sand at metal/mold interface may reach very high temperatures, close to that of steel. So, liquid steel has a tendency to penetrate a finite distance into the mold or core. The result is either a steel-sand crust attached to the casting or fin type attachments resulting from local or crevice penetration into the mold. This problem may lead to starting from rough surface to sand burn on to sand burn-in and finally the metal penetration. This defect is characterized by metal intrusion into the mold wall upto thickness varying from a few mils to as much as 200 mm (8") in extreme cases. Heavier steel sections usually have a thick layer of adherent material than light sections. In general such defects occur at localised hot-spots produced by re-entrant angles, internal cores, or a projection into the mold cavity.

Metal penetration may occur in soft rammed spots or at hot-spots. This kind is basically known as "Mechanical Penetration". But in dense molds also, metal penetration may occur, which is through the mechanism of "Chemical reaction penetration" or "Vapour state penetration". The next mechanism of metal penetration is known as "Subsequential penetration". Factors which have been found to affect metal penetration are :

1. The longer molten steel is held in contact with sand, the greater the chance for metal penetration.
2. Increased metal pressure increases metal penetration. Since metal normally does not penetrate sand because of surface-tension effects, the effect of pressure is to overcome the normal resistance offered by surface tension. High metal pressures are, of course, directly related to the differential metal pressure head (height) that exits in the molten steel.
3. High pouring temperatures lead to penetration.
4. Increased penetration occurs with increased grain size or poor ramming of the sand.
5. Mold washes tend to prevent penetration.

6. Increased penetration occurs with increased carbon content.
7. Large metal mass compared with a small sand mass increases the penetration.
8. Moisture or other gases in the sand may react with the sand or metal to open channels for penetration.

Mould washes are probably effective in reducing metal penetration by closing the pores in the sand mold and also by reducing the amount of oxidation that occurs on the metal surface.

Besides the use of mold or core washes, areas that are troublesome from the standpoint of metal penetration can be improved by using zircon (zirconium silicate) sand mixtures as a facing material. Zircon sands are highly refractory, have a high heat conductivity and a high density, and are not wet by metals. These factors are all instrumental in reducing metal penetration. Because of the cost, Zircon sand is used only in severe cases where the cost is justified.

8.2.2. Burn-On

Burn-on is differentiated from metal penetration by the fact that it is related to oxidation of the casting surface. This results in penetration of the sand by the molten oxide which wets the sand grains. Progressive oxidation at the metal surface develops a pitted condition with encrustations formed by the liquid-oxide penetration into the sand clinging to these areas because of the rough casting surface created by the oxidation. Although burn-on may have the appearance of metal penetration, it is essentially oxide penetration into the casting surface (coupled with oxide penetration into the sand) rather than metal penetration.

8.2.3. Ceroxides

Closely related to burn-on is the accumulation of nonmetallic particles that normally occur on the cope surface of steel castings. These particles, low melting mixtures of oxides of silicon, manganese, iron, and aluminium of variable composition, are called ceroxides. They solidify as a glassy material frequently intermingled with sand grains. They are attributable to the following factors :

1. *Pouring practice.* Bottom pouring causes fewer ceroxides than lip pouring.
2. *Composition of ladel linings.* A high alumina lining results in less ceroxide on the casting surface than a silica ladle lining.

3. *Gating practice*. A streamlined gating practice produces less ceroxide than other methods tried, although a whirl gate is quite effective.
4. *Mold atmosphere*. If the mold atmosphere is made more reducing, ceroxide formation is reduced, but erosion of the mold may increase.
5. *Composition of the steel*. Increasing manganese in the steel increases the amount of ceroxide formed, and aluminium may react with fire-clay refractories to form a viscous product which may erode from the refractory lining surfaces and enter the mold cavity.

8.2.4. Core and Mold Washes

The greater difficulty in achieving good surface finish and avoiding metal penetration in steel castings suggests the use of mold or core washes to a greater degree than with other casting alloys. The function of mold or core washes are :

1. To prevent metal penetration
2. To eliminate cutting (sand erosion)
3. To prevent scabbing
4. To give smooth casting appearance
5. To obtain clean casting surface
6. To modify the mold atmosphere
7. To assist in cleaning the casting

Much higher casting temperatures are involved in case of steel castings compared to other alloys. The choice of coating refractories includes zircon, silica chamotte and in special cases magnesite and olivine. Zircon is the most popular general purpose coating for steel. It is a versatile material for all section thicknesses of steel castings. Because of its high refractoriness, it can be used on surfaces of silica, olivine and chamotte.

The two steel alloy ranges for which zircon is not completely acceptable are the high chromium-nickel stainless steels and the high manganese austenitic steels. The latter suffer burn-on due to the combination of manganese oxide and silica to form a slag which fuses at the pouring temperature. A coating based on magnesite and free from silica has given the most satisfactory results, particularly on moulds and cores of silica sand. Olivine sands for moulds and cores are an improvement over silica sand.

Coatings for steel castings are often subjected to much high drying temperatures in the core ovens then those used in the production of castings from lower melting point alloys. This causes problems which are not experienced in the latter case. Also the coatings are applied more thickly in steel casting practice and must have a stronger resistance to cracking and spalling. These contingencies are catered for by modification to the binder system and control of the grading of the refractories.

Table I. Typical Coating Materials for the Mold and Cores of Steel Casting

Casting Type	*Liquid*	*Refractory*	*Other Materials*
Steel	25 gal water	150 1b silica flour	12 1b western bentonite
Steel	Water as required	100 parts magnesite	20 parts western bentonite, 20 parts cereal, 0.15% sodium benzoate
Steel	3-4 gal water	8 qt. silica flour	1 qt. western bentonite, 0.5 gal water-soluble resin
Brass, bronze	15 parts water	11 parts plumbago	1 part western bentonite

8.2.5. Hot Tears

Steel castings begin to freeze very soon after being poured because of the high freezing temperature range of steel. In many cases, this freezing begins before the cores have been heated to the point where they acquire collapsibility. The combination of a relatively rigid core and a weak contracting metal (because of being in plastic temperature range) often results in the formation of cracks in the casting, referred to as HOT-TEARS. This problem can be quite acute in certain steel castings and therefore, the causes of hot-tear formation have been the subject of considerable study.

The problem of Hot-tears is associated with both metallurgical variables and sand variables. The commonly acceptable view is that this phenomenon occurs during the last stages of freezing of steel, when the solid metal grains are still surrounded by a thin film of liquid. The tearing occurs when the natural construction of the steel is inhibited either by the geometry of the casting or by the mold. Hot-tearing is more likely to occur if the casting contains local areas that freeze at a later period than the balance of the surrounding metal. These so called "Hot-spots" which remain fluid larger than the surrounding metal serve as a focal point for the concentration of contractual strains. It can be said that the probability and intensity of tearing are related to core composition and properties as follows :

(*i*) Strong, hard cores and molds that collapse slowly under heat are more likely to cause Hot-tears than weak, easily collapsible cores or molds.

(*ii*) Increased core density increases the intensity (length) of hot-tearing.

Among the metallurgical variables, elements such as sulphur and phosphorus are known to aggravate the hot-tear formation. In one investigation, aluminium deoxidation was found to give more resistance to hot-tearing than Si deoxidation. But metallurgical variables are of lesser significance than the mold and core variables.

8.3. SOLIDIFICATION FACTORS

8.3.1. General

It is well known that steel, like other metals, has certain typical characteristics and therefore the aspect of gating and risering of steel casting must be correctly handled. Careful attention has to be paid to the following factors in the actual manufacture of steel castings :

(*i*) Fluidity of liquid steel

(*ii*) The problem of "Hot-tear" formation.

(*iii*) Proper design of gating and feeding systems with necessary provision for providing tapers, pads, chills etc. The gating system should be so designed as to avoid problems like mold erosion, slag entrapment, cold shuts or premature solidification, etc. Also, feeders of adequate dimensions and numbers have to be provided such that a totally sound casting is obtained. For this purpose, the "Feeding distance" of feeders has to be calculated in advance. In this connection provision for padding sections and chills to gain directional solidification, use of exothermic riser sleeves and antipiping compounds etc. are important factors.

In the following sections, some of these important factors and recent developments related with them are enumerated.

8.3.2. Fluidity

A spiral type fluidity test is used to describe the fluidity of molten cast steel. Effect of melting variables and composition on the fluidity of steel is shown in Fig. 8.1.

In pouring commercial steels limited to certain analysis ranges it appears that temperatures rather than analysis may be viewed as the major factor which should be controlled to obtain the best conditions. This is not to imply that chemical-composition variables,

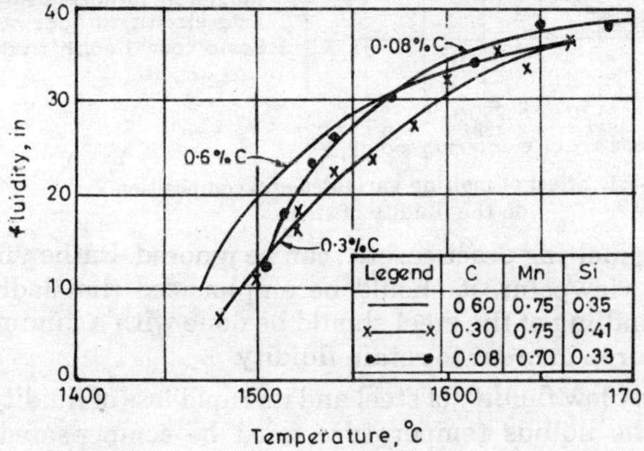

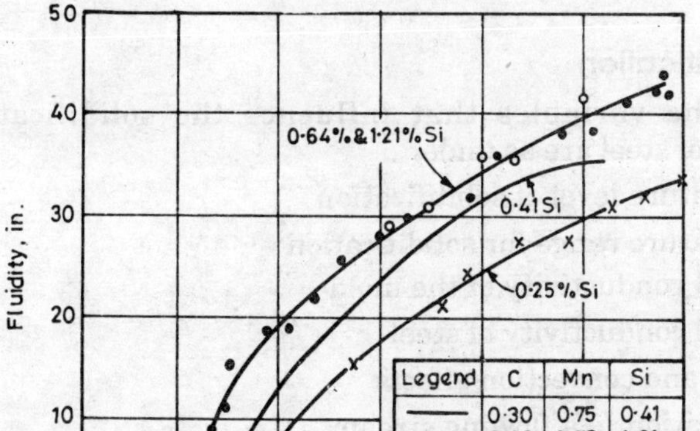

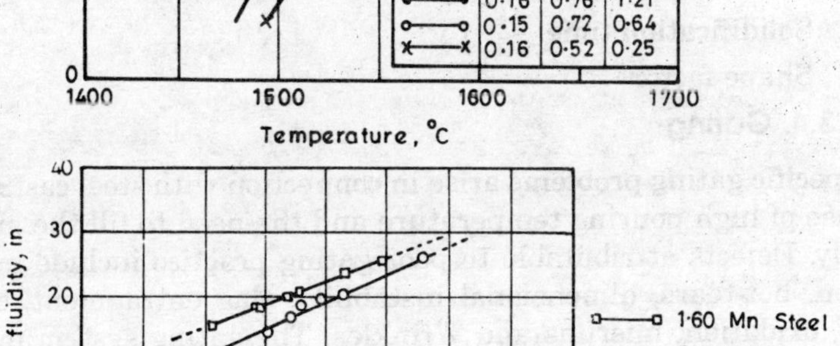

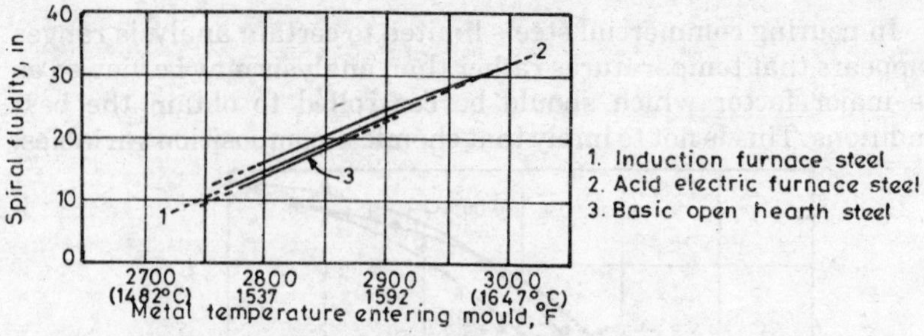

Fig. 8.1. Effect of melting variables and composition on the fluidity of steel.

mold conditions, melting practice, etc., can be ignored. Rather, from an engineering viewpoint, it should be emphasized that ladling, pouring, and handling of the steel should be done with a minimum of temperature drop in order to retain fluidity.

The relatively low fluidity of steel and its rapid loss of fluidity as it approaches the liquids temperature must be compensated by gating design.

8.3.3. Solidification

Some of the variables that influence the solidification characteristics of steel are as under :

1. Temperature level of solidification
2. Temperature range for solidification
3. Thermal conductivity of the mold
4. Thermal conductivity of steel
5. Gravity and convection effects
6. Solidification in a flowing stream
7. Solidification time
8. Shape factor

8.3.4. Gating

Specific gating problems arise in connection with steel castings because of high pouring temperature and the need to fill the mold quickly. Defects attributable to poor gating practice include mold erosion, hot-tears, dimensional instability, slag entrapment, cold shuts, oxidation, misruns and Wrinkles. The gating system must therefore be designed not only with respect to filling the mold, but also with the objective of avoiding these defects.

8.3.5. Feeding

The following factors are of utmost importance with regard to the feeding practice :

(*i*) Riser size and feeding distance, the use of "shape-factor method" for determining riser dimensions and their placement are vital.

(*ii*) Riser necks.

(*iii*) Padding sections to gain directional solidification.

(*iv*) Use of chills to gain directional solidification.

(*v*) Use of exothermic riser sleeves.

(*vi*) Use of knock-off risers.

Developments pertaining to only the use of "Exothermic Riser Sleeves" is discussed in the following section.

8.3.6. Exothermic Riser Sleeves

As is well known, the exothermic materials can be used to reduce the size of feeders. For example, a $3\frac{1}{2}$ inch riser with $1\frac{1}{2}$ inch sleeve of exothermic material is about as effective in feeding a casting as a 7" uninsulated riser. This aspect is demonstrated in Fig. 8.2.

One precaution that should be emphasized is that, when using exothermic materials, the riser cavity should not be filled too high. A rule of thumb which has been given is that the riser should be no higher than its diameters. Some authorities have advocated heights equal to one-third to one-half the diameter. If riser is filled too high

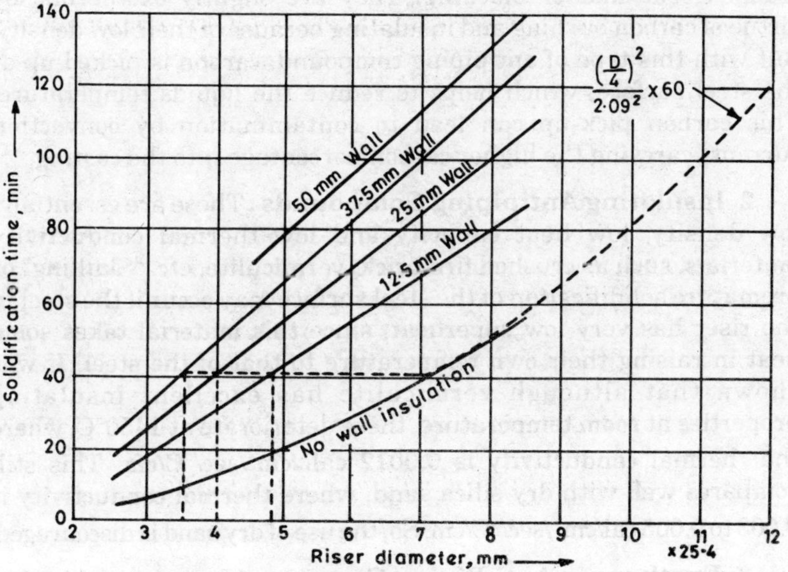

Fig. 8.2. Combinations of riser diameters and sleeve walls which match solidification times of uninsulated risers.

freezing occurs in the lower portion of the riser before the casting has freezen. Thus the molten metal at the top of the riser cannot feed the casting and under-riser shrinkage occurs.

Broadly speaking, there are three main types of exothermic material used in steel foundries :

1. *Surface additions* : these include anti-piping compounds and thermit materials.
2. Moldable exothermic materials for lining risers.
3. Moldable exothermic material to replace metal padding.

8.3.6.1. Surface additions

It is estimated that the radiant heat losses from the surface of a open riser are approx. 25% of the useful heat in riser when no cover is used. Any saving of this heat would lead to more efficient feeding. For maximum efficiency, it is essential that anti-piping compounds are used when exothermic sleeves line the risers, since the % loss of heat by radiation from the surface is going to be more in this case. There are three main types of anti-piping compounds :

(*i*) Carbonaceous (*ii*) Insulating (*iii*) Exothermic

1. **Carbonaceous Antipiping Compounds :** are essentially crushed coke and or blacking. They are slightly exothermic by virtue of carbon burning and insulating because of their low density. But with this type of antipiping compound, carbon is picked-up at the steel surface, which helps to reduce the liquids temperature. This carbon pick-up can lead to contamination by convection currents carrying the higher carbon percentage into the casting.

2. **Insulating Antipiping Compounds :** These are essentially low density, low heat capacity and low thermal conductivity materials, such as crushed fire-brick, vermiculite, etc. "Skulling" or premature solidification of the steel surface may occur if the steel in the riser has very low superheat, since this material takes some heat in raising their own temperature to that of the steel. It was shown that although vermiculite has excellent insulating properties at room temperature, these deteriorate at 1450°C, where the thermal conductivity is 0.0012 cals/cm^2/sec/°C/cm. This still compares well with dry silica sand, where thermal conductivity is 0.003 to 0.005 cal/cm^2/sec/°C/cm. So, the use of dry sand is discouraged.

3. **Exothermic Anti-Piping Compounds :** are useful in that the heat emitted by their reaction delays the "skulling" of low

superheat steel riser and that the sintered residue, which bridges the riser cavity, serves as an insulator by virtue of its low thermal conductivity. Ferrux 101 and Ferrux 40 (Proprietary anti-piping compound) are especially effective in this regard. Coverage of anti-piping compound should be sufficiently thick, at least equal to 1/10th of the riser diameter. This will mean a consumption of 2 – 3.5 kgs/t of the steel casting.

Use of THERMIT MATERIALS is another development. It is basically a feeding aid. The casting is headed as for sand risers, with head volume ranging from 50 to 100% of the casting volume, depending on the casting geometry. These heads are then deliberately short poured to a height equal to 1/3 of the riser diameter with a minimum height of 3". 'Anti-piping compound' is added and then depending upon the riser diameter, a waiting period is necessary to allow the initial feeding, to occur. On diameter upto 5", the waiting period is negligible and the thermit can be added within a minute or so of the termination of pouring. Fig. 8.3 gives an indication of the required waiting time and Fig. 8.4 gives a guide to the weight of materials required. Yields of upto 80% have been achieved with this practice.

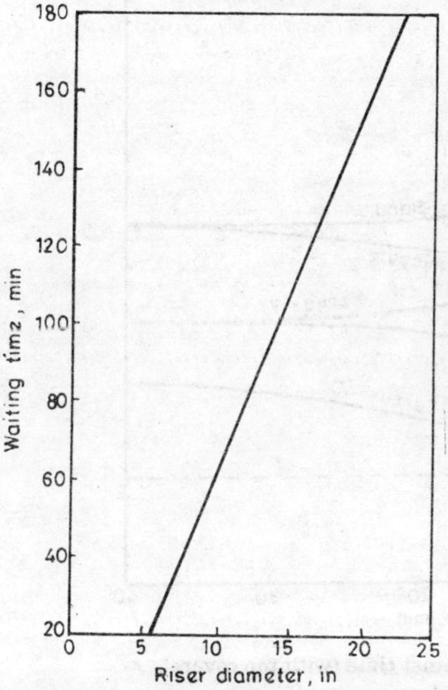

Fig. 8.3. Waiting time for "Kalmex" or "Thermexo" additions.

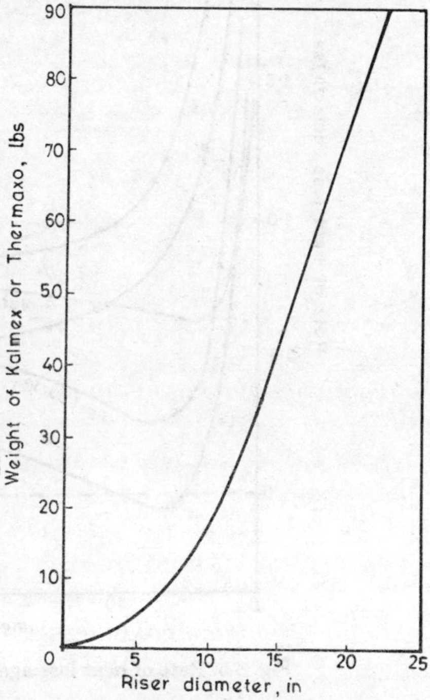

Fig. 8.4. Weight of "Kalmex" or "Thermexo" additions.

8.3.6.2. Moldable Exothermics

Most of the useful heat loss from a riser is through the walls and any reduction in this improves the efficiency of the riser. Exothermics are available in powder form and as prefabricated sleeves so that riser cavities can be lined with exothermic materials either by ramming up in situ or by insertion of a complete sleeve. There are many grades of exothermics, classified by their sensitiveness (time to react in contact with molten steel) so that the fastest would be used on smaller diameter riser and the slowest on large risers.

Except on very small diameter risers, the exothermic riser does not raise the temperature of the steel in the riser, but is sufficient to prevent premature freezing of the steel and then the incandescent residue insulates against rapid heat losses. Investigations at Foseco International showed that the heat loss

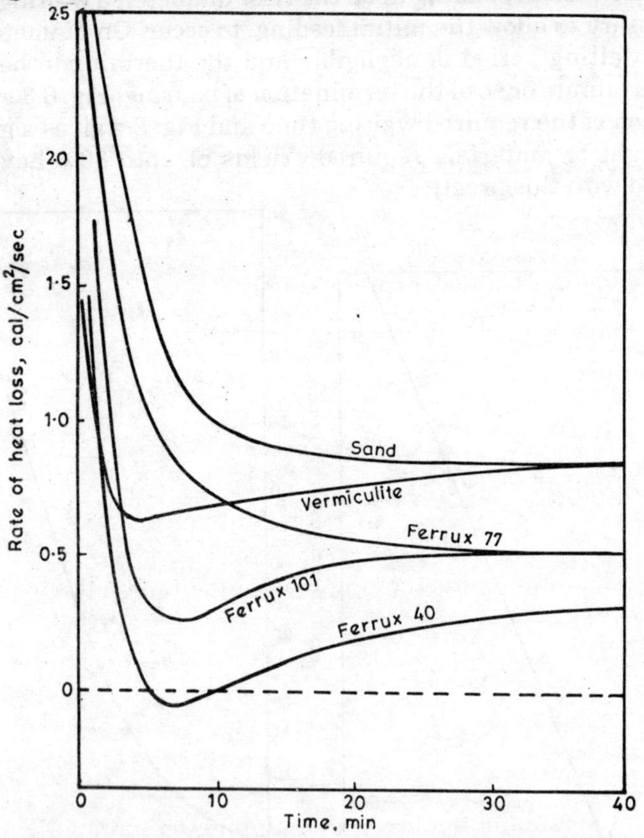

Fig. 8.5. Rate of heat loss against time (with top covers).

through a thickness of medium sensitivity exothermic was 0.58 cal/cm^2/sec, whereas through similar thickness of silicon sand the heat loss was 1.2 cal/cm^2/sec. The freezing time of an exothermic riser can thus be extended to twice that of sand riser of equal diameter. Since exothermics liberate quantities of gas when burning, and in turn need atmospheric oxygen to complete the oxidation of aluminium, adequate venting is required at the interface of the sleeves and backing sand and one quarter inch diameter vents every 2 in. should prove adequate for this purpose.

The advantage of exothermics is due to the fact that the heat generated by the reaction satisfies its own thermal capacity and so there is no drain of useful heat from the steel in the riser.

As the time of solidification increases as with larger dia risers, the relative importance of the two properties namely thermal capacity and thermal conductivity of the surrounding material

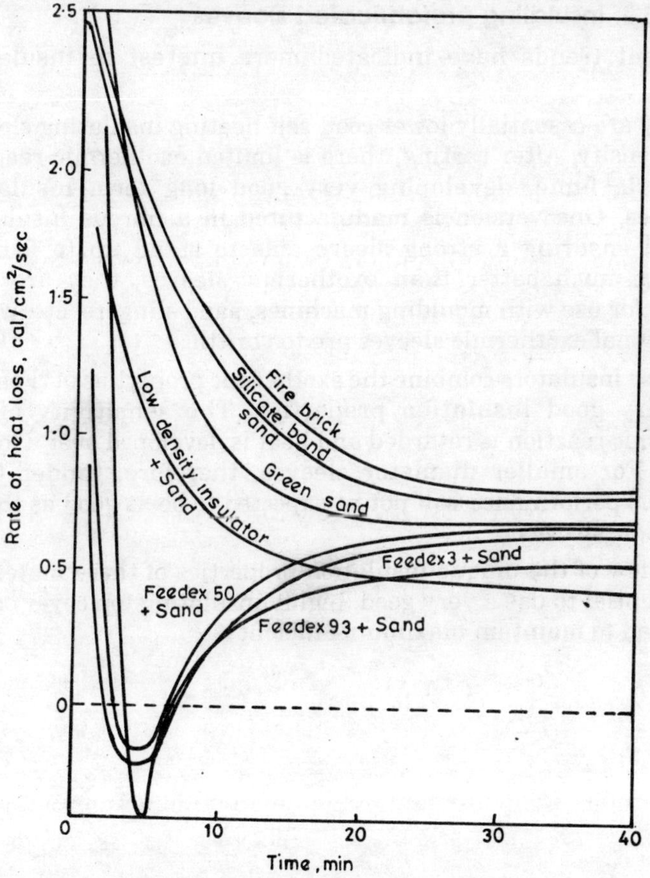

Fig. 8.6. Rate of heat loss against time (with side linings).

changes and the low thermal conductivity plays the major part in improving the feeding. Exothermics are still useful upto say 36" dia riser since the incandescent residue is insulating, but beyond this diameter, it is better to employ a more suitable insulator than exothermic, or resort to a thermit material.

It can be seen from the results of Fig. 8.5 and 8.6 that no matter what material is added to the top of the feeder, or what material is used to line the riser (as shown in Fig. 8.6), there is an immediate heat loss corresponding to thermal capacity requirements. Then depending upon the characteristics of the material this rate reduces rapidly and eventually heat may be added to the riser metal (where the graph goes below the zero line) and then the rate of heat loss becomes fairly steady. This rate of heat loss depends on the thermal conductivity of the materials used.

8.3.6.3. Insulating prefabricated sleeves

Recent trends have indicated more interest in insulating sleeves.

They are essentially lower cost, self-heating insulating sleeves of low density. After casting, there is limited exothermic reaction with little fume, developing very good long term insulating properties. One version is manufactured in a fibrous insulating material ensuring a strong sleeve able to stand up to foundry handling much better than exothermic sleeves; they are very suitable for use with moulding machines, sand slingers, etc. where conventional exothermic sleeves are too brittle.

These insulators combine the exothermic properties of "Feedex" with very good insulation properties. The sensitivity of the exothermic reaction is retarded and heat is developed over a longer period. For smaller diameter sleeves, therefore, (under 5 in. diameter) performance will not be expected to be as good as that of exothermic sleeves.

In view of the unique insulation properties of these materials, it is essential to use a very good, highly insulating top cover on the riser head to maintain maximum efficiency.

NINE

Energy Optimization and Pollution Control

9.1. INTRODUCTION

The different processes used for extraction and refining of metals from their primary and secondary sources and their adoption for use of mankind are highly energy-intensive. Necessity of efficient utilisation of this costly and scarce service, especially after the major impact of "Energy-Crisis" was felt in early 1970's, has resulted in rigorous attempts of "Energy-Auditing" and large scale adoption of "energy-conservation measures". Study of "Energy-balance" at each stage of the process and for each equipment/process in use, is necessary for identifying the major "energy sinks". Adoption of proper "energy-saving" and "energy-recycling devices/techniques" and incorporation of high-productivity and efficient production techniques have in many cases resulted in most economic and optimum figures of specific energy consumption. Similarly increasing awareness of environment and ecology in common masses has necessitated proper "Environment Impact Assessment" of different human, social and industrial activities prior to their approval and licensing or actual adoption. Adequate pollution control measures must be adopted by both the existing and the new industries so that ambient concentrations of various pollutants in atmospheric air, water and land/soil do not exceed the prescribed safe limits. In the following, we shall deal with these two important aspects of industrial activity, in relation to Mini-Steel Plant practice, especially the Electric Arc Furnace steel-making.

9.2. ENERGY OPTIMIZATION

9.2.1. General Treatment

Consequent to universal disapproval of Bessemer Convertor and open-hearth processes, due essentially to their inherent techno-economic limitations, three routes viz., blast furnace-basic oxygen furnace (BF-BOF), direct reduction-electric arc furnace (DR-EAF) and scrap-charged electric arc furnace (EAF) are only used in modern steel-making technology. Of these, the first BF-BOF route is highly energy intensive and is suitable only for large-capacity Integrated Iron and Steel Works, due essentially to the high capital-cost component. Of the three steel-making processes viz. BOF, EAF and Open-hearth processes, the first is most energy efficient, followed by EAF with OH process having the lowest thermal efficiency. This is reflected by the data presented in Table-1, based on energy-balance assumptions indicated therein.

Table 1. Energy balance of different routes of steelmaking

S.No.	*Route*	*Energy balance*	*Heat utilisation (pct.)*
1.	Basic Oxygen Furnace (BOF)	50% Hot Metal + 50% Reaction Heat	90%
2.	Open Hearth Furnace (OH)	75% provided energy + 25% Reaction heat	35%
3.	Electric Arc Furnace (EAF)	85% provided energy + 15% Reaction heat	75%

Data on cost-analysis of a typical scrap-based EAF process indicated the following distribution :

Labour - 6.5%, Refractories - 6.5%, Electrodes - 6.5%,

Energy - 18.5% and Raw Materials - 62%

which reflect to the significance of proper energy-savings and energy-optimisation for cost-reduction in this steel-making practice.

Because of favourable economics and technological factors, the EAF process has shown spectacular rapid growth as reflected by the share of EAF production of world steel, which was mere 1.3% of total steel production in 1930 and rose steadily to 8.3% in 1950, 14.3% in 1970, 20% in 1979 with an anticipated 25% by 1990. According to one estimate in a UNIDO publication, the world crude-steel production in beginning of 1980's was 55% by BOF and 22% by EAF melting scrap and D. R. Iron, which is expected to rise by 2000 A.D. to BOF - 57% and EAF - 41%.

The electrical energy to be used in the EAF process may be generated either in Hydel stations, in Nuclear Power Plants or in Thermal Power Plants, based either on solid, liquid or gaseous fuels viz. coal, lignite, fuel oil, natural gas, coal gas etc. These sources operate at different overall average efficiencies of power generation viz. 25% for Thermal stations, 35% for Nuclear power plants and 70% for Hydel stations. Value of the energy conversion efficiency also contributes to the variations in the figures of energy-balance of different electric-steel furnace operations.

The furnace efficiency is defined as the ratio of arc power to the circuit-power. Its actual value depends upon several factors, such as (*i*) the furnace size (*ii*) rate of power input, showing a maxima in power vs. current plot (*iii*) continuity of working (*iv*) electrical characteristics of the furnace and supply-electrical efficiency and power factor decreasing with circuit current (*v*) furnace-brickwork and amount of cooling water, and, (*vi*) steel-making process viz. all scrap charge or scrap + D.R. Iron charge, extent of finishing needed, use of gaseous oxygen for melting and/or refining etc.

Energy required for steel making furnace equals the sum of heat needed for actual steel-making and the heat to make-up losses of all types. According to one estimate, the overall energy consumption varies from 600-700 kWh/tonne excluding generating losses at the station and transmission-losses to the furnace, equivalent to 500-540 kWh/tonne with D.C. supply depending upon the furnace size and charge composition or equivalent to 400-425 kWh/tonne in a UHP furnace compared to 525 kWh/tonne for conventional melting furnace.

The various input-items for energy in the EAF steel-making include, (*i*) supplied energy as electrical power (*ii*) reaction heat, (*iii*) heat resulting from combustion of electrode material (carbon) to CO, and, (*iv*) heat duc to combustion of CO to CO_2 within thc furnace. In furnaces using gaseous oxygen for assisted-melting and/or refining, this exothermic reaction source also contributes to additional heat input. On the expenditure side, the two main useful items are heat in molten steel and molten slag. A small proportion of heat is used to meet the requirements of endothermic oxidation of impurities by iron-ore. However, a major portion of energy is lost as electrical power, thermal losses with exhaust gases and cooling water. Electrical losses amount to 5-10%. These depend upon the transformer rating and furnace practice and are reduced by use of large cross-sectional conductors and use of on-line computers for energy-optimisation acting as greatest energy saving devices. Waste-gases heat loss, also amounting to 5-10% and resulting due to the extraction of furnace gases for cleaning, varies with the fume-extraction practice and furnace pressure. Water-cooling to protect critical components of furnace such as electrode-holders, doors, refractories in walls, roof etc. by incorporation of the water-cooling panels etc., to protect the lining also, may amount to heat losses of similar dimensions. Radiation-and conduction-losses through the refractory-lining, roof, side-walls, open-doors etc. are not easy to determine and increase progressively with the progress of campaign as the refractory lining wears. Non-tilting, bottom-pouring furnaces reduce heat loss due to this item.

The energy-consumption data on DR-EAF and scrap-based EAF processes depend upon a variety of independent and dependent factors, totalling 26 in number. Data presented in Table-2, refer to the theoretical analysis, by different energy-requirements of steel-making from ore for the different routes. As will be observed, in general, the BF-BOF route requires 18.5-22.5 GJ/tonne of steel with 20 GJ/tonne on an average, and DR-EAF route may show 10% or more energy consumption. However, suitable modifications and optimization of plant practice may reverse this trend.

Table 2. Energy required to make liquid steel from ore

S. No. of Source of data	*BF-BOF route (GJ / tonne)*	*EAF route (GJ / tonne)*	*DR-EAF route (GJ / tonne)*	*Remarks*
1.	21.85	13.33	20.88	(*i*) Scrap rated at 7.4 GJ/tonne (*ii*) DR-EAF route uses gas-based DR process & includes 20% scrap.
2.	18.46	6.79	—	Scrap zero-rated
3.	36.62	19.92	40.04	All three figures refer to finished steel.
4.	20.66	8.62	22.96	(*i*) Model Plant data (*ii*) Scrap zero-rated
5.	22.51	7.26	27.70	Scrap zero-rated

Values of energy requirements of scrap-based EAF vary appreciably, due essentially to the different assumptions of energy-ratings of the scrap used. According to one estimate, the specific energy consumption for 100% scrap-charged EAF is 3700 kWh/tonne of steel (1 kWh = 3.6 MJ), which may amount to about 8 GJ/tonne based on scrap-energy equivalent of 7.4 GJ/tonne (energy already spent on scrap). For DR-EAF process, this figure for scrap-based EAF, increases with an increase in percentage of DR-Iron in feed to the furnace at constant degree of metallisation, becoming 7000 kWh/tonne at 50% scrap and rest sponge-iron. Increased degree of metallisation at same composition of charge reduces energy consumption, essentially due to the lower slag-volume with such feed. Energy equivalence of DR-Iron or the pre-reduced pellets is assumed to be less than 7.4 GJ/tonne, decreasing with degree of metallisation.

Typical Heat balance data for a scrap-charged EAF are presented in Table-3. As will be observed, more than 3/4th requirements, is supplied as electrical energy and more than 1/3rd of the supplied energy is lost in thermal losses, electrical losses and in slag.

Table 3. Average Heat/Energy Balance of Electric Arc Furnace-Scrap Melting

S.No.	*Energy / Heat Input*		*S.No.*	*Energy / Heat Output*	
	Item	*Percentage*		*Item*	*Percentage*
1.	Electrical Energy	77.9	1.	Thermal losses	24.3
2.	Reaction Heat	9.4	2.	Electrical Losses	6.9
3.	Electrode Combustion to CO		3.	Ore-reduction	0.7
4.	Combustion of CO to CO_2	10.9	4.	Heat content of slag	7.0
			5.	Heat content of metal	61.1
	Total	100.0		Total	100.0

It must, however, be noted that these figures in a heat-balance vary appreciably with the size of the furnace, type of feed charged, the steel-making practice, power-input rate, continuity of working, electrical characteristics of the furnace and power-supply, furnace brickwork and the amount of water-cooling. Large size furnaces having small surface area/volume ratio, low metallurgical load in the charge requiring less flux and thus resulting in lesser slag-volume, use of preheated charge and faster melting and refining techniques, thus reducing the overall "heat-time", as also adoption of heat-saving and waste-heat recovery, recycling and/or reuse techniques, result in reduced net losses and higher overall thermal- and electrical-efficiencies of the EAF process and also higher productivity of the furnace and consequently low specific-energy consumption figures. Heat-balances of three present day common modes of EAF operation are presented in Table-4, which clearly reflects to the superiority of oxy-fuel burners as additional source of energy to the furnace over the classical operation, with the operation using gaseous oxygen to assist melting and refining in between the two. As already pointed out, care must, however, be exercised, while analysing data from different sources, on such comparisons, due to variation in the equivalences between electrical and thermal forms of energy from different sources.

Table 4. Comparison of Energy Balances of Three Operating Procedures of Electric Arc Furnace.

S.No.	*Operational parameter*	*Classical operation*	*Operation with oxygen*	*Operation with oxy-fuel burners*
1.	Energy, kWh/tonne	537.5	502	450
2.	Coke, kg/tonne	5	10	17.5
3.	Oil, kg/tonne	—	—	2.5
4.	Oxygen, Nm^3/tonne		10	25
5.	Electrodes, kg/tonne	5.4	5	3.7
Total energy, Gcal/tonne		1.557	1.515	1.456

Equivalence basis - 2500 kcal = 1 kWh; 8000 kcal/kg coke; 9000 kcal/litre oil; 32,000 kcal/kg; Electrode and 2000 kcal/Nm^3 oxygen.

Effect of size of furnace on the efficient utilisation of energy in EAF is reflected by the data presented in Tables-5 & 6, respectively for a 3-tonne and a 120-tonne furnace. Small furnaces, in addition to high capital cost for same equivalent capacity, are also energy-wise less efficient due to increased thermal losses, as pointed out earlier also. It has been estimated that specific energy consumption for a 5-tonne and a 25-tonne furnace are 800 and 450-500 kWh/tonne respectively. Thus for production of 50 tonnes of steel, *(i)* Ten 5-tonne furnaces will require 40,000 kWh power but, *(ii)* Two 25-tonne furnaces only 25,000 kWh power, reflecting to a major saving in power cost in production of steel using large size furnaces.

Table 5. Typical Energy distribution in a 3-tonne Basic Arc Furnace

S. No.	*Heat - Item*	*Percentage value*
1.	Useful Heat	60
2.	Heat flow through walls, roof and bottom	16
3.	Heat flow through openings	4
4.	Loss to water on cooling	5
5.	Loss to escaping gases	8
6.	Electrical losses	7
	Total	100

Table 6. Heat Balance of a 120-tonne furnace

	Heat Item	*MWh*	*kWh/tonne*	*Percentage*
A. Input :				
1.	Transformer Primary Input Power	67.6	564	76.6
2.	Metal Reaction	10.1	84	11.5
3.	Coke combustion	4.2	36	4.8
4.	Electrode Combustion	6.3	52	7.1
	Total	88.2	736	100.0
B. Output :				
1.	Heat content of			
	Metal	46.1	384.6	52.3
	Slag	9.2	77.2	10.5
	Electrodes	4.4	36.9	5.0
2.	Heat to			
	Walls	5.9	49.3	6.7
	Waste gases	7.8	64.9	8.8
	Roof	1.8	14.7	2.0
	Hearth	0.7	5.9	0.8
3.	Water cooling			
	Elbow	3.3	27.5	3.7
	Shell	1.2	10.3	1.4
	Roof	0.6	5.4	0.7
4.	Electrical Losses	7.2	59.8	8.1
	Total	88.2	736.5	100.0

According to another estimate, the useful heat increases from a mere 54% value of the total energy input from all sources including also the exothermic reaction heat, for a 5-tonne furnace to over 70% for furnaces of 18-120 tonne capacity for high power inputs used during melting. This reflects to the superiority of MIDI Steel Plants over Mini-or Micro-Steel Plants based on EAF steel-making.

9.2.2. Methods adopted for Energy Conservation and Optimisation

For improving energy efficiency of EAF and reducing the specific energy consumption, several modifications have, therefore, been made in the recent years in the conventional EAF practice based on 100% scrap charge. Important among these include use of,

(*i*) **High-Power (HP) and Ultra High-Power (UHP) Furnaces** - current density has been increased from 18 A/cm^2 to 20 A/cm^2 (= 40 kA for a 500 mm dia carbon electrode) and even upto 24 A/cm^2 for impregnated high quality graphite electrodes thereby appreciably reducing the heat time and thus increasing "productivity" of the furnaces without any appreciable loss of refractory life or "availability" of the furnace.

(*ii*) **Water-cooled Panels** - reduce refractory consumption and corresponding thermal losses, improve furnace lining life and hence "availability" and "productivity" and also result in appreciable energy-conservation and reduced energy costs of steel-making.

(*iii*) **Oxy-fuel burners for assisting melting** - placed symmetrically downwards and away from electrodes to aid melting and thus reduce overall "heat-time" - results in appreciable decrease in heat-time and hence specific energy consumption by minimising heat losses and increase of productivity of the furnace – all resulting in reduced cost of production.

(*iv*) **Continuous charging of furnace** - of pre-reduced pellets, DR-Iron, preheated scrap and granulated B.F. iron has made it possible to charge this metal feed slowly and continuously to minimize roof-swinging and associated heat losses and also the overloading of the furnace, during melting operation, thus reducing the overall heat time and hence specific energy consumption of the process. Use of DR-Iron or pre-reduced pellets provides also an alternative cheap and efficient steel making path from ore.

(*v*) **Use of split-shells or replaceable shells** - the former for furnaces to reduce relining time and also improve furnace availability, and, the latter with one shell in operation and the other under relining stage.

(*vi*) **Scrap-preheating** - the exhaust gases from the furnace leave at about 1000°C and even in the dust-laiden condition, can preheat the scrap-feed kept in charging baskets. A preheat from 25°C to 200°C is equivalent to reduction of energy consumption by 34.4 kWh/tonne, to 300°C by 50 kWh/tonne and upto 350-500°C from 64.6 to 95.6 kWh/tonne respectively. This preheating of scrap uses the sensible heat of exhaust gases or waste gases prior to gas cleaning for fume extraction.

(*vii*) **Rapid process** - process-modification to result in simultaneous refining, for phosphorus removal by addition of patented fluxes suitable for phosphorus removal and melting of charge, appreciably shortens the heat time thereby improving the productivity of existing units and reducing their specific energy consumption.

(*viii*) **Computer-process control** - Computers have been used, based on adoption of "Thermal" and "Metallurgical" models and "Con-cast-EAF co-ordination". Taking appropriate corrective action at proper time with the help of computers have also been used for energy-optimisation, as pointed out earlier and thus have immensely improved process efficiency and economics.

The furnace productivity over a campaign is also characterised by two counteracting requirements needing a workable compromise viz. the shortest possible melting time and the lowest and most-evenly distributed refractory wear. Also furnaces in the "melt-down period" alone are fit for maximum power control, for which several approaches making use of computer viz. the furnace-control, maximum-control, coarse-control, precision-control and priority-control techniques have been adopted.

(*ix*) **Ladle Metallurgy** - Adoption of ladle metallurgy, with induction-stirring, arc-heating, degassing etc., for either refining or finishing the heat or both transfers the major load from the furnace to ladle and thus has appreciably contributed to energy-conservation and process-economics in EAF steel-making.

(*x*) **Bottom Pouring** - Use of bottom-tapping, instead of conventional tilting furnace have dual advantages of minimising energy consumption as also reduced wear and tear of refractory - due to no necessity of lifting and swinging the roof.

Thus we see that an analysis of different sinks for the energy supplied to the furnace has resulted in minimizing losses to improve energy-efficiency of EAF process of steel-making.

9.3. POLLUTION CONTROL

9.3.1. Pollution Problem - General Treatment

Environmental pollution, in the broadest sense may be defined as the undesirable changes, brought about by physical, chemical or biological factors and resulting essentially from human activities, in the natural quality of the environment, consisting of atmospheric air, land or surface soils and surface waters - the latter present as streams or reservoirs. Large scale industrial activity leads to both "fugitive" or "diffuse" emissions and "stack" or "point" emissions of some toxic and obnoxious gases, vapours, liquids and/or solids, called pollutants, into atmosphere, land or water, creating serious ecological and lethal or sub-lethal health problems, reflected in, (*i*) irritation of sensual and respiratory organs, chronic diseases and even fatalities among human being and both land-and aquatic-animals due to either inhalation, ingestion or absorption through skin-contact of these hazardous pollutants, (*ii*) mild to severe damage to plants resulting into poor yields and even total destruction of vegetation and other plant life, and, (*iii*) corrosion and/or erosion of structurals built of metallic and other constructional materials. Extent of damage in each case depends upon the extent of exposure, as measured by the time of exposure and concentration and toxicity of pollutants and susceptibility of the subject or the individual exposed.

A normal individual breathes approximately 20,000 times a day and may thus intake from 15.7 to 18.2 kg of air per day depending upon the state of his/her health. (16 kg on an average equivalent to approximately 5 million litres of air per annum), which is much greater than his/her daily intakes of food and water and hence the gravity of health-hazards associated with air-pollution. Five major primary air-pollutants, resulting over 90% air-pollution, include (*i*) carbon monoxide (CO), (*ii*) oxides (SO_x) of sulphur viz. SO_2, SO_3 etc., (*iii*) oxides (NO_x) of nitrogen viz. N_2O, NO, NO_2, N_2O_5 etc., (*iv*) hydrocarbons (HC), and, (*v*) particulates (SPM). Secondary air-pollutants, include a variety of other organic and inorganic chemicals and their photo-chemical, oxidation, hydration or other reaction-products, generated in atmosphere. "SMOG" resulting from interaction of water-vapour or atmospheric-fog with smoke or by other photo-chemical reactions, impairs visibility and causes serious health-problems.

(*i*) **Water pollution :** results from solids in waters and effluents from industries, human settlements, mine waters, waste-materials from ocean-going vehicles, pesticides, insecticides etc. from agricultural land, sewage, "acid-rains" etc. The pollutants - both organic and inorganic in nature, may be present in water as suspended solids (SS), dissolved solids (DS), liquids or gases or as immiscible liquids such as oils. Trace-metals impurities of different toxicities in ionic-form constitute the major problem of soluble pollutants.

(*ii*) **Land-or soil-pollution :** results from "Dust-fall" or "fall-out" and "rain-out" or "wash-out" effects of pollutants from atmospheric air, spraying of chemicals at agricultural sites or due to solid-wastes or liquid effluents discharged on land from industries, human settlements and sewerage systems etc.

(*iii*) **Noise-pollution :** results from vibrations-both in 'audible' and 'inaudible' ranges, generated from leaking effluents or machinery under operation. Its adverse effects include irritation, nervousness, hyper-tension, muscle-spasm, mental exhaustion etc. and may also temporarily or permanently damage the 'hearing-capability' of the exposed victim.

(*iv*) **Thermal pollution :** results from acute temperature and/or humidity changes of the environment and depends upon the ambient air temperature and velocity, mean radiation temperature and humidity of the air. It may cause from "uncomfortable to unsafe" working conditions culminating into excessive stress and even "burns" of the workmen/victims, exposed to such conditions.

(*v*) **Visual pollution :** results either from "surface-excavations", due to mining, or "hoarding" in cities, and affect adversely, due either to "degradation of land-scape" or "increased traffic and similar hazards" due to high-rise structurcs obstructing view.

9.3.2. Control of Pollution

For efficient control, the pollution problem should be properly identified its sources and effects in relation to each stage in the processing flowsheet adopted in plant or industry under scrutiny. A proper scientific "cause-effect-remedy analysis" with respect to each of the potential pollutants involved is necessary to develop adequate awareness of "pollution-hazards". Abatement to "Zero-pollution levels" being impossible in actual practice, two major approaches are used for effective and economic control of pollution.

(*i*) **Legislation against pollution :** to act as "major deterrant" for fear of penalty likely to be imposed on default or negligence on the part of owner, management or workman. "Quality standards" for water and air in relation to each potential pollutant must be framed and different "levels" of effects of exposure to them in relation to their harmful effects/reactions designated. The different industries must be identified and classified based on their pollution-potentials to frame a "generalised policy of pollution-control" and also the "monitoring and control measures" to be adopted for each.

(*ii*) **Adoption of scientific methods :** these include, (*a*) identification of major pollutants and their sources for specific case under scrutiny (*b*) monitoring of pollution at different times and places within and in the vicinity of the industry, and, (*c*) selection of suitable method and technology in the light of (*a*) and (*b*) for control of pollution.

The different techniques adopted for pollution-control are classified as, (*i*) Source-control or "process-specified measures" – for reduction, limitation or elimination of the source of pollutant, and, (*ii*) pollutant removal preferably incorporating a "non-waste technology" or the "Equipment and plant specific measures". Wherever possible, cost-reduction methods including (*i*) recovery of wastes as by-products, and, (*ii*) segregation and blending of wastes to facilitate treatment, recycle and reuse should be adopted.

For **Air-Pollution Abatement,** the following approaches are used :

(*i*) Collection of particulate matter-solids or even liquids from dust- or aerosol-laiden gaseous emissions are separated making use of settling chambers and flues, inertial collectors, cyclones and multi-cyclone separators, fabric filters or electrostate precipitators or by the collecting action of liquids, generally water, employing wetting and entrainment in liquid streams, from which these are subsequently separated prior to discharge of effluents. Appliances used for this washing of particulates, include simple-gravity or cyclonic-spray towers, packed-bed scrubbers, jet-scrubbers,vortex-scrubbers, venturi-scrubbers and other specially designed scrubbers.

(*ii*) Collection of gaseous pollutants in aqueous medium using water or other suitably selected solvents or chemically reactive fluids employing absorption, adsorption on solid surfaces or chemical reaction for collection by either "throw-away" or "by-product recovery" methods. For this purpose, contact is made either in film absorbers using packed- or fluidised-columns, jet scrubbers in spray-columns or bubble-end drop-absorbers in plate columns, bubble columns, rotating disc columns or nozzle spray-columns. Thermal incinerators of combustible pollutant gases *e.g.* Hydrocarbons, Hydrogen, Carbon monoxide etc. is practiced to convert them to harmless products of combustion. Adsorption of pollutant gas or vapour uses suitable solid adsorbent-either activated charcoal or metallic adsorbers or chemicals, which is usually regenerated and recycled for reuse. Odour-control is achieved by absorption of chemicals or contaminants causing such odours.

For **water-pollution abatement,** different physical, physico-chemical and chemical methods of treatment for separation, decomposition or destruction of pollutants present in "waste-waters" or polluted "process-water" are used. The common methods adopted include, (*i*) flocculation or coagulation followed by sedimentation in thickeners, or settling tanks/ponds and filtration, (*ii*) froth-flotation for suspended oily-liquids, and, (*iii*) waste-water treatment methods such as lime-soda process requiring sludge-disposal facility also, Ionic Exchange, Reverse Osmosis, Chemical treatment, Bio-chemical treatment and special treatments for complete destruction of extremely toxic pollutants such as cyanides etc.

Solid waste disposal methods include : (*i*) combustion in Thermal Incinerators for combustible wastes, (*ii*) 'Back-filling' *e.g.* in mines and other 'land-reclaimation' operations, and, (*iii*) underground or under-water dumping in sealed and/or protected containers, if necessary, *e.g.* for radio-active wastes.

In case of **Noise-Pollution Control,** no person should be exposed to noise-levels above 90 dB (A)-leg (8 hrs.). With high noise levels, "hearing protectors" must be provided and several statutory provisions must be met by employers/management. "Direct-sound" from machines or leaking effluents and "Indirect-sound" reverberated in the premises or a result of "air-borne" or "structure-borne" transmission is controlled either by, (*i*) Absorption in suitable absorbers e.g. fibre-glass, wade, rock-wool or heavy drapes for converting "sound energy" into "heat energy" which are incorporated in roofs, side walls, floor or as suspended curtains in the building, (*ii*) enclosures made of absorber materials to contain sound within them, and, (*iii*) reduction of sound at source by proper design and operation of machinery and use of 'silencers, wherever possible, isolation of vibrations generating machinery, damping of vibrating components, proper design and selection of sizes and shapes to minimize vibrations and "force-pulse" modifications to reduce the intensity of vibrations.

Thermal pollution abatement is achieved by, (*i*) proper ventilation and air-or water-cooling of working areas, (*ii*) use of heat-radiations absorbing curtains of asbestos or felt-matting, saturated with water for "evaporative-cooling" between the heat-source and the workmen, (*iii*) use of "protective sheets or screens" at the open-doors of furnaces-air or water-cooled and made of metallic materials, generally aluminium with glass-observation windows.

9.3.3. Environmental Problems in Small-Scale Steelmaking in EAF

The main environmental problems in small-scale steel making or in Mini Steel Plant practice with special reference to Electric Arc Furnace (EAF) steel making include,

(1) **Particulate (SPM) and gaseous emissions** - from the furnace and associated equipment/operations - constitute major problem and requires special control and containment or collection equipment and practice.

(2) **Water-treatment** - either for recirculation in cooling water system for electrode clamps and arms, roof rings wall-panels, doors etc. and also in rolling mills and reheating furnaces of Mini-Steel Plants or for the dirty effluent from wet collection systems for dust and fume (SPM), prior to its discharge in water streams/reservoirs or on land and may necessitate use of clarifying equipment such as thickeners, settling tanks/ponds, filters, pump etc.

(3) **Disposal of solid waste** - including (*i*) dust and fume from primary-and secondary-dry or wet gas/air-cleaning systems, (*ii*) slag, produced at 50-75 kg per tonne of steel made from scrap charge and more with increased use of pre-reduced pellets or direct-reduced iron in furnace charge, and, (*iii*) scrap-may be recirculated in the plant, which is possible for dust and fume also after agglomeration or used for land-reclaimation by filling the slag and un-usable scrap, dust and fume also.

Recycling of solid-wastes is limited due to, (*i*) presence of tramp-elements (*ii*) very fine size, difficult to handle and transport, (*iii*) oil-contamination and, (*iv*) variable composition of the waste and its consequent adverse effects on process- and quality-control.

(4) **Noise and vibration** - Arc furnace operation is associated with "Noise" produced both "inside" *i.e.* from arc, and, "outside" the furnace or plant *i.e.* from auxiliary services & equipment. The normal noise levels in EAF operation may be 115-120 dB (A) - with maximum at the "Melt-down" stage (cf. safe noise level). No specific arrangement is generally made, except either use of air-conditioned control-pulpits for men and occasional attenuation by absorption in the plant. Lately a successful approach suggested is to contain noise by placing the entire furnace in an accoustic-cell, which is a very expensive preposition and somewhat inconvenient operationally, in such an eventuality, for most melting and refining cycle, the furnace is enclosed and the operator works in clean, comfortable and safe condition.

The major emissions from the EAF step include gases and particulates including dust and fume. Typical gaseous constituents emitted are presented in Table-7 and include N_2 and O_2 from infiltrated air, CO and CO_2 from combustion of electrode and steel making - refining reaction of C-boil, water vapour from moist solid charge, SO_2, SO_3 and vapourised hydrocarbon from oily or greasy scrap. Besides few others may be emitted under specific conditions of high temperature, reducing atmosphere or while using contaminated charge.

Table 7. Constituents present in gases during melting of steel

Common constituents	*Possible under certain conditions*
Nitrogen	Ozone
Carbon dioxide	Nitrogen oxides, NO_x
Oxygen	Hydrogen
Carbon monoxide	Hydrogen sulphide
Water vapour	HCl and HF gases
Sulphur dioxide	SiF_4 vapours
Sulphur trioxide	Ammonia
Vapourized Hydro-carbons	Acetylene

Dust originates from the solid charge materials *e.g.*, scrap, DR Iron, flux, ore etc. Fume originates from volatilisation of iron and tramp-elements at furnace temperature, oxidation in vapour state and subsequent condensation and is generally spherical in shape. It may also show flocculating tendency and vary in size from extremely fine to about 90% of –400 mesh (–37 micron) size. The composition of dust and fume depend upon the quality of scrap and other components of the furnace charge, the type of steel made and operating conditions of the furnace. While melting high-grade scrap, fume may consist of over 90% iron oxides but with dirtier-scrap with residual-tramp elements, appreciable concentrations of Cu, Zn, Pb, Mn, S, P, Na, K etc., may also be present, with maximum concentrations of Cu and Zn upto 25% and 5% respectively. Quantity of dust and fume also varies appreciably from 0.25-3.5 grains/Cu ft. of ventilation air from EAF, which is equivalent to 4-30 lbs/ton of metal poured or approx. 1-2% of steel made, and depends upon, (*i*) melting rate, (*ii*) quality of scrap and other raw materials charged, (*iii*) vapour pressure of tramp elements present as minor constituents of the charge, (*iv*) temperature of operation, (*v*) degree of oxidation affected, (*vi*) air-infiltration or O_2-lancing in the furnace, and, (*vii*) rate of gaseous evolution from the furnace. Lowest emission rates are achieved with clean scrap, uniform coarse-sized additions, low melting rates, low gas temperature and low gas-extraction rates.

Typical relevant data on dust and fume for different types of charges and steel-melting practices are presented in Tables-8 and 9.

Table 8. Typical Analysis of EAF dust - scrap melting

Item	*Wt. Percent*
Total Fe	38.0
Zn	3.3
Pb	1.4
Na_2O	0.25
K_2O	0.20
S	0.25
C	1.00

Table 9. Typical data on charge, emissions and characteristics of dust and fume (composite dust at ESP from 3 furnaces)

Typical charge		*Typical dust and fume emission (shell evacuation & ESP collection)*	*Typical characteristics of dust and fume*	
Fluxes, carbon & ore	–5%	SiO_2 – 2%	A. Particle size Distribution	
Turnings & borings	–7%	Al_2O_3 – 3%	Size in microns	Weight percent
Home scrap	–20%	Iron - 25%	0–5	71.9
No. 2 baled scrap	–25%	Lime - 6%	5–10	8.3
Miscellaneous	–43%	MgO - 2%	10–20	6.0
Scrap (auto etc.)		ZnO - 37%	20–44	7.5
		CuO - 0.2%	>44	6.3
		MnO - 4%		
		P_2O_5 – 0.2%		
		SO_3 – 3.0%		
Total	100%		B. Sp. gr. of dust & fume = 3.93	

9.3.4. Method of Dust and Fume control in EAF Plants

The actual system used in any specific case depends upon the quality and quantity of dust and fume to be extracted *e.g.* for the use of electrostatic precipitators. conductivity of particulates is an important criteria of selection.

In basic oxygen (BOF) steel making, generally two approaches have been adopted for gas-cleaning, viz., (*i*) full-combustion system, in which CO of the gases is completely burnt to CO_2 prior to discharge into atmosphere, and, (*ii*) suppressed-combustion system, in which CO-rich converter gases are used as a by-product fuel after gas-cleaning.

Two approaches have been used for dust and fume collection from EAF exhaust gases, viz., (*i*) Primary Extraction system or "direct evacuation" from the furnace shell from a separate opening in roof to capture fume and dust escaping from the furnace directly, and, (*ii*) Secondary Extraction system or "secondary evacuation", designed to capture dust and fume from all other operations, equipment and places outside the furnace *e.g.* in the metal pouring and charging arrangements or while slagging or tapping the furnace or in secondary-steel making or ladle-treatment facility etc.

Four commonly adopted methods of particulate collection include,

(*i*) General shop-ventilation-by natural or mechanically induced draught by exhaust fans installed at roof of the plant.

(*ii*) Canopy Hoods, located above or under the crane-ways in suitable areas of important activity such as slagging, tapping, ladle- treatment etc.

(*iii*) Roof-mounted close-fitting hoods or local exhaust hoods of either, (*a*) stationary or fixed type, or (*b*) swivel or movable type. For these, efficiency is adversely affected by cross-draft inside the plant, and,

(*iv*) Direct extraction or evacuation from the furnace shell through separate or fourth hole in the furnace roof/5th hole if continuous charging is adopted through the fourth hole.

The first two methods are also obviously less efficient due to cross-draft and external atmospheric conditions and hence need much higher blower powers and capacities. The fourth system is most efficient and is usually adopted in recent furnaces.

All these systems are schematically presented in Figs. 9.1, 9.2 and 9.3, which do not need any further comments.

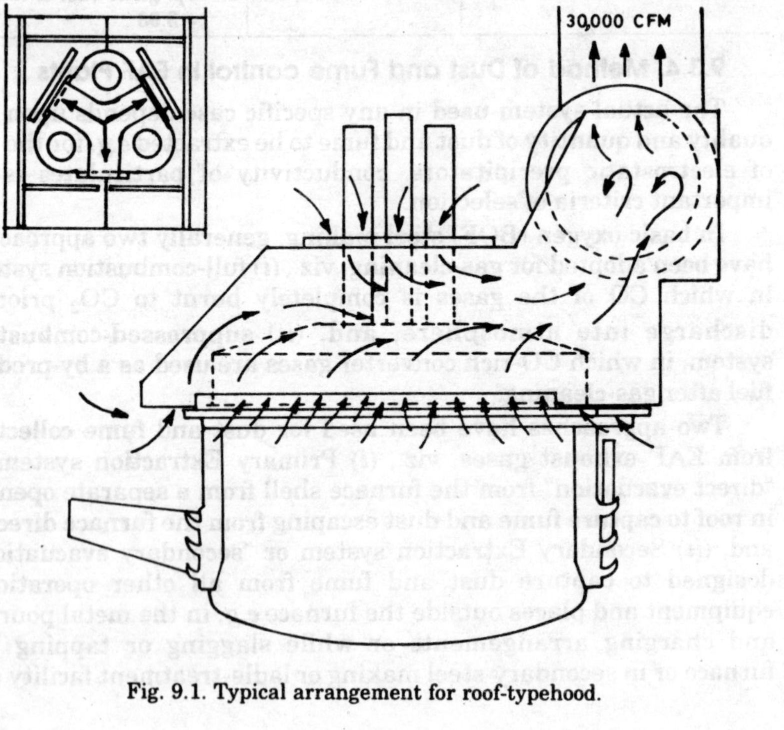

Fig. 9.1. Typical arrangement for roof-typehood.

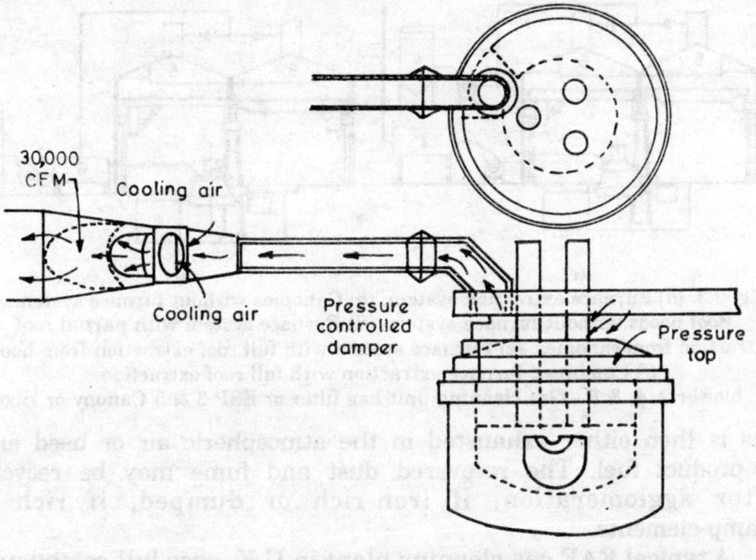

Fig. 9.2. Method of fume control by shell evacuation or direct shell top.

The dirty gases from the furnace or the plant, laiden with dust and fume, are then cleaned either in Bag filters in dry state or in Electrostatic separators or ESPs in dry or wet state, or in spray-towers, venturi scrubbers etc. in the wet gas cleaning systems. The cleaned

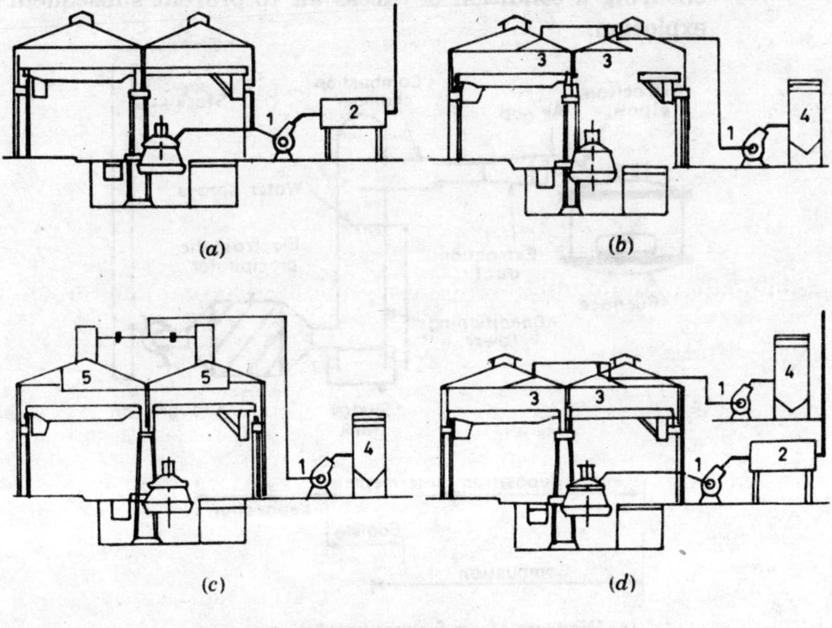

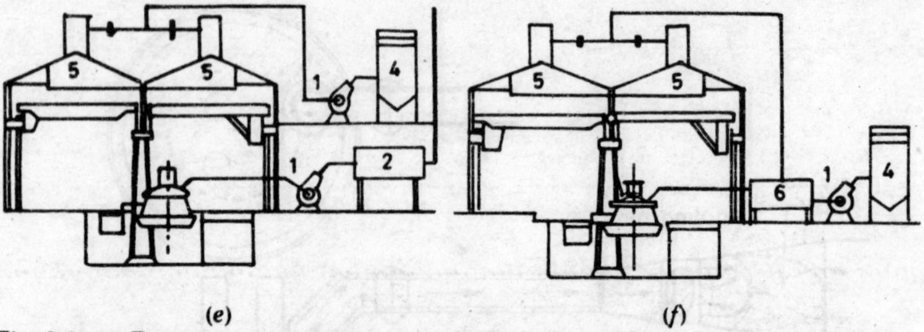

Fig. 9.3. (*a*) Furnace extraction system, (*b*) Canopies without furnace system, (*c*) Roof hoods without furnace system, (*d*) Furnace system with partial roof extraction from canopies, (*e*) Furnace system with full roof extraction from hoods, (*f*) Combined furnace extraction with full roof extraction.
1. blower 2, 4, & 6 - Gas cleaning unit bag filter or ESP 3 & 5 Canopy or Hood.

gas is then either exhausted in the atmospheric air or used as a by-product fuel. The recovered dust and fume may be recycled after agglomeration, if iron-rich or dumped, if rich in tramp-elements.

A typical EAF gas-cleaning plant in U.K. uses full-combustion system, water-cooling or gases for humidification and wet ESP for dust and fume collection. The basic elements of this system, shown in Fig. 9.4, include,

(*i*) A horizontal combustion chamber, designed using isothermal modelling, to burn the waste gases from the fume ensuring a condition of excess air to prevent subsequent explosion.

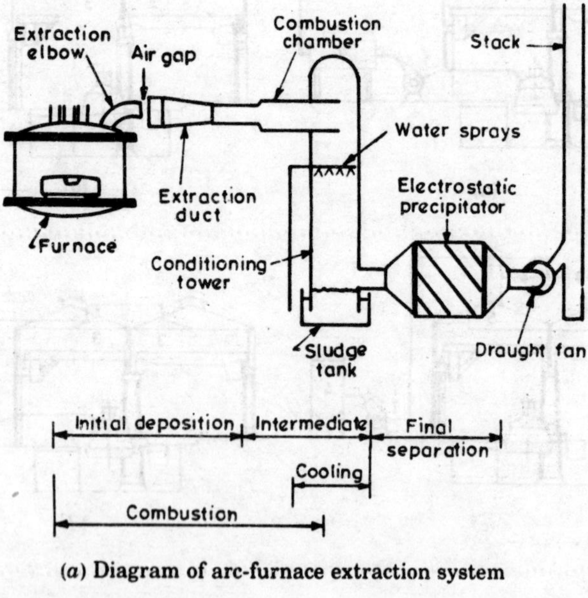

(*a*) Diagram of arc-furnace extraction system

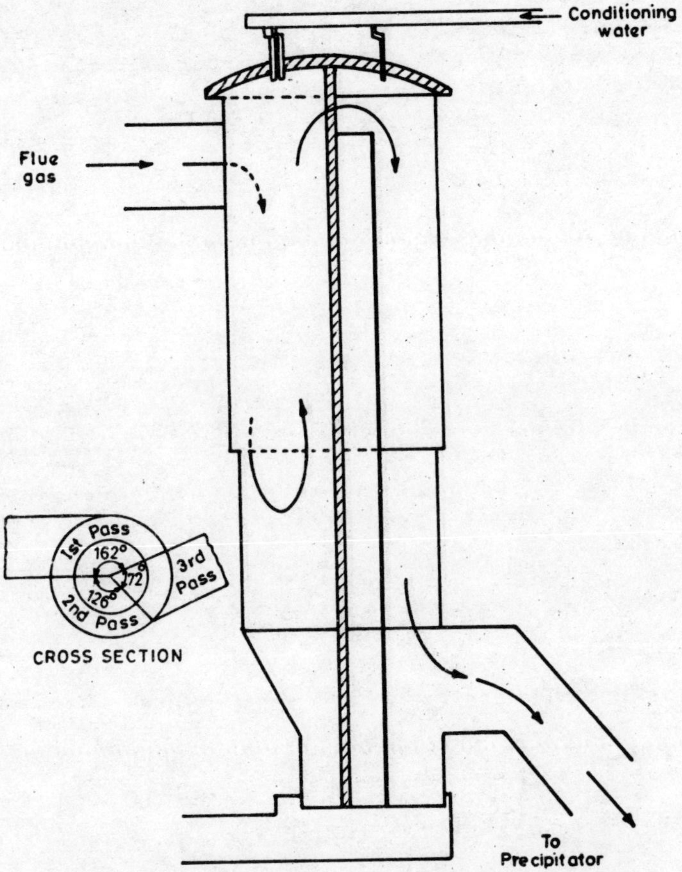

(*b*) Schematic drawing of conditioning chamber.
Fig. 9.4. Typical arrangement of Arc Furnace Extraction System for full-combustion water scrubbing and ESP gas-cleaning system.

(*ii*) a multi-pass cooling and humidification chamber.

(*iii*) wet electrostatic precipitators, leading to a fan and a stack, operating at extraction volume of 16.5 Nm^3/s for five 120/180 tonne furnaces.

The object of conditioning chamber of the E.A.F. extraction system is to cool the effluent gases from the furnace to below 100°C and saturate them with water vapour, any unsaturation leading to caking of the damp fume on precipitator electrodes, not easily removed by regular flushing of the precipitator and resulting failure of the E.S.P. dust-collection system. Mathematical modelling of humidification chamber was used to overcome this difficulty for efficient plant operation.

PART-2

ONE

Advances in Electric Arc Furnace Steel Making Practice for the Production of Tonnage and Quality Steels

1.0. INTRODUCTION

A number of developments have taken place in recent times in Electric Arc Furnace (EAF) steel making practice with a view to achieving faster melting rates, greater efficiency of power utilisation, and the production of quality steels. In 1987, EAF steel making accounted for about 27% of world steel production and it grew at an annual average rate of 5% between 1970 & 1987. Projections suggest that EAF steel making could reach 40% of total raw steel production by the year 2000, based on the continual availability of cheap scrap and moderate increase in the price of electrical power.

Technological developments of the EAF process in the 1960s and early 1970s were based essentially on increasing its tonnage capability and maximising its power input/ton steel produced. In these days, it was realised by larger steel works that EAF could be a viable alternative to open hearth furnace for the tonnage production of carbon steels. At the onset of 1980s greater emphasis was laid on the production of quality steels using EAFs and process optimisation.

In modern EAF plants, the traditional role of EAF steel making has now been split into three separate steps :

(*a*) Scrap preheating in a separate vessel.

(*b*) Scrap melting and

(*c*) Refining of molten metal in an array of secondary (or ladle refining) processes.

Transferring the major refining phase out of the EAF to another treatment ladle can save vital minutes of production. Besides becoming an efficient scrap melter capable of very high production rates and precise control, a modern EAF can readily be adapted to use a wide range of raw materials. The above combination (traditionally known as duplexing) can lead to the production of quality steels in an efficient and economic manner.

A host of new measures and technologies have been introduced in modern EAF plants to improve its overall performance, which are described in following sections.

1.1. CHANGES IN FURNACE DESIGN AND PROCESS EQUIPMENT

1.1.1. UHP Transformers for Rapid Melting

An increase in the size of EAFs coupled with proportionate increase in the capacities of high power (HP) or ultra high power (UHP) transformers to feed these furnaces have led to considerable increase in the productivity (t/hr.) of EAFs. This change in place of regular power transformers not only expedites melting but also it results in more stable arc in terms of uniform and consistent power input. The UHP transformers are best utilised while making plain carbon tonnage steels. It is because, while making high quality complex alloy steels, the reducing period is longer and the power of transformer is utilised less effectively. Therefore, EAFs equipped with UHP transformers should be assigned only the task of rapid melting and oxidising refining and the entire operation of desulphurisation, deoxidation, alloying, composition and temperature adjustment or degassing etc. should be transferred to another equipment of secondary steel making *e.g.* Ladle furnace (LF). Thus, UHP EAFs function most efficiently in a duplexing operation. Transformer ratings of EAFs following double slag practice for the manufacture of alloy steels and a single slag practice for the manufacture of plain carbon tonnage steels are given in the following table :

Nominal Capacity of furnace, t	*Double Slag Practice*		*Single Slag Practice*	
	Standard rating, KVA	*Specific Power KVA/t*	*UHP rating, KVA*	*Specific Power KVA/t*
40	12,500	312.5	22,000	550
50	15,000	300	30,000	600
100	25,000	250	50,000	500
120	35,000	291	70,000	583
150	40,000	266	80,000	533

Because of the use of water cooled panels (WC panels), most furnaces above 50t are equipped with transformer specific power in the range of 600 to 900 KVA/t and secondary voltages in the range of 600 to 900 V are now quite common. Performance data of four different UHP furnaces is summarised in the following table :

Item	*Units*	*1*	*2*	*3*	*4*
Shell Diameter	mm	3350	4100	5200	6700
Tap weight	t	22-20	27	57	140-145
Transformer rating	MW	13	20	34	80
Max. melt-down power	MW	9.5	13	25-26	45-53
Approx. power factor	—	0.70 0.75	—	—	0.67 0.69
Melt-down Currents	KA	3 × 26—27	3 × 30—42	45-50	—
Electrode dia	mm	350	400	500	600
Electrode length	mm	1830	1830	—	—
Energy consumption	KWH/t	510-530	—	490	510-540
Electrode consumption	kg/t	5	—	5.5	4-4.75
Top-to-tap time	hr-min.	—	1.30 to 1.55	2.15	2.20 to 2.30

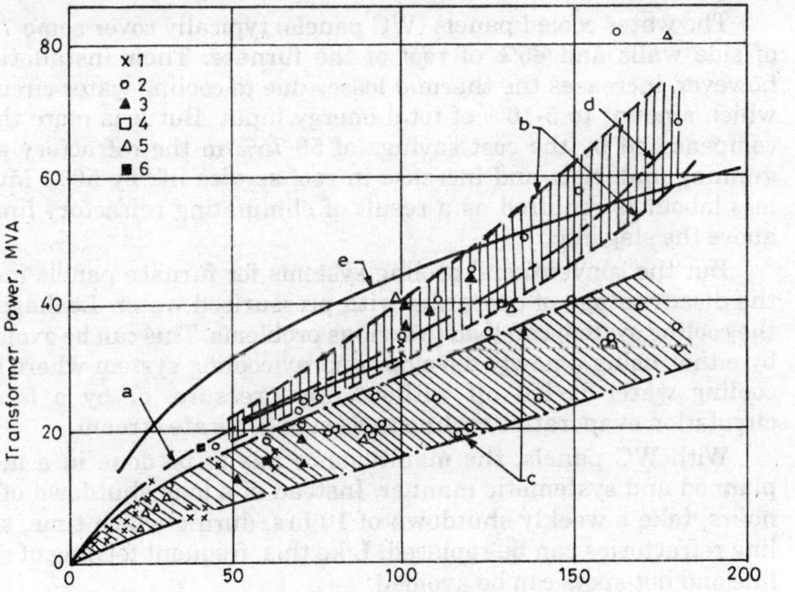

Fig. 1.1. Transformer power versus furnace capacity
1. USSR; 2. USA; 3. FRG; 4. France; 5. Great Britain; 6. Japan; (*a*) recommended by VNIIETO; (*b*) recommended by DEMAG; (*c*) region embracing the majority of existing furnaces; (*d*) recommended in the USA; (*e*) recommended by V.A. Romenets and A.M. Leontyev.

Fig. 1.1 compares the power of transformers of furnaces employed in the USSR, USA, FRG, and other countries. In modern practice, there is a general trend to increase transformer power. The significance of UHP has gradually shifted to the concept of "ultra high productivity". This has led to increased productivity to the tune of nearly 50 t/hr. However, under Indian conditions, the specific output of EAFs is nearly 10 t/hr.

1.1.2. Water Cooled Furnace Panels (WC Panels)

After the melt down stage, the tap voltage is decreased to decrease power input. The objective is to protect the side-walls and roof refractory lining from the reflected arcs. A higher voltage tap and therefore longer arcs have to be avoided during the reducing period. Usually, hot spots would occur on the opposite side of these electrodes. And the repairing of hot-spots by fettling and gunning is trouble some and costly and ultimately cause loss of productivity. So, quick heats are possible only by installing water-cooled side walls and roof. With W.C. panels, the furnace can be operated continuously at high power input (500-1000 KVA/t) and high efficiency without any danger to side walls and roof and under these conditions, the transformer (UHP) is fully exploited.

The water cooled panels (WC panels) typically cover some 70% of side walls and 85% of roof of the furnace. Their installation, however, increases the thermal losses due to cooling water circuits which amount to 5-10% of total energy input. But it is more than compensated by the cost savings of 50-75% in the refractory and gunning materials and increase in roof service life by 50%. Much less labour is required as a result of eliminating refractory lining above the slag line.

But the conventional cooling systems for furnace panels incur the disadvantage of operating with pressurised water. Leakage in the cooling system can lead to serious problems. This can be avoided by either using a newly developed spray cooling system where the cooling water is led off without any pressure, or by a forced circulation evaporative cooling system to generate stream.

With WC panels, the maintenance has to be done in a more planned and systematic manner. Instead of a long shutdown of 40 hours, take a weekly shutdown of 10 hrs, during which time, slag line refractories can be replaced. Like this, frequent fettling of slag line and hot-spots can be avoided.

By operating the furnace continuously, the following advantages are obtained :

(1) Drastic reduction in electrode consumption. (From usual 5-5.5 kg/t down to 2.5-3.5 kg/t: nearly 50% reduction).

(2) More scrap can be accommodated in the furnace (~20% or more liquid tapped after installing WC panels).

(3) Power consumption decreases by ~20% down to 400-500 KWH/t.

(4) Productivity increases and naturally the cost of production decreases. Productivity has been reported to increase upto 45% mainly because of a decrease in tap to tap time (ranging between 60 to 90 minutes per heat).

(5) Reduction in refractory consumption both for roof and side walls.

Two more recent developments have enhanced the use of WC panels and roofs :

(1) Electrode Controllers (2) Foamy slags.

1.1.3. Electrode Technology

The electrode cost has been observed to be about 6-8% of total production cost in Electric Arc Furnace steel making. This percentage would vary for different capacities of EAFs, and for different qualities of steels produced.

Besides the quality of graphite electrodes, a number of operating parameters influence the specific consumption of graphite electrodes/t of liquid metal produced. These values are usually in the range of 5 to 9 Kgs/t liquid metal produced. Fig. 1.2 shows, how this figure is generally related with capacity of EAFs.

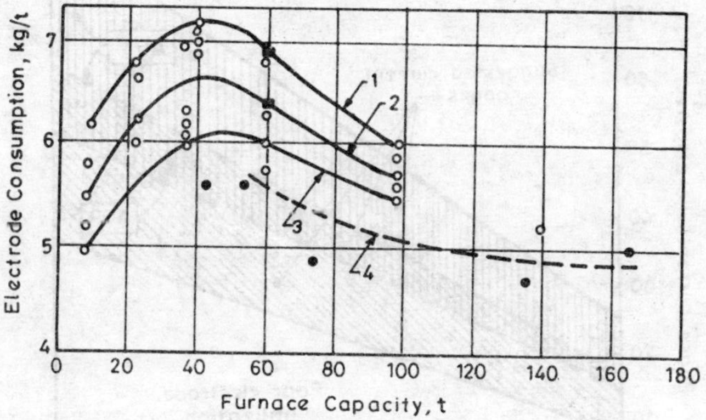

Fig. 1.2. Consumption of electrodes in various electric furnaces.
1, 2, 3 - Soviet furnaces (1963, 1965, 1967); 4- foreign furnaces.

With the advent of HP and UHP furnaces and with power levels approaching 1000KVA/t, the current density necessary to be carried by the electrodes is much beyond the safe level of operation for regular power electrodes (RPE). Special electrodes meeting the need of HP & UHP furnaces are respectively termed as HP and UHP electrodes. These electrodes are required to have homogeneous properties to :

(*a*) Operate at high current density level - low resistivity.
(*b*) Have, low coefficient of thermal expansion.
(*c*) Have high thermal shock resistance.
(*d*) Have high mechanical strength.

But an electrode with low coefficient of thermal expansion will normally have poor strength. Also, high strength electrodes will have poor thermal shock resistance. Therefore, a judicious blending of these properties is necessary to make it match the specific requirements. The Calcined Petroleum Coke (CPC) required for HP/UHP electrodes is Needle Coke, which has low CTE. This Needle coke is not manufactured in India at present and is required to be imported.

Besides this, binder pitch/impregnation pitch of a certain quality required in the process has also to be imported. Both these factors increase the cost of manufacture of HP and UHP grade electrodes. Fig. 1.3 shows the current carrying capacities of regular and premium grade electrodes :

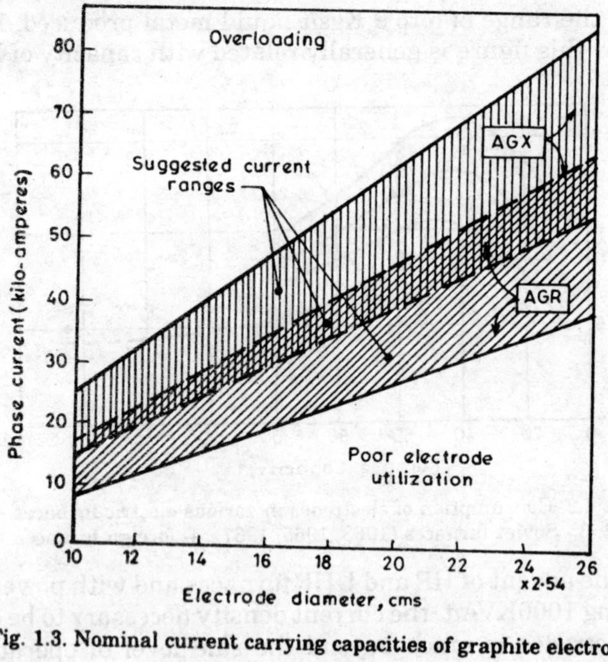

Fig. 1.3. Nominal current carrying capacities of graphite electrodes.

It is well known that electrodes are consumed in practice due to following factors.

(*a*) Consumption due to oxidation - 40-50%

(*b*) Consumption at tip (Tip loss) - 35-45%

(*c*) Consumption due to breakage/ - 10-15% stub end losses

Consumption due to oxidation is dependent upon the surface temperature of electrodes and on the velocity, turbulence and oxygen content of gases passing over the electrode. A very low furnace pressure and extended duration of processing period (tap to tap time) increase oxidation losses. Therefore, Kg of electrode consumed/t of LMP (liquid metal produced) is high. So, the furnace has to be operated carefully to minimise electrode consumption.

Tip loss is directly related to the electrode current and rate of power input. Short arcs and high current results in reduced refractory wear but in the process, greater electrode tip consumption occurs. Also, dipping of electrode tip in the bath for carbon pick-up also results in greater tip loss. Fig. 1.4 shows relationship of Tip Loss with mean electrode current.

Breakage and stub end losses can be reduced by preventing scrap falls, avoiding contact with non conducting materials (which cause wandering of arc), minimising thermal shock, rigid electrode arms and furnace masts and efficient operation of the electrode regulators. It can, therefore, be concluded that electrode usage has a important bearing on electrode consumption.

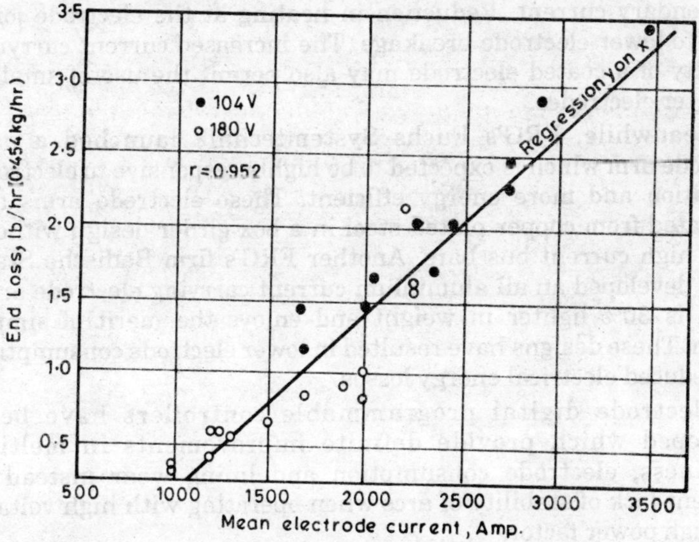

Fig. 1.4. Rate of tip loss versus electrode current.

New Developments

Two major developments have taken place :

(1) Water Cooled electrodes.

(2) Coated electrodes.

Introduction of water cooled electrodes has reduced the average electrode usage from 4.5 to 3 Kg/t of liquid metal produced (LMP). These electrodes can run cooler with reduced oxidation losses at high temperature and electrode diameter can often be made smaller thus reducing the surface oxidation. But these water cooled systems require considerable modification in equipment, maintenance, cooling water volume and meltshop space. Electrodes can also be cooled externally by sprayed water. This produces a cooling water film on the surface of graphite electrode, which minimizes oxidation loss. This has been claimed to reduce the electrode consumption by 10-15%.

The use of coated electrodes has reportedly lowered electrode consumption by 20-30% together with less than normal breakage. Typically, an electrically conductive electrode coating consists of three layers of arc sprayed aluminium, followed by a layer of sprayed refractory component, and lastly, a fusing arc treatment. This coating provides a protective layer highly impermeable to oxygen during operation. Sidewall oxidation should be delayed until the coated electrode is deep within the EAF and the high temperature has destroyed the coating. Coatings also provide an electrically conductive outer skin to reduce the current density within the electrode. As these coatings are nearly 40 times more conductive than the graphite, they could carry a significant portion of secondary current. Reduction in heating at the electrode joint tends to lower electrode breakage. The increased current carrying capacity of a coated electrode may also permit the use of smaller diameter electrodes.

Meanwhile, FRG's Fuchs Systemtechnik launched a new electrode arm which is expected to be highly responsive to electrode regulation and more energy efficient. These electrode arms are fabricated from copper plated steel in a box girder design without using high current bus bars. Another FRG's firm Badische Stahl Engg. developed an all aluminium current carrying electrode arm, which is 30% lighter in weight and enjoys the merit of simple design. These designs have resulted in lower electrode consumption and reduced electrical energy losses.

Electrode digital programmable controllers have been developed which provide definite improvements in melting steadiness, electrode consumption and lining wear instead of inherent lack of stability of arcs when operating with high voltage and high power factor.

1.1.4. Oxy-fuel burners and Scrap pre-heaters

In EAF cold zones between the three phases occur due to which the charge in these areas does not melt down quickly and uniformly. To overcome this problem and for achieving efficient, rapid melting, oil/gas oxy-fuel burners can be employed. For a small to medium size EAF, a single burner introduced through the slag door can effectively reach all the three "Cold-spots". Use of nearly 10-30 m^3 of oxygen/t during melting is almost a standard practice. The thermal yield of burners is improved by injecting tar or even coal into the flame, which basically increases flame radiation. But sufficient suction of exhaust gases from inside of furnace is necessary to activate the combustion inside the furnace.

The "Split-Fire burner System" developed by an American firm is expected to be more efficient and cost effective than the conventional door mounted system. It provides two oxy-fuel flames from a single burner block, firing broadly across the cold spots to roof ring and are fired downwards tangentially through the furnace cold spots. All these kinds of burners, however, present lots of maintenance problems. Table below shows comparative results of three different companies who have installed Oxy-fuel systems :

	Company A	*Company B*	*Company C*
Mounting location	Slag door	Sidewall	Roof
Number of burners	1	3	3
Heat size, t	20	110	70
Transformer capacity MVA	7.5	28	25
Sidewalls	Silica	Water cooled	Water cooled
Roof refractory	Silica	Alumina	Alumina
Primary scrap	Turnings	Mix	Structural
Firing rate, Gjh^{-1}	5	76	41
Results			
Production increase, %	10	15	15
Power saving $KWht^{-1}$	16	40	25
Electrode savings, kgt^{-1}	0.7	0.2	—

M.B. Wells and F.A. Vonesh : Iron Steelmakers, 1986, 13(11), 13-22

As expected, the production increase attained with a three burner system is higher than that with a single door mounted burner.

Several other variants of burner systems have been introduced recently. For example, a premixed combustion burner installed at the Bethlehem Steel's sheelton works is capable of high velocity flames (1000 $m.s.^{-1}$) and high energy input (50 GJh^{-1}). Its flame can cut a trough into the scrap burden with no rebound or impingement on the refractory walls. Another new burner system known as the "Pyreton system" was introduced by the American Combustion Company. It consists of water cooled burner with separate introduction of air, oxygen and fuel into the combustion chamber (Fig. 1.5). Individual mixing ratios can be adjusted to

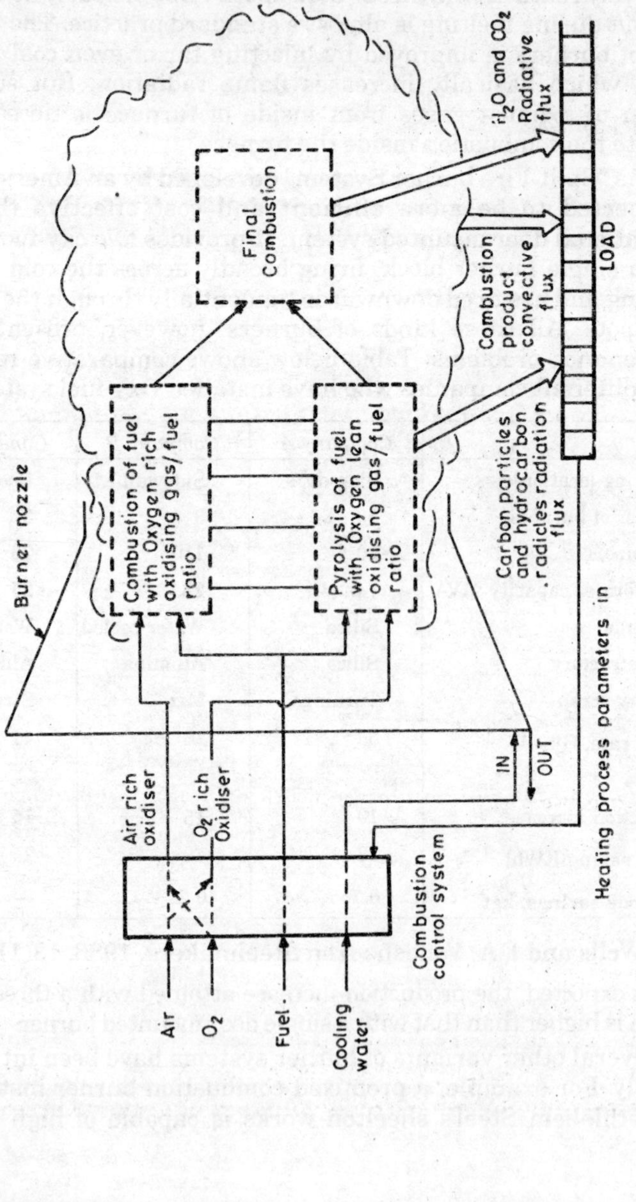

Fig. 1.5. Pyretron-combustion and heating process flow diagram.

provide a very broad range of flame characteristics. Test data have shown a 12% improvement in productivity with lower nitrogen pick up, and minimum maintenance problems.

Scrap heating has emerged as the most effective way of recovering waste heat from exhaust gases and recycling this energy into the charge. Conventional scrap preheaters may recoup upto 10% of the heat loss from the exhaust gas and reduce the energy requirement for the EAF process by 20-25 KWH/t of steel tapped (HMP). Additional advantages of scrap preheater include faster melting cycle, increased yields, lower electrode and refractory cost and the possibility of concentrating Zinc in the EAF dusts.

Oxy-fuel burners coupled with scrap preheating arrangement as shown in Fig. 1.5 would help in reducing power and electrode consumption substantially and increase productivity by more than 50%. Scrap can be easily preheated to 300–400 °C by the scheme of things shown in Fig. 1.6. The accompanying table also shows the combined advantage of using WC panels, Oxy-fuel burners and scrap preheating arrangement.

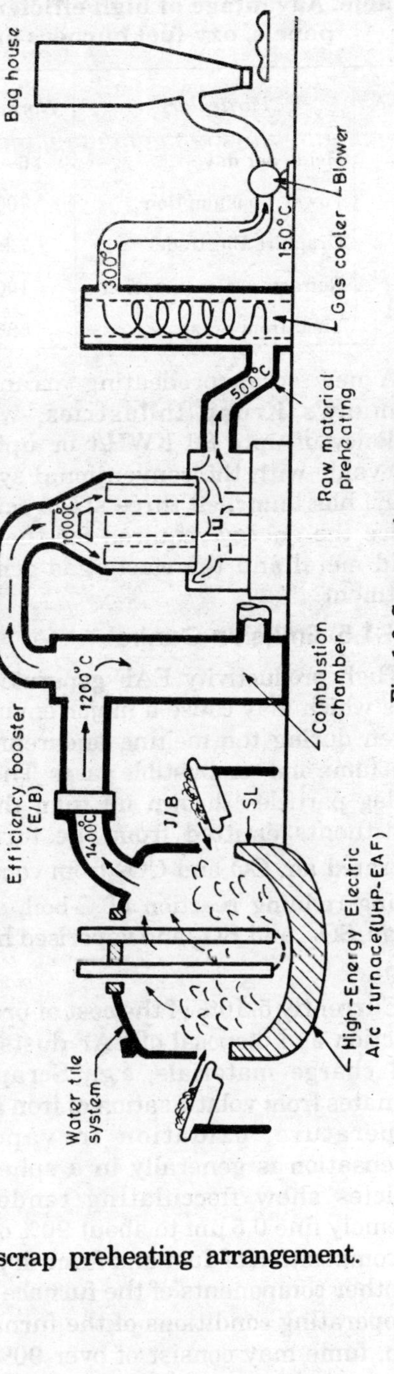

Fig. 1.6. Scrap preheating arrangement.

Table. Advantage of high efficiency operation using water panels, oxy-fuel burners and scrap preheating

S. No.	*Particulars*	*Normal operation*	*High Efficiency Operation*
1.	Heats per day	6	10
2.	Power Consumption	700 KWH/ton	500 KWH/ton
3.	Graphite Electrode	7 kg/ton	4 kg/ton
4.	Refractory Consumption	100	30
5.	Yield from Scrap	88%	99%

A new scrap preheating arrangement has been devised by Germany's Krupp Industries, which shows greater energy efficiency of upto 80 KWH/t or upto 2.5 times higher than that achievable with the conventional systems. Japan's Nippon Kokan (NKK) has launched three stage scrap preheater which is said to reduce the energy required for the EAF process by 50 KWH/t of liquid metal and the waste gas generated requires no additional treatment.

1.1.5. Emission Control

High productivity EAF generaters large volume of exhaust off gases which may cause a major environmental problem. Injection of oxygen during the melting and refining period produces both iron oxide fume and combustible gases. These emissions are supplemented by slag particles thrown up from the foamy slag. Typical gaseous constituents emitted from the furnace include N_2 and O_2 from infiltrated air, CO and CO_2 from combustion of electrodes and steel making refining reaction of C-boil, water vapour from moist solid charge, SO_2, and SO_3 and vaporised hydrocarbons from oily or greasy scrap.

Currently 5-10% of the cost of production is associated with the collection and disposal of EAF dusts. The dust originates from the solid charge materials, *e.g.*, Scrap, DRI, flux, Ore etc. Fume originates from volatilisation of iron and tramp elements at furnace temperature, oxidation in vapour state and subsequent condensation is generally in a spherical shape. These condensed particles show flocculating tendency and vary in size from extremely fine 0.5 μm to about 90% of (-) 400 mesh (-37 μm) in size. The composition of dust and fume depends upon the quality of scrap and other components of the furnace charge, the type of steel made and operating conditions of the furnace. While melting high grade scrap, fume may consist of over 90% iron oxides, but with dirtier scrap with residual tramp elements, appreciable concentrations of Cu, Zn, Pb, Mn, Sn, S, P, Na, K etc. may also be present with maximum concentrations of Cu and Zn upto 25% and 5% respectively. In some cases, Zn may go upto 10-15% and lead may be present in the range of 3-6% along with cadmium and chromium, which are known as hazardous compounds. Quantity of dust and fume also vary appreciably from 0.25 to 3.5 grains/cu. ft. of ventilated air from EAF (equivalent to 4-30 lbs/t of metal poured or approx. 1-2% of steel made). Dumping of this dust is becoming unacceptable, as it creates pollution and environmental problems. A number of processes have been developed to process these dusts.

The EAF baghouse dust can alternatively be recycled back to the furnace. Two common recycling methods are available. The first recycling option employs a pellitizer by which EAF dusts are agglomerated as macropellets and charged into the EAF. The merits of pelletising recycling are its lower capital cost, less dust generation, and little adverse effects on steel or slag chemistry.

The second recycling option is also known as the direct injection technique where baghouse dust and screened coke breeze are fed separately into a high speed screw conveyer followed by a water spray. The screw conveyor is used to densify the micropellets formed and these are then screened and stored for 1-3 days to gain adequate green strength. A good material balance between slag metal and dust was reported which indicated that nearly all of the Zinc, Lead and Cadmium, injected as micropellets returned to the dust during recycling.

Many improved pollution abatement techniques have been developed. The customary approach is to use a combination of overhead canopy hoods to capture charging and tapping emissions and a fourth hole for direct fume extraction during furnace melting. An alternative development is the emergence of a clean house enclosure or "dog house" system which not only takes care of fume emission, but offers some abatement of noise levels by 20-30 dB during melt-down. It has also cut energy usage by 10-15 KWH/t, as it requires less fan capacity than a conventional EAF with fourth hole and canopy.

1.2. DEVELOPMENTS IN PROCESS TECHNOLOGY

1.2.1. Raw-material Charging

Charging practices affect both the EAF productivity and the material consumption rate. Faster melting requires an optimum scrap mix for each heat. Scrap chemistries, densities and the basket loading sequence must be considered. Development of a computerised charging programme is essential for proper control. For an optimal EAF operation, the recommended procedure is to charge two-thirds of the scrap in the first basket, and the remainder one-third in the second basket which replaces the volume of scrap melted. This gives maximum furnace filling rate and enables maximum energy utilisation for melting scrap, with the lowest heat losses, by ensuring minimum exposure of arc radiation to the WC panels. Continuous feeding of fragmented scrap via the roof during melting limits the charging basket to one and reduces the power off times and heat losses caused by furnace roof opening. Also, pneumatic injection of lime during melting improves both, dephosphorisation and desulphurisation and eliminates the need to charge lime in the basket. Automatic lime and alloy charging facilities became indispensible, if productivity has to improve.

The use of DRI (or HBI) and its charging system and related aspects are discussed in subsequent sections.

In Italy, densification and cleaning of scrap is being practised and the plants have been equipped with shredders or shears with magnetic separators or vibrating screens in the yard.

1.2.2. Oxygen Lancing

In the past 10 years, Oxygen lancing has become an integral part of the modern steel making facilities. Conventional use of water-cooled oxygen lances through sidewalls is effective only for decarburisation in the molten bath and may increase the electrode consumption if used improperly. To overcome these drawbacks, West Germany's Badische Stahl Engineering designed and built an "Oxygen lance manipulator" to cut scrap and to inject oxygen into the bath for decarburisation and into the slag to assist foaming. This lance manipulator consists of two independently controllable lance mechanisms mounted as one unit, and are located at the end of gantry arm. It reportedly reduces the melting cycle by 12%, lowers the electrical power usage by 60 KWH/t or (13%) with less labour intensive and improved working conditions.

Tonnage oxygen has long been used in EAFs for the production of stainless steel heats from all S.S. Scrap. The use of oxygen enables much higher temperatures to be attained of the order of 1800°C and these conditions are known to favour the oxidation of carbon without undue loss of chromium. It is known that there is a definite Cr/C ratio for every temperature and the higher the temperature, the higher is the value of this ratio. Most of the chromium oxidised can be recovered from the slag by reduction with ferrosilicon, aluminium-silicon or ferrosilicon chrome, the latter being commonly favoured as being the most economical method of adding both silicon and low carbon chromium provided always that an addition of chromium is required. Oxygen injection is carried out in two stages with vigorous rabbling interposed between the two stages. This is necessitated by the layering effect due to the rapid and complete elimination of carbon from the upper layers of the bath, while the layers close to the bottom are relatively un-decarburised. Vigorous rabbling prior to further oxygen injection is, therefore, necessary. Also, oxygen blowing is known to flush out dissolved hydrogen from the melt. Chromium recoveries in the range of 90 to 94% have been recorded with carbon contents as low as 0.05%. The power consumption for heats larger than 10 tonnes is in the range of 500 - 550 KWH/t, which reflects a saving of about 200 KWH/t over straight melting and refining practice. This of course more than repays the additional cost of oxygen.

The oxygen blowing technique referred to above has now been overtaken by more modern processes like AOD, VOD and CLU, which are discussed in detail in Chapter-3.

1.2.3. Furnace Powder Injection and Foamy Slags

The main objective of injecting powder materials into an EAF is to create a favourable metal bath condition for metallurgical reactions. A number of powder materials are injected into the slag/metal system to achieve different objectives; and some of these materials are listed in the following table ;

Material	*Functions*
Calcium carriers (CaC, CaSi)	Desulphurisation, deoxidation
Reactive powders (lime, flue dust, fluorspar, iron oxide)	Desulphurisation, Dephosphorisation
Mixtures (Scale-limestone, coal limestone, scale-coal)	Foaming slag, improvement of reaction kinetics
Lime powder, flue-dust, oxygen and Argon carriers.	Decarburisation

Injection of carbon (anthracite or coke breeze) and oxygen into a steel bath produces large volume of carbon-monoxide. This gas foams the slag from its normal thickness of 100 mm to 300 mm and it envelopes the electrode tips and submerges the arc. Notable reduction in both the energy and electrode consumption can be achieved through the use of quasi-submerged arc operating practice with a foamy slag.

As the inside of the furnace is kept oxidising in nature during the melting period to activate the combustion within the furnace, it is necessary to inject carbon to the molten steel after the melt down by means of carbon injector in order to achieve desired meltdown carbon content.

The disadvantage of oxygen assisted melting is that it produces extremely oxidised molten steel bath. With carbon injection, it is possible to decarburise the steel and avoid overoxidised bath at the end of melting period. This results in ideal melt down chemistry.

If the molten steel is extremely over oxidised, the refining period is extended considerably and over oxidised bath results in lower yield and increased consumption of ferro alloys. If powdered carbon is injected into the molten bath, it is possible to get out of extremely oxidised condition.

If carbon injection is done at the end of oxidising period, deoxidation is done along with recarburisation. In this manner molten steel enters from the oxidising period to reducing period with lower oxygen content. Carbon is ideal deoxidiser at it can reduce the oxygen content without leaving behind any products of deoxidation in the bath. Steel makers have been practicing electrode dipping for producing cleaner steel. Carbon injection performs the same operation at lower cost. So, carbon injection leads to better steel with lower inclusion ratings. Lower oxygen results in increasing scrap yield and lower consumption of deoxidisers. When it is desired to decrease the excess oxygen without increasing carbon, the carbon injection must be done in a predetermined manner towards the end of oxidising period. Thus, oxidising and refining periods can be made overlapping, with the overall reduction in tap to tap time. Since both oxygen and FeO contents of slag can be reduced simultaneously with recarburisation, it becomes very easy to switch over to single slag practice.

In order to achieve efficient S and P removal, reactive slag powders (*e.g.* flue dust lime/fluorspar/iron oxide) can be injected into the molten bath via a side wall entry submerged lance. Using this approach, it is possible to remove substantial amounts of sulphur and phosphorus. However, the final levels of S and P attainable are dependent on the quality of scrap being used. Injection facilities installed at two LTV steel works in North America have shown that in furnace dephosphorisation (to < 0.003% P) and desulphurisation (to around 0.016% S) could be achieved. In addition, by injecting an appropriate combination of Oxygen and Argon gases with solids (lime and flue dust), melts with a Carbon level of 0.020% C could be attained during the usual 15 min. refining period. Powder injection also exposes fresh, colder liquid to the hot arc thereby homogenising the bath with respect to temperature gradients existing in the melt.

In another investigation using twin lance for parallel injection of oxygen and lime fluorspar mixture, it has been reported over a three year trial period that injection of 30 $Kg.t^{-1}$ of lime fluorspar mixture in the ratio of 6 : 1 gave an effective phosphorus removal in the range 68-75% with simultaneous removal of some 30-50% of sulphur. Steels were produced with 0.005 P and 0.2-0.4%C.

1.2.4. Slag-Free Tapping (EBT) and Hot-heel Practice

On account of the oxidising meltdown as well as operation with a foaming slag upto shortly before tapping, both the bath as well as the slag are strongly enriched with Oxygen. With the change-over to reducing ladle metallurgy, as with the oxygen convertor or EAFs, it is necessary to separate the iron oxide rich furnace slag from the steel to the fullest possible extent. Prevention of oxidising slag from entering the ladle is essential for the efficient operation of subsequent ladle refining and continuous casting. The slag can either be retained in the furnace on tapping or be removed from the ladle after tapping. The deep level or siphon tap was already introduced at the end of 1970's whilst the "eccentric bottom tap" was a further development of the central bottom tap and was first introduced at the beginning of 1983. Various slag separating techniques are shown in Fig. 1.7.

The emergence of eccentric bottom tapping (EBT) has made a major contribution in achieving a "slag free trapping" condition.

Slag separation on tapping

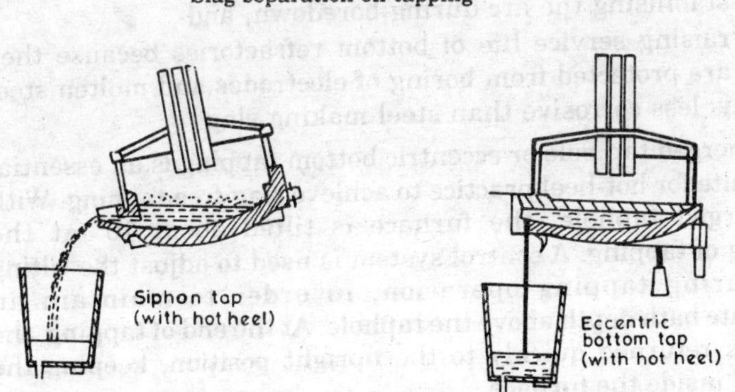

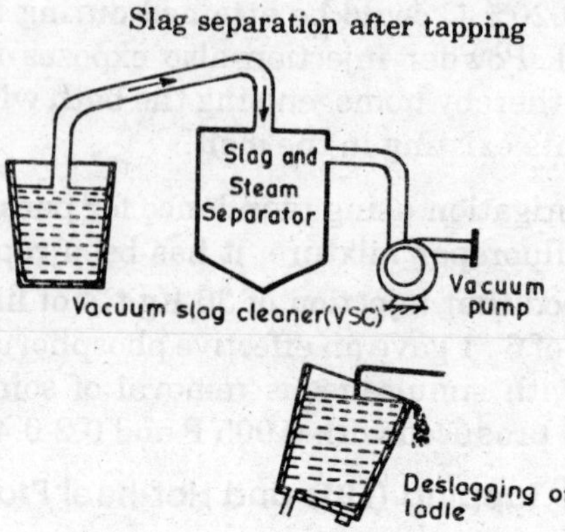

Fig. 1.7. Methods of slag separation on tapping or after tapping.

Additional benefits of EBT lie in the extension of its water cooled panels, lower temperature loss, and reduced gas pick-up by molten steel during tapping. With EBT, the tapping time is so short that neither alloying nor desulphurisation agents can be so effectively added as they can be with the siphon tap. The hot oxide rich slag remains in the furnace and provides excellent initial conditions for the dephosphorisation of the subsequent heat. The ladle is protected against contact with oxide rich slags. So, the life of ladle is improved. On tapping, undesirable reactions with alloy additives in the ladle are avoided. For slag free tapping, a liquid residue (hot-heel) amounting to some 15-20% of molten steel is retained within the furnace hearth with both methods (Siphon tap and EBT). Uptill 1986, more than 30 furnaces were provided with EBT system. The practice of retaining a "hot-heel" in the furnace hearth has following advantages :

1. providing efficient melting and refining for the next heat.
2. stabilising the arc during boredown, and
3. raising service life of bottom refractories because they are protected from boring of electrodes and molten steel is less corrosive than steel making slag.

Submerged tap hole or eccentric bottom tapping is an essential prerequisite for hot-heel practice to achieve slag-free tapping. With a submerged taphole, the furnace is tilted about 10° at the beginning of tapping. A control system is used to adjust the tilting angle during tapping operation, in order to maintain an appropriate bath depth above the taphole. At the end of tapping, the furnace is returned quickly to the upright position, keeping the "hot-heel" inside the furnace.

Fig. 1.8 shows a schematic view of the EBT furnace set up and Fig. 1.9 shows the refractory lining of EBT furnace bottom belonging to Bedische Staklwerke AG, FRB. (BSW). The bottom sole of 40/48 MVA 5.5. m dia UHP are furnace of Badische Stahlwerke, FRG, is shown in Fig. 1.8. This furnace is fitted with W.C. walls and roof and provided with additional Oxy-fuel burners. The bottom sole is flat to guarantee slag free tapping. The sole casing is lined with two layers of magnesite bricks for safety, topped by a 356 mm layer of dolomite bricks and rammed with a 100 mm layer of magnesite mass. The tapping nozzle has a life of 250 heats. The replacement of nozzle takes two hours. By using the eccentric bottom tapping, the life of the refractory material has been increased to 500 heats and it corresponds to full one month operation without any intermediate repairs.

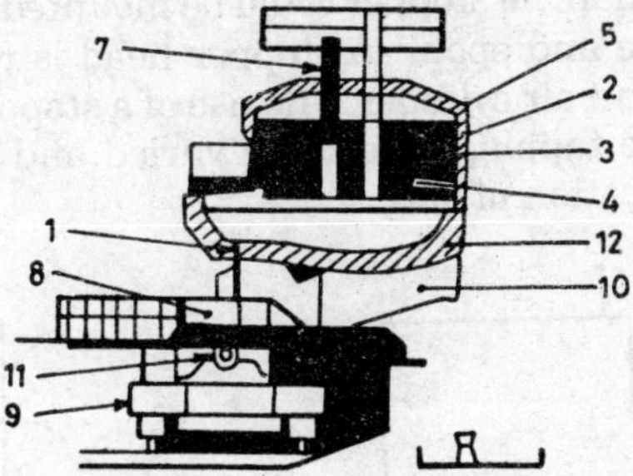

1. Bottom Taphole
2. Shell
3. Steel WC Panels
4. Copper WC Panels
5. Water Cooled Roof
6. Roof Panels
7. Electrode Sealing Ring
8. Ladle
9. Ladle Car
10. Rocker
11. Swivelling Plateform
12. Bottom

Fig. 1.8. Eccentric Bottom Tapping Schematic.

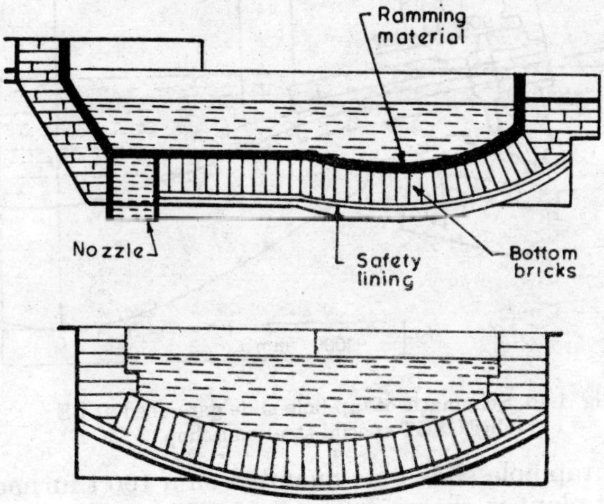

Fig. 1.9. Refractory Lining of EBT Bottom.

In case of the above plant (BSW), the tap to tap time could be reduced from 71.5 minute to 68.5 minutes. This is due to a reduction of the tapping time by 1 min., of the tapping temperature by 15°C and of the fettling time between the heats. This lowered temperature and the shorter tap to tap and fettling times have led to energy savings of 15 KWH/t. Also, refractory brick consumption has been reduced from 0.4 kg/t to 0.1 kg/t (saving of 0.3 kg/t) and fettling mass consumption from 2.7 kg/t to 1.2 kg/t (saving of 1.5 kg/t). Also, the consumption of ferroalloys has been reduced by 5%.

Another technique which employs "interruption of stream" involves the use of slide gates and mechanical stoppers and reladling. During tapping, the weight of steel in the ladle is monitored and the tap-hole slide gate is activated at the first sight of slag. Occurrence of slag can be detected either visually or electromagnetically. Alternatively, a plug type mechanical stopper could be mounted outside the furnace above the taphole and spout. A stopper head is rammed into the taphole by activating an air cylinder. The use of a stopper head as a tapping valve allows the tapping rate to be varied and the tap hole to be sealed on first indication of slag.

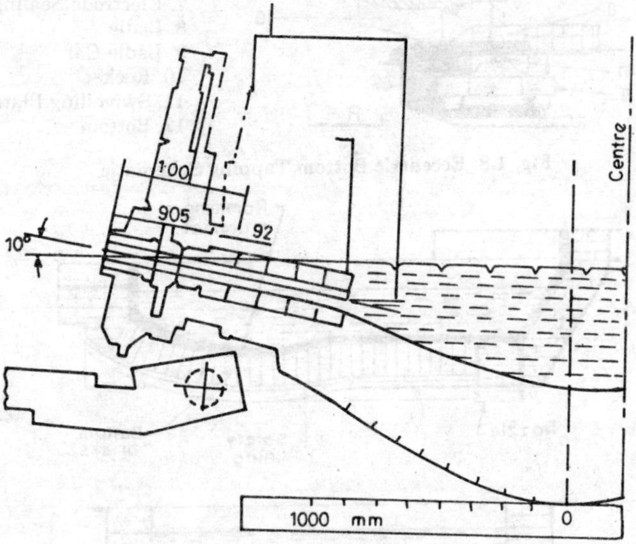

Fig. 1.10. Schematic of tap-hole slide gate used at LE CREUSOT special steel meltshop.

A typical tap-hole slide-gate mounted on a 100 t furnace of Le Creusot steel plant is shown in Fig. 1.10. This system has shown almost total reliability. Since the slide gate is above the liquid level in the bath, so it can be opened during melting. On tapping, the gate is closed by the operator when slag begins to appear. An estimated carry-over of only 200 kg is experienced, leaving behind 1.5 t of slag in the furnace. Previously, the above plant employed reladling to separate slag from liquid steel effectively. Slide gate practice gave excellent results metallurgically (steel containing $P < 45$ ppm, $S < 10$ ppm, $H < 1.5$ ppm and $N < 45$ ppm regularly produced). A new slide gate is installed every 25 heats, needing only 10 minutes of furnace off power.

1.2.5. Submerged Inert Gas Stirring

Attempts have been made to intensify the heat transfer within the molten bath and the molten bath and unmelted scrap, leading to faster melting rates. Submerged inert gas stirring has been conducted in a 18 t EAF by Union Carbide Corp., USA. Argon & Nitrogen were injected through three tuyers placed in the furnace bottom as shown in Fig. 1.11. These injection sites basically represent the locations of cold-spots within the furnace. Improvements of 5-6% in total energy efficiency for melting, faster melting rates and greater alloy recoveries have been reported with total gas flow rates of 0.19 $N.M^3.min^{-1}$. The process has demonstrated a nitrogen dissolution rate of 7%, lower than those generally encountered in a conventional steel making.

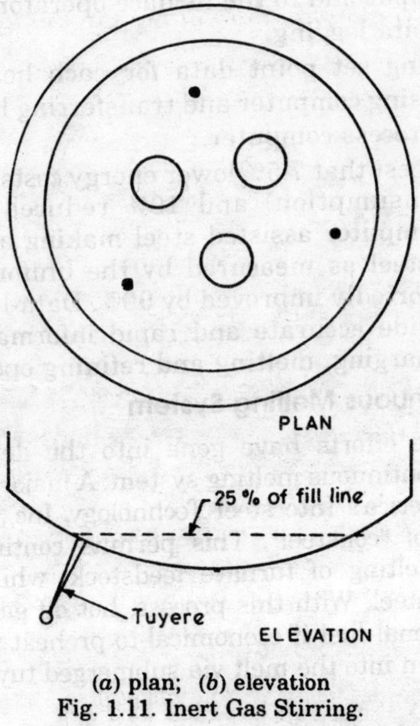

(*a*) plan; (*b*) elevation
Fig. 1.11. Inert Gas Stirring.

1.2.6. Process Automation (Computer Assisted Steel Making or CAS)

CAS is the key to achieving optimised furnace performance, and to keep down raw material costs and energy consumption. This system is normally designed to provide the advice, action, and information necessary to enable the entire meltshop to be controlled automatically while still allowing manual operation at the operator's discretion. Reliable process measurements are vital for the EAF process control system. America's UHP International Inc. has introduced a new electronic metering system for the EAF to display simultaneously 25 parameters affecting optimum delivery and the use of electrical energy. These displayed signals are available as input data for computer control and automation.

The computer in EAF process control performs the following main functions :

(*i*) monitoring the meldown operation achieved by using direct on line measurement.

(*ii*) controlling the thermal balance and energy balance through electrical power input and oxyfuel burners.

(*iii*) controlling the alloy additions, oxygen input and flux additions through metallurgical calculations.

(*iv*) communicating with the scrap yard operators for scrap data input and to the furnace operators with melt state, melt data logging.

(*v*) receiving set point data for each heat from the data processing computer and transferring heat records to the data process computer.

Results suggest that 7.5% lower energy costs (electrical energy and oxygen consumption) and 10% reduced electrode usage attributed to computer assisted steel making are realistic. Also, the quality of steel as measured by the uniformity of chemical composition reportedly improved by 90%. Data-logging and report generation provide accurate and rapid information for effective scheduling of charging, melting and refining operations.

1.2.7. Continuous Melting System

Considerable efforts have gone into the developments of an improved EAF continuous melting system. A major stride in this area has been by America's Intersteel Technology, Inc., which introduced the technology of "consteel". This permits continuous preheating, charging and melting of furnace feedstock, while simultaneously refining molten steel. With this process, hot off gases from EAF are used with additional fuel if economical-to preheat the scrap. Oxygen and coal are blown into the melt via submerged tuyers to assist rapid melting. Molten steel is tapped periodically from the bottom while maintaining full power most of the time.

If the charge is preheated with furnace off gases to 500°C, energy requirement would drop to around 380 $KWH.t^{-1}$, as against 470 $KWH.t^{-1}$, with 100% cold scrap practice. The production rate could be much higher, 2t $MVAh^{-1}$ as against 1.1 t $MVAh^{-1}$ for a conventional EAF. If the scrap preheating temperature is raised to 900°C using combined furnace off gases with ~30 $N.m^3.t^{-1}$ natural gas, a sharp decline in energy requirement to 280 $KWH.t^{-1}$ could be anticipated. Noise level, fume emission and baghouse load are reduced as both the bath and arc are under an almost continuous cover of foamy slag. With an array of secondary steel making (or ladle refining) processes, the system could produce heat every 45-55 minute.

1.2.8. Use of DRI (or HBI) in EAFs

1.2.8.1. General

The availability of a homogeneous high quality feed stock for steel making in the form of DRI and HBI has helped to replace or substitute steel scrap in EAFs. But DRI differs from scrap both in physical and chemical characteristics and these differences have an impact on melting techniques depending upon the practice employed. Sponge iron offers following advantages :

1. Sponge iron has a chemically known homogeneous and uniform composition and that enables the steel maker to predict his end point analysis.
2. It has very low impurity level and the tramp elements like Ca, Sn, Zn, Cr, Ni, As, Sb, Pb are almost absent, thus acting as a purifier to the "dirty" scrap and increasing productivity by reducing refining requirements.
3. Metallurgical reaction kinetics based on mixing and heat transfer are greatly enhanced due to vigorous carbon boil for reaction between FeO and carbon.
4. The uniform size of sponge iron (DRI or HBI) lends itself to conveyor transportation and continuous charging.
5. Absence of impurities can also have significant economic implications, not the least of which is the possibility of diluting low cost steel scrap to produce an equivalent quality steel. On the other hand, the utilisation of sponge iron to replace poorer quality scrap will assist in meeting the more stringent quality requirements of the market.
6. With the advent of HBI, the steel maker has a dense material that does not create "iceberg" formation problems, melts faster, has controlled gangue, does not

reoxidise easily, occupies less storage space, and is easy to handle.

1.2.8.2. Effect on Heat Time

The objective of charging DRI in EAF is to derive optimum benefit from the special characteristics of DRI. This aim is achieved by continuous feeding of DRI in EAF at particular rate, such that refining is carried out simultaneously. The main advantage of this process is that at the end of the charging period, the steel obtained is ready for tapping. But in case of 100% scrap heats, relatively more time is spent in melting several charges of scrap (4 to 5 bucket charges of bundled scrap) till the furnace is completely filled and thereafter a refining period of 15-60 minutes is absolutely essential to meet final analysis specifications. A decrease in total heat time

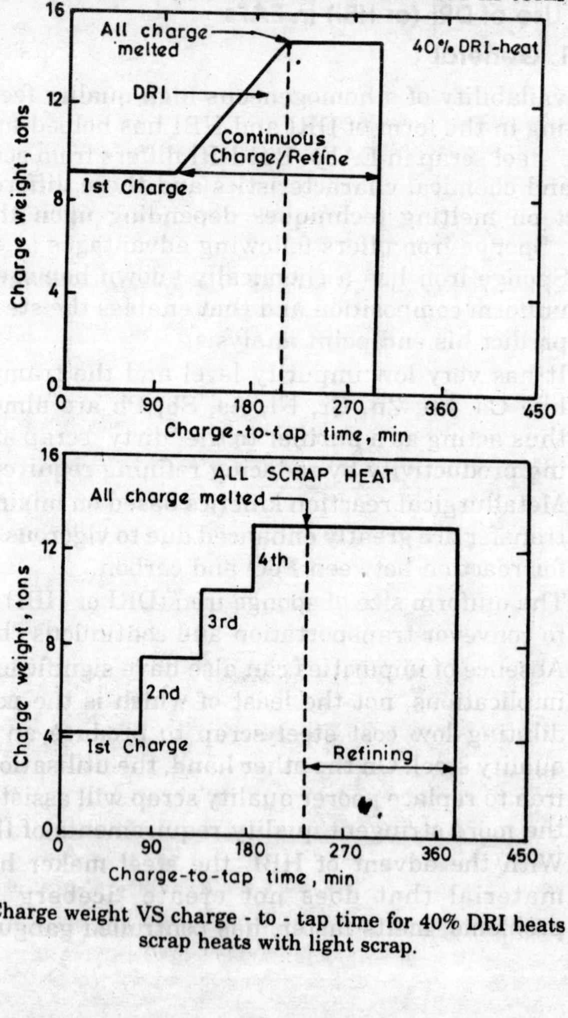

Fig. 1.12. Charge weight VS charge - to - tap time for 40% DRI heats and 100% scrap heats with light scrap.

for a 40% DRI heat in comparison to 100% scrap heat using light bundled scrap is shown in Fig. 1.12. (A gain of 21.3% in heat time is obtained for 40% DRI heat).

1.2.8.3. Effect on Energy Consumption

Compared to 100% scrap charge, the use of DRI requires additional energy for melting. This is because of the need to melt and separate the gangue of DRI by slagging. In general, the energy consumption increases steadily with increasing proportions of sponge iron (or DRI) in the charge mix due to higher slag volume. Also, the degree of metallisation of DRI affects energy consumption. Low metallisation (Or higher percentage of gangue in DRI) increases energy consumption and adversely affects the economics of steel production due to :

(*i*) additional energy consumed due to larger slag volumes and extra reduction of (FeO), and

(*ii*) longer heat durations, lower metallic yield and increased consumption of electrodes and refractories.

A degree of metallisation higher than 92-93% is necessary for economic reasons. Effect of sponge iron (DRI) proportion and gangue content on energy consumption is shown in Fig. 1.13.

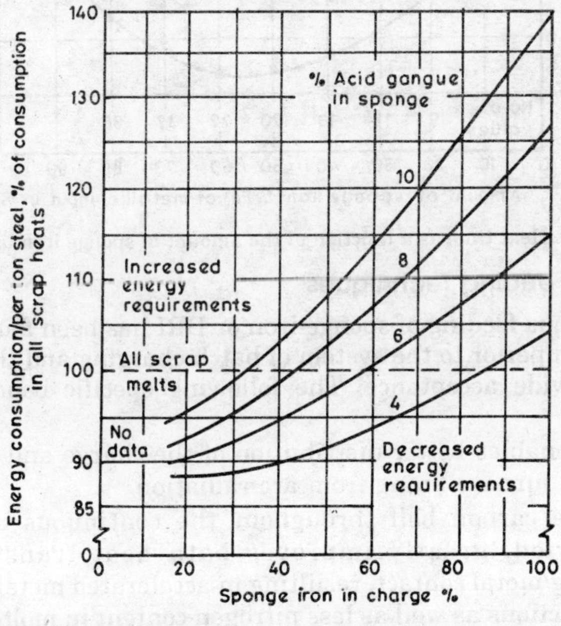

Fig. 1.13. Effect of sponge iron proportion and gangue content on energy consumption.

As mentioned earlier, due to continuous charging of DRI, the heat time and furnace productivity improve (Fig. 1.12). This is mainly because of simultaneous melting and refining taking place. Upto 30-40% DRI in the charge, there is no significant increase in power consumption, but with higher proportions of DRI, the power consumption increases, depending upon the gangue content and its nature. (Fig. 1.13). As shown in Fig. 1.14, the optimum seems to be around 50-60% of DRI in the charge. But a total evaluation has to be made based on energy consumed, heat time, quality of the DRI (gangue content, DOM and its cost), refractory and electrode consumption etc., before deciding on charge proportions for individual plants.

It has been found that with 70% DRI usage, the productivity is up by 21%. Below 30% DRI usage, no increase in productivity is observed. Between 30% and 80% DRI, the gain in productivity is about 1 ton/hr/each 10% DRI increase. The increase in productivity is mainly due to the shortening of conventional refining period as a result of overlapping of melting and refining periods (Fig. 1.14).

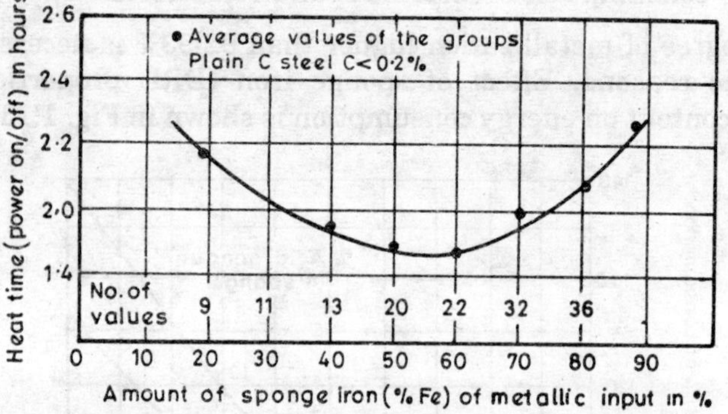

Fig. 1.14. Heat time as a function of the amount of sponge iron used.

1.2.8.4. Feeding Techniques

Continuous feeding of sponge iron or DRI has been found to be immensely superior to the system of batch charging and therefore, has found wide acceptance. The following specific benefits are noteworthy :

1. It enables better distribution of the charge and protects the furnace lining from arc radiation.
2. The carbon boil throughout the continuous charging period, greatly improves bath heat transfer and slag/metal contact, resulting in accelerated metallurgical reactions as well as less nitrogen content in molten steel.
3. There is combined melting and refining reaction while continuous feeding is in progress.

4. The lower electrical losses resulting from continuous charging produce 10-14% additional useful power for melting.
5. Heat losses are eliminated.
6. There is less "power off" time and "full power" can be applied during a larger portion of heat cycle.
7. Comparatively stable arc is obtained with reduction in noise levels.

Above effects continue to increase furnace productivity and decrease both electrode and power consumption compared to batch charging. The optimum benefits are generally believed to be attained when the melting of DRI takes place at the slag/metal interface. Also optimum melting results are possible with controlled feed rate of DRI, while feed rate is dependent on its composition, bath temperature etc. For most furnaces, a feed rate of 27-35 kg/Min/MW of applied power is maintained. When greater percentage of DRI is to be used in the EAF charge, it is necessary and advantageous to use the continuous feeding technique. During

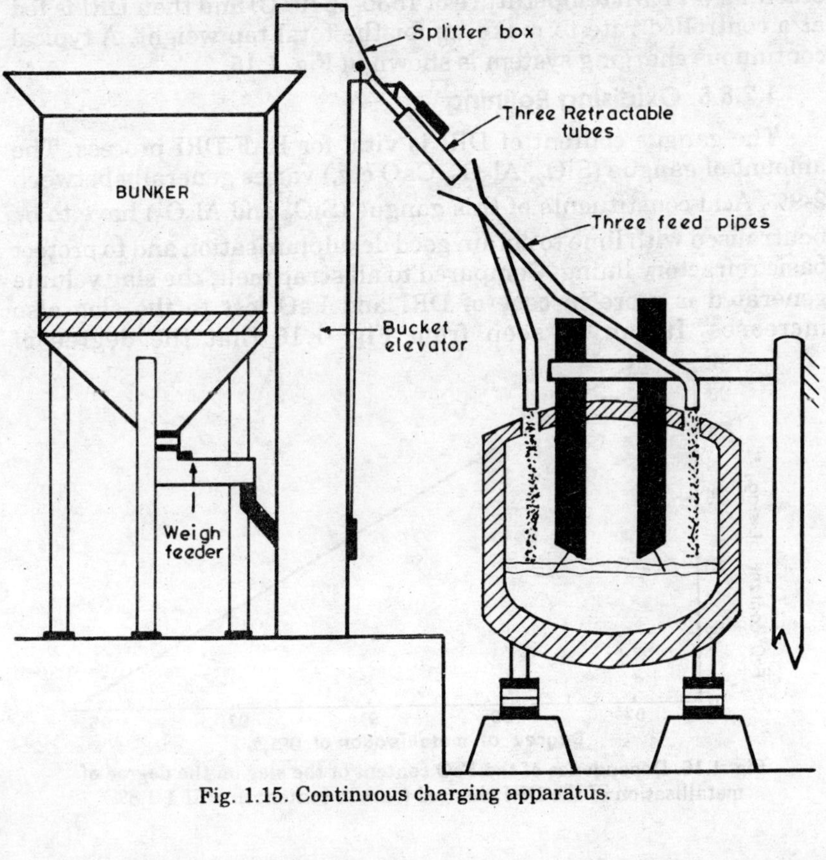

Fig. 1.15. Continuous charging apparatus.

continuous feeding of DRI, full power supply is maintained and the bath composition and temperature is adjusted such that at the end of charging, the heat is ready for final adjustment and tapping.

Three different charging systems are normally employed :

1. System employing three-holes on the roof for charging the DRI between the electrodes and the side-walls.
2. System having single-hole at the centre of the roof and charging the DRI into electrode delta region.
3. System having single-hole at the roof between the electrode and side wall and charging the DRI between the electrode and side wall.

Continuous charging through a single-hole is advantageous for small EAFs and charging of DRI briquettes (HBI). In case of relatively larger EAFs, and for melting of unbriquetted DRI, it is advisable to apply three-hole system. The usual practice is to charge a scrap bucket amounting to 20-50% of total charge initially. This scrap is melted partially to create a molten pool (hot-heel) which is essential (at bath temperature of 1530-1570° C) and then DRI is fed at a controlled rate to make up for the total tap weight. A typical continuous charging system is shown in Fig. 1.15.

1.2.8.5. Oxidising Refining

The gangue content of DRI is vital for EAF-DRI process. The amount of gangue (SiO_2, Al_2O_3, CaO etc.) varies generally between 2-9%. Acid constituents of this gangue (SiO_2 and Al_2O_3) have to be neutralised with lime to obtain good desulphurisation and to protect basic refractory lining. Compared to all scrap melt, the slag volume generated is more in case of DRI and FeO lost to the slag also increases. It can be seen from Fig. 1.16 that the degree of

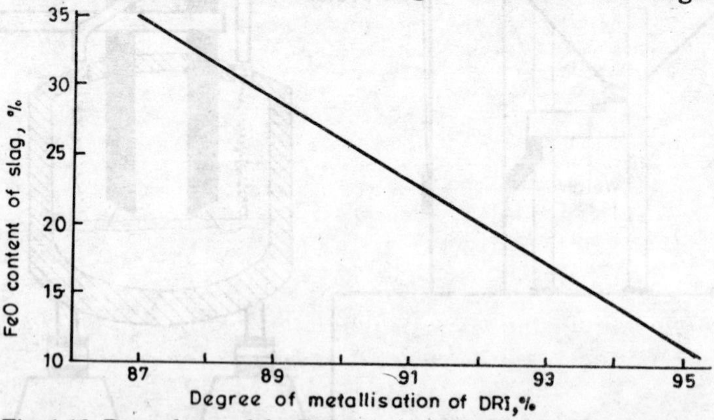

Fig. 1.16. Dependence of the FeO content of the slag on the degree of metallisation of the DRI charged (carbon content in DRI 1-1.8%)

metallisation (DOM) of DRI controls the (FeO) content of slag. So, from steel making point of view, the gangue in DRI should be as low as possible to achieve low power and flux consumption. The energy consumption of 40% DRI heats with 8% gangue in DRI, the power consumption goes up by 5%. The theoretical requirement of lime for 40% DRI heats to maintain an average basicity (CaO/SiO_2) of 2, is about 3 kg/t of liquid steel and slag quantity is about 115 kg/t of liquid steel. Corresponding values of the above for 100% scrap heats are 30 kg/t and average 70 kg/t liquid steel respectively. And for every 50 kg of slag produced per ton of steel results in increase in power consumption by about 50 KWH.

The iron oxide (FeO) and carbon contained in DRI have important roles to play is refining of steel in EAF. The refining of steel requires an active bath which is brought about by the reaction of the carbon contained in the metallic charge and the FeO in the slag. Active carbon boil results in improved heat transfer and slag/metal reactions during melt-down and thus improved refining efficiency. During this reaction of FeO with carbon, approximately half of the iron is recovered in the steel bath and the balance goes to the slag (in the form of FeO). It has been found that for this the DRI should possess a minimum of 0.6% carbon. Ideal oxygen content in DRI (as FeO) has been related with the rating of furnace capacity and to power input to ensure proper boil during continuous feeding as

$$\text{Ideal \% oxygen (in DRI)} = 0.475 \times \frac{\text{Furnace capacity, t}}{\text{Furnace power rating, MW}}$$

at a feed rate of 28 kg/min/MW of DRI.

There is approximately 9-10% FeO and varying carbon contents present in DRI depending upon the DR process employed. Coal based DRI has a carbon content of around 0.1% whereas gas based DRI has a carbon content in excess of 1.0%. A typical HBI employed at Mukand steel, Bombay contains ; carbon - 1.2% min, Sulphur - 0.015% max, phosphorus = 0.040% max, total gangue = 6% max and the DOM of this HBI is 93%. The size of HBI is approx. 30 mm × 60 mm × 90 mm. It has already been stressed that the DOM must lie in the range between 82 and 95%. Also, the sulphur content in the DRI must be less than 0.04% and preferably below 0.035% to avoid time consuming and expensive refining operations involving high basicity reducing slags. But the dephosphorisation efficiency is good as a result of low bath temperatures during the DRI feeding period and as a consequence of the residual oxygen contained in the DRI.

1.2.8.6. Tramp Elements and the Quality of Finished Steel

The use of DRI or HBI in EAFs has greatly helped to decrease the amount of deleterious tramp elements like copper, cobalt, tin, arsenic, antimony, nickel and molybdenum from the finished steel. These tramp elements are obtained basically from the steel scrap and are concentrated in the finished steel upon processing in the EAF. Some deleterious elements present in scrap are partially eradicated as a result of oxidising refining and are lost to slag, but quite a few contaminants still remain in liquid steel, lending undesirable physical properties to the finished steel. For the production of quality steels, therefore, the quality of scrap must be controlled. However, the level of tramp elements in most of the available scrap is in the range of 0.5 to 1.0%. Under this condition, sponge iron or DRI plays a vital role as a dilutant of tramp elements to obtain desired quality steel. Fig. 1.17 clearly indicates the effect of using DRI/HBI in the charge mix to decrease the menace of tramp elements.

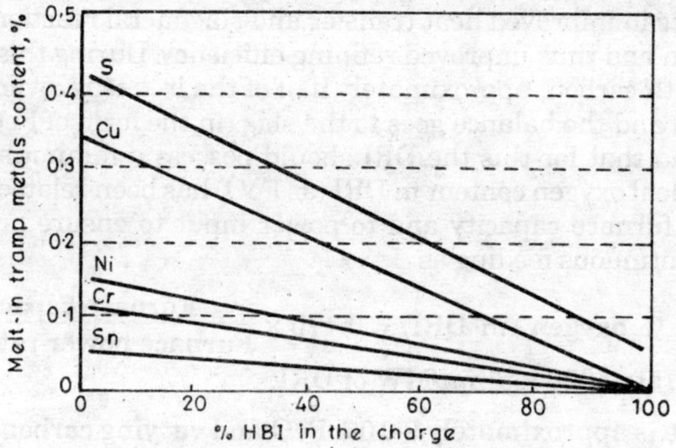

Fig. 1.17. Effect of HBI on melt-in tramp element content.

In general, the presence of tramp elements mentioned above have been found to induce undesirable resistance to deformation, hot shortness and a number of other metallurgical defects. But the mechanical properties and formability of steels produced with 80% DRI and 20% scrap have been comparatively much better and the low S and P obtained at the melt out stage even eliminated the need for desulphurisation. The practice at Mukand Steel has shown that using the above charge mix, good quality hot rolled and cold rolled products can be obtained.

The practice at Mukand Steel has shown how a certain high quality steel specifying a "P" level of 0.015% or another steel grade for which desulphurising with refining slags may be difficult, can be suitably and conveniently produced only by employing high purity feedstock like DRI or HBI, upto a proportion of 40% in the charge.

In rod steel manufacturing, as the carbon content increases, the drawability reduces as we go for finer wire drawing. In such cases, the drawability can be improved with controlled tramp element levels to 0.25 max. For cast steel products, very close tolerances for tramp element are not specified. Even ISI specification has permissible limit of tramp elements upto 1.0% max. But the experience at Mukand Steel has shown that the tramp level in cast steel becomes very significant for the products of special applications *e.g.* wear resistance steel castings of intricate shape and geometry used for earth moving equipment or used under dynamic forces. Such castings have shown cracking tendency when the tramp element exceeds 0.25%. It was found that the cracking tendency is reduced drastically when the tramp elements are controlled well below 0.2%.

For certain cast products which are used at sub-zero temperature, the impact value of steel must be high. Investigations at Mukand Steel have shown that the impact properties of steel can be improved by controlling tramp elements to a limit of 0.2% max and by decreasing S and P levels down to 0.015%. For example, the steel containing S and P at the level of 0.035%, exhibited an impact value of 0.4 kg.cm. For the same steel when the S and P was brought down to a level of 0.015%, the impact value improved to 3.0 kg.cm. In order to control both the tramp elements as well as the S and P levels, DRI or HBI as a dilutant is a viable alternative.

Further trends could seek steel with tight specifications on micro alloying (V, Nb, MO) and residual elements (Al, Ca, O, Cu, Ni, Ti, Mn, Co, As, Sn and Sb) which are improved from the considerations of welding integrity, corrosion protection and other factors. The use of high purity feed stock like DRI or HBI and advances in secondary steel making techniques have served to achieve these requirements. The combined process of EAF - secondary steel making - continuous casting is expected to become the dominant process for the production of high grade steels in the future.

1.3. ALTERNATIVE DEVELOPMENTS

1.3.1. General

The conventional EAF steel making process has firmly established itself as a highly efficient melting unit for the steel industry. However, despite all the improvements already made in a modern A.C. EAF, it suffers from the following drawbacks :

1. It suffers from arc instability, noise generation, utility system disturbance (or flicker effect) and excessive furnace side-wall deterioration.
2. It must absorb substantial losses of thermal efficiency and transmission incurred during electrical power generation.
3. It uses consumable graphite electrodes to conduct electricity to the scrap and this increases energy costs by nearly 30%.

Efforts to overcome these limitations have prompted the development of the DC arc furnace, and both the DC and AC plasma furnaces. Also, the impact of high electricity charges and the sharp rise in the cost of graphite electrodes have generated an intensive search for alternative forms of energy to melt scrap. The current technology is to consider the use of coal and the plasma torch to melt scrap. To stand any chance of replacing the conventional EAF, these alternatives must not only have an advantage over the existing technology, but should also be as equally efficient and reliable.

1.3.2. DC Arc Furnace

These furnaces are normally characterised by a centrally positioned graphite electrode which acts as the cathode. DC arc generation provides more efficient heat transfer, and arc heat is focussed towards the anode. Unlike an AC arc, a higher portion of DC arc heat (72% as against 65% from the AC arc heat) is directly delivered to the melt. Also, the DC arc is burned vertically whereas the AC arc burns at an angle and is forced by the electromagnetic field towards the furnace wall, which in turn, causes considerable hot-spots. This favourable behaviour of the DC arc has enabled it to operate at a higher voltage and consequently a longer arc for efficient melting of scrap.

Tests carried out with single electrode DC arc furnaces in Europe and the USA have confirmed the following desirable effects :

1. Lower electrode consumption of less that 2 kg t^{-1} (about half that of the conventional AC EAF).
2. Less noise emission (around 5-10 dB).
3. Reduced utility system disturbances in the form of flicker (*i.e.* half the value of the comparable three phase AC EAF).
4. Drop in refractory wear (< 4 kg t^{-1}).

5. Better temperature distribution, and improved bath stirring caused by electrodynamic forces.

The 62t DC arc furnace of 48 MW (60 MW maximum) rating has been operating at Societe Metallurgique de 1' Escaut (SME), France, since 1986. In DC arc furnace, the bottom electrode is the main problem area. The SME furnace is equipped with three hearth electrodes, to match the three top electrodes, in order to reduce current density in the 5 m diameter furnace, and still achieve high melting rates. The hearth electrodes are steel billets which are water cooled outside the furnace, and have a life of 150 heats. One big advantage of three electrode d.c. arrangement has been that W.C. panels above the slag-line have been replaced by ordinary refractory lining, which has proved to be more economical.

A current density of 40,000 A/electrode is employed during operation, which means a total of 120 KA. This current density is achievable only with more than one electrode. The electrodes are ordinary quality carbon with a diameter of 500 mm and the present consumption is around 2.3-2.5 kg/t of liquid steel. Attempt has been made to bring this figure down to 1.8 kg/t with smoother operation of furnace. The best productivity has been 73 min. tap-to-tap for 75 t charge, but the objective was to reach greater than 50 t/hr productivity. Average tap-to-tap time has been around 80 min at a power consumption of 520 KWH/t, although a record 460 KWH/t had been reached with oxygen blowing at 16 NM^3/t. The predicted advantage of lower noise had been fulfilled, with a maximum of 100 dB recorded inside the doghouse enclosure. Also, a higher tapping temperature of 1680°C achievable with the furnace eliminates the need for additional heating in LF refining process. The refractory consumption on an average has been 5 kg t^{-1}. SME DC furnace also generated 30% less flicker as compared with the conventional AC furnace.

Fig. 1.18 shows a 115 t capacity EBT DC arc furnace of 95 MVA rating. The shell dia of this furnace is 6.6 m, with a shell height of 4.1 m and is provided with a 700 to 800 mm dia electrode. The furnace can tap 85 t liquid metal leaving behind a hot-heel of 30 t. The electrode (Cathode) can carry a maximum of 120 KA and the furnace is provided with a 4 bottom electrodes (anode), which are spray water cooled. Details of bottom electrodes are shown in Fig. 1.19. This furnace was built by Mannesman-Demag Huttentechnik.

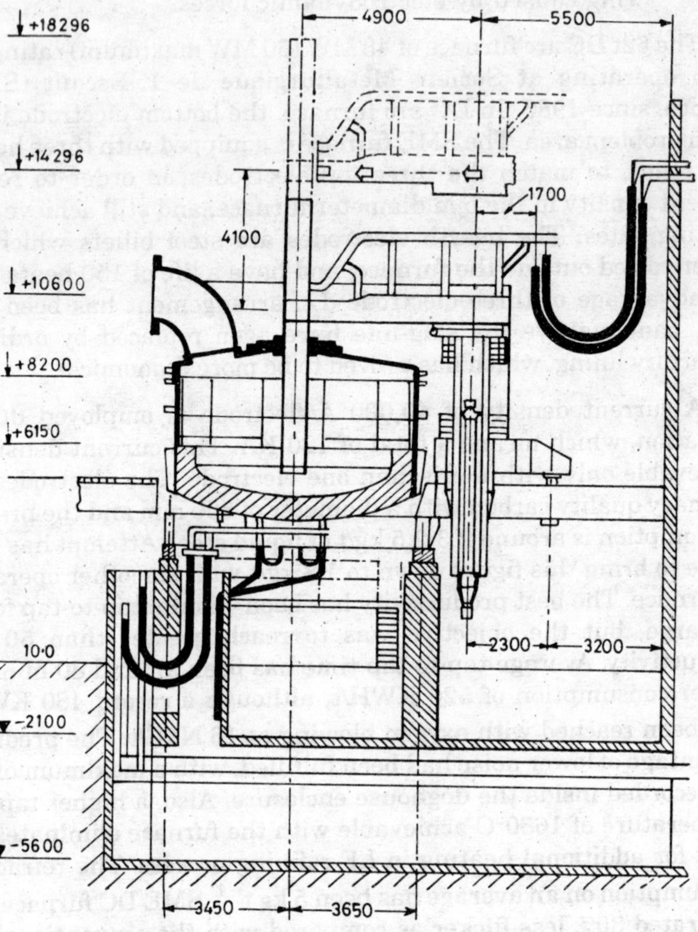

Fig. 1.18. 115t, EBT DC Arc Furnace.

1.3.3. Plasma Furnace

In the past decade, plasma furnace steel making has been the subject of intensive development. It dispenses with the conventional electrode and instead energy is transmitted through the plasma arc by a plasma torch. The DC plasma furnace features a single water cooled bottom electrode (anode) and 3 or 4 plasma torches connected as cathodes and introduced diagonally through the furnace walls. In 1986, a three phase AC installation with a rating of 20 MVA was built, and is being run as 10t melting furnace. Plasma arc heating is

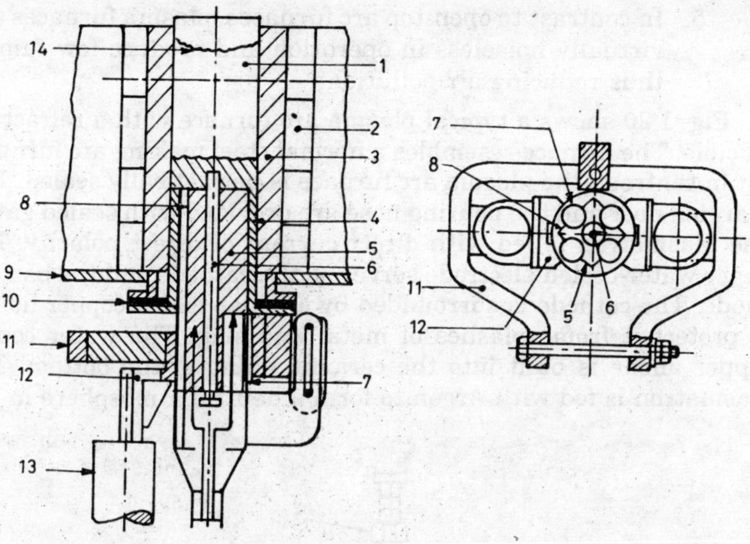

1. Steel Cylinder 2. Electrode sealing ring 3. Ramming mix 4. Steel shell 5. Copper cylinder 6. Tension bolt 7. Cooling chamber 8. Thermocouple 9. Shell bottom 10. Insulation 11. Clamping unit 12. Contact pad 13. High-current cable, Water cooled 14. Solidus line

Fig. 1.19. DC bottom electrode (Anode).

used in modern practice for both melting of metals in refractory lined furnaces and remelting in water cooled moulds. Melting in plasma arc furnaces with ceramic crucibles offers following advantages :

1. It avoids contamination of the metal with carbon of the electrodes and with hydrogen from the furnace atmosphere.
2. The plasma jet may be composed of any mixture of gases which also can be used for metal alloying (such as Nitrogen); melting in a plasma-arc furnace with a burner fed with an inert gas is equivalent of melting in a vacuum furnace.
3. Plasma arc furnace can be used for melting metals with volatile components, since the pressure above the metal is substantial and impedes evaporation.
4. The rate of melting is rather high due to the high concentration of energy, high speed of the plasma, quick heat transfer to the materials being heated and relatively low loss of heat by radiation.

5. In contrast to open-top arc furnaces, plasma furnaces are virtually noiseless in operation and produce few fumes, thus reducing air pollution.

Fig. 1.20 shows a typical plasma arc furnace with a refractory crucible. The furnace resembles a normal steel making arc furnace, but in contrast, the plasma arc furnace is hermetically sealed. The charging door and the pouring nose are provided with sealed gates. The plasmatron is fed with direct current of direct polarity. The upper water-cooled electrode serves as the cathode and the bath as anode. The cathode is surrounded by a water cooled copper nozzle to protect it from splashes of metal and slag. The water cooled copper anode is built into the ceramic lining of the bottom. The plasmatron is fed with Argon to form a neutral atmosphere in the

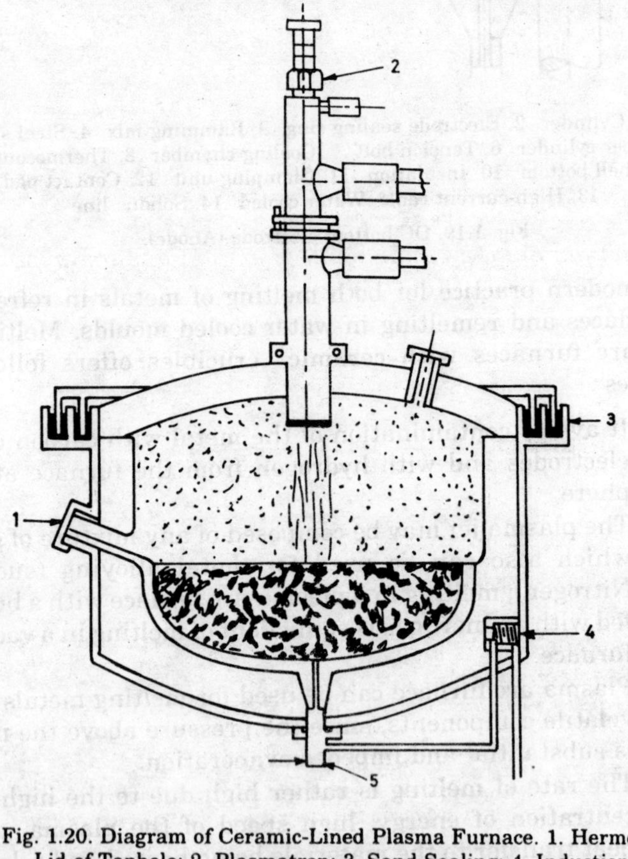

Fig. 1.20. Diagram of Ceramic-Lined Plasma Furnace. 1. Hermetic Lid of Taphole; 2. Plasmatron; 3. Sand Sealing; 4. Induction Coils for Metal Stirring; 5. Bottom Electrode.

furnace. The length of the arc is greater than in common arc furnaces, so that the nozzle can be made countersunk in the roof and thus prevented from damage by the charge. Melting takes place mainly around the anode. For a furnace of 1 t capacity, the melting time is around 2 hrs, argon consumption is 5.1-5.4 m^3/hr, voltage 177 V, current 990-2500 A and the consumption of energy is around 1225 KWH/t.

A few kinds of plasma torches used in practice are shown in Fig. 1.21 (*a*, *b* and *c*). In a fully sealed plasma arc furnace with ceramic crucible and the use of pure argon, the conditions of melting approach those in vacuum induction furnaces.

In plasma arc melting, alloying additions are almost completely assimilated by the metal : Mn by 96%, V by 95%, C by 86% and Ti by 85-90%. The metal can well be deoxidised with carbon due to the low partial pressure of CO in furnace atmosphere. After melting in a plasma arc furnace, the metal has only 0.0008-0.0025% oxygen, 0.0001-0.0002% hydrogen and 0.001-0.003% nitrogen. Melting of stainless, ball bearing and heat-resistant steels has shown that

(*a*) Oxygen content is reduced from 0.004/0.006% to 0.002/0.004%.

(*b*) Non-metallic inclusions are considerably reduced and

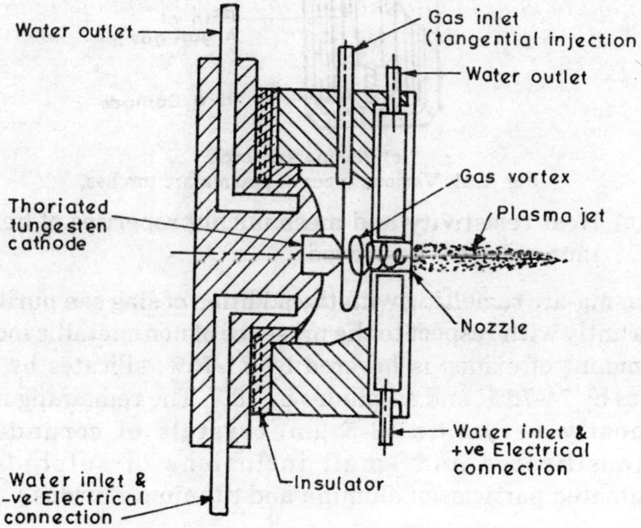

(*a*) The cross-section of a plasma torch using a gas-vortex stabilized arc.

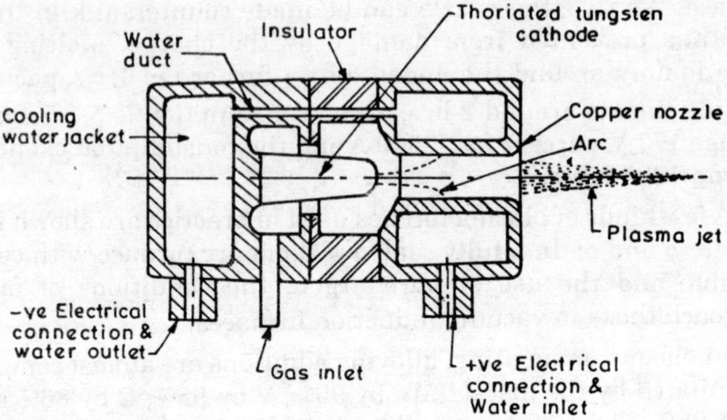

(*b*) The cross-section of a plasma torch using a gas-sheath stabilized arc.

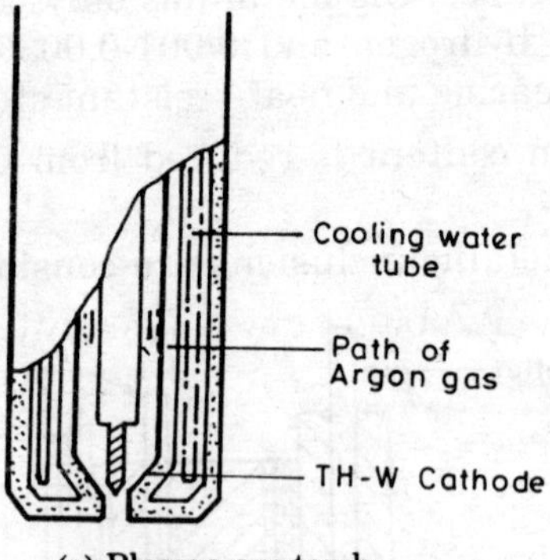

(*c*) Plasma arc torch

Fig. 1.21. Various types of plasma arc torches.

(*c*) Heat resistivity and mechanical properties of heat resistant steel are improved.

Plasma-arc remelting with the addition of slag can purify a steel significantly with respect to the presence of non-metallic inclusions. The amount of oxides is lowered by 71-75%, silicates by 67-71%, globules by 74-76%, and sulphides by 78%. The remaining inclusion are mostly very fine (3-5 μm) crystals of corundum and aluminosilicates and small inclusions of sulphides with disseminated particles of alumina and titanium nitrides.

The total amount of inclusions diminish from a level of 0.0067 to 0.0018%. Precision ball-bearings of high reliability made of plasma-arc remelted steel had a longer life than those made of electro slag remelted steel.

Major drawbacks of this process are the excessive refractory wear resulting from intensive power loading and the hearth requiring more maintenance owing to the bottom electrode. Additional problems remaining to be resolved involve upgrading the plasma torch from the present of 5-8 KA to 25 KA, and the unsatisfactory life of nozzles and cathodes.

1.3.4. EAF with Bottom Fuels Injection

This new approach of coal injection with combined blowing of oxygen into the EAF (known as the K-ES process) has been developed by a joint team of Japan's Tokyo Steel manufacturing Co and the Klockner CRA technology group of Germany. It involves injection of coal fines and other pulverised materials, as well as

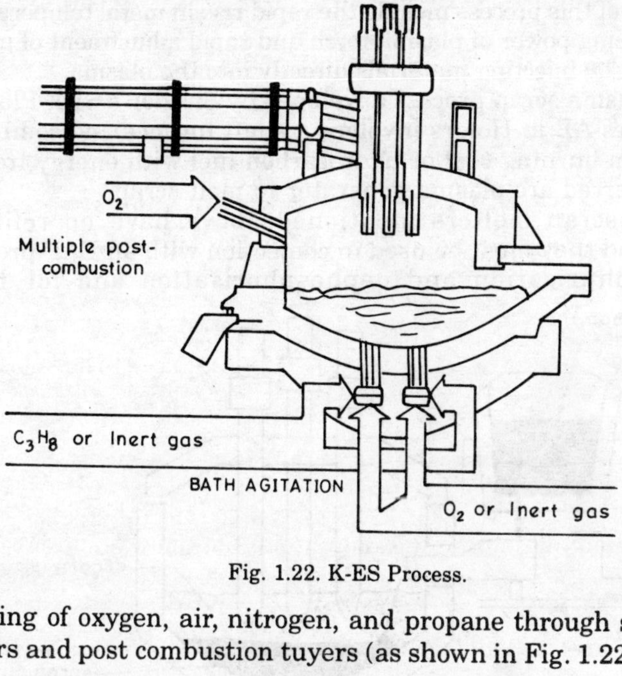

Fig. 1.22. K-ES Process.

blowing of oxygen, air, nitrogen, and propane through submerged tuyers and post combustion tuyers (as shown in Fig. 1.22).

Efficient post-combustion of process gases within the furnace, and a rapid melting process through oxidation and accelerated bath agitation, have been achieved with effective heat-transfer to the scrap and the melt. If this process is operated with oil-oxygen burners, but without coal injection, it requires electrical power in the range of 410 KWH/t liquid, steel. Power requirement reportedly drops to 300 KWH.t^{-1}., when the process is operated with coal additions of 22-30 $kg.t^{-1}$ (at a coal efficiency of 4.5 $KWH.kg^{-1}$) with a melting time of 48-50 min, and productivity raised by over 33%.

1.3.5. Scrap Melters

Owing to the rapid rise in electricity charges, techniques that can replace part of the electrical energy used for melting scrap with less expensive fuels are of great interest. Many scrap melters have been proposed. Among the more important proposals are the plasma fired cupola and the Plasma scrap process. In the plasma fired cupola sponsored by America's Westinghouse and the Electric Power Research Institute, a non-transferred arc plasma torch was installed at the tuyer zone of a cupola. This plasma energy replaces much of the energy of coke which would, otherwise supply the melting power. Metal chips (loose borings and other machining wastes) which normally required briquetting are injected into the plasma and melted quickly. Additional advantages of this process include the rapid rise in metal temperature with increasing power of plasma torch and rapid adjustment of metal composition by injecting materials directly into the plasma.

The plasma scrap process developed by Sweden's SKF Plasma Technologies AB in Hofors involves a shaft furnace and combines energy from burning coal or hydrocarbon fuel with energy from a non-transferred arc plasma generator to melt scrap.

These scrap melters mentioned above have no refining capacity and they must be used in conjection with an EAF process for desulphurisation and dephosphorisation and for final

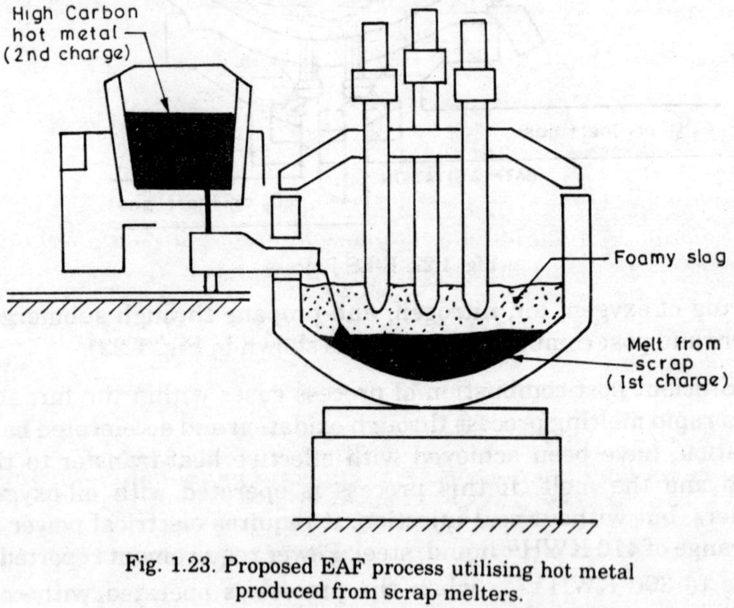

Fig. 1.23. Proposed EAF process utilising hot metal produced from scrap melters.

adjustment of the carbon levels. Fig. 1.23 shows a proposed process to utilise the hot metal produced from scrap melters, with simultaneous melting of steel scrap within the EAF.

This set-up could both increase the productive capacity as well as improve the steel quality.

1.4. THE PRODUCTION OF QUALITY STEELS

In modern times, the specifications set particularly for high alloy and special steels are becoming more stringent. Such steels generally will have tight specifications on microalloying (V, Nb, Mo) and residual elements (Al, Ca, O, Cu, Ni, Ti, Mn) and would be expected to contain negligible quantities of gases and non-metallic inclusions. Besides, these steels will be expected to conform to tight chemistry with regard to particularly sulphur and phosphorous in addition to other elements usually present in steel. Structurally such steels will be expected to be very homogeneous in their properties. Such quality steels are in great demand in aviation industry, radio-electronics, automobile industry, instrument making, space engineering for construction of nuclear power plants, gas mains in polar regions, thermal power plants, defence applications, shipbuilding and numerous other applications.

The rigid or stringent specifications laid down for "Quality Steels" meant for critical applications can not be easily met by "primary processes of steelmaking" like EAF, OH, LD, KALDO, Ajax, Rotor, LDAC etc. Even if an attempt is made to meet the specifications using the primary processes of steelmaking, it would usually result in excessively long periods of processing, extra cost of production and under utilisation of expensive equipment. The modern strategy is therefore, to employ special process techniques along with primary processes to produce steels of more stringent specifications. The special process techniques are generally termed as "Secondary steelmaking techniques". The combination of primary processes and the special process techniques may be termed as "Duplexing". The basic idea behind such a duplexing operation is to carry out normal melting and oxidising refining in the primary processes of steelmaking as quickly as possible and then transfer the more tedious and lengthy operations of deoxidation, alloying, composition and tapping temperature adjustment and desulphurisation etc. to the special process technique (or the secondary unit). This combination results in the production of quality steels in a most efficient and economic manner. This combination also results in optimum utilisation of both the units (Primary as well as secondary steel making units) in a proper manner. Secondary steelmaking sometimes is also termed as "ladle metallurgy", 'off-furnace treatment', 'ladle-refining' or simply 'secondary metallurgy'.

Secondary steel making processes are provided with all or some of the following facilities :

1. Vacuum degassing of liq. steel.
2. Inert gas flushing/stirring.
3. Intermixing of the metal with a specially prepared slag or alloy.
4. Blowing powdered materials into the metal.
5. Oxygen blowing under vacuum.
6. Induction stirring.
7. Graphite electrode heating.
8. Alloy additions under vacuum.

Secondary steel making processes can achieve the following well laid out objectives :

1. Homogenisation of the melt with respect to the composition and temperature.
2. Alloying and controlling the metal composition strictly according to the specifications.
3. Deoxidation.
4. Modification of non-metallic inclusions.
5. Removal of non-metallic inclusions.
6. Desulphurisation.
7. Degassing with respect to the dissolved gases like Hydrogen and Nitrogen.
8. Decreasing the concentration of unwanted elements.
9. Grain control in the final metal.
10. Precise tapping temperatures control.
11. Extensive decarburisation without loss of Cr under vacuum.

Liquid metal treatment can be carried out in one of the following kind of equipment :

1. In a conventional teeming ladle lined with fireclay brick and provided with vertical stopper.
2. In a ladle with a gate type stopper and the refractory lining made of basic or high alumina refractories.
3. In a capped ladle.

4. In a ladle provided with means for blowing in an inert gas or gas-powder mixture through the bottom.
5. In a ladle type plant provided with a cover (roof) through which electrodes are inserted to heat-up the metal during the treatment.
6. In a plant of the convertor type in which the metal is blown with oxygen, argon or nitrogen or steam.

The common teeming ladle, which was hitherto employed for separation of slag from metal & convenience of metal casting into ingot moulds, has now become an additional refining unit in modern steel plants, in which a number of operations such as removal of dissolved gases and non-metallic inclusions, deoxidation, heating to the desired teeming temperature, temperature and composition homogenisation etc. are carried out; while the primary processes of steelmaking such as LD, OH or EAFs serve the sole purpose of preparing a pre-refined metal through speedy melting and oxidising refining (removal of C, Si, Mn and P). The pre-refined steel is then finally treated and refined and finished to the specifications in a secondary unit. A number of secondary steelmaking techniques are now available for the production of quality steels. The important ones are : ASEA-SKF process or Ladle furnace (LF), VAD, ESR, VOD, AOD, CLU, LD-PHOB process, VODC and UDDACON process. These processes are described in detail in subsequent sections (Chapter 3).

TWO

Continuous Casting of Steels

2.1. INTRODUCTION

The first patent on the idea of continuous casting of steels was obtained by Sir Henry Bessemer in the year 1856. Later, several new processes were patented *e.g.* Evanon (Russia) 1905, Duter (Germany) 1909 and Junghans (Germany) 1933. During 1930-40, continuous casting machine for non-ferrous metals was adopted as a production method. In 1943, Junghans demonstrated the new process by casting steel from a small capacity Bessemer converter into continuous strands on a vertical machine. Since 1943, the process has shown phenomenal growth, to the extent that continuous casting machines for billets and blooms are considered an integral part of Mini Steel Plants today. Besides, this process has been extensively employed in integrated steel plants for the casting of large slab sections, which are ultimately rolled down to plates and sheets. In 1948, the number of machines was only one and the production only 0.28×10^6 t/annum. In 1968, the number of machines rose to 243 and the production increased to 35.8×10^6 t/annum. The growth of continuous casting machine is identical to that of LD. Steel making. It is because the continuous casting process offers many advantages over the conventional casting pit practice with nearly identical or slightly superior cast and rolled product quality. It does not mean, however, that the continuous casting process is free of practical problems. By 1969, about 5% of the world's steel was cast by continuous casting machines and in 1979, this figure was 18%. It is estimated that in 1985, this figure was 41%.

Among the major producers of steel today, South Korea heads the list with about 95% of the steel being continuously cast. In Japan, the ratio is around 93%, while in the EEC countries as a whole, the ratio is about 88%. USA, however, is lagging behind with a ratio of 56% while the USSR & Eastern Europe are further behind with only 17% & 18% respectively. This ratio in India is around 25% & progress is being made steadily.

Continuous casting started in India with the commissioning of a billet caster at Mukand Iron & Steel Works and at the Canara Works in 1965. Since then, more than 80 more continuous casting machines have been set up or are under installation. The corporate plant of SAIL forecasts that by the year 2000 AD, the integrated steel plants are expected to produce about 17 mt of crude steel, of which 98% will be through the continuous casting route.

2.2. PRINCIPLE

Fig. 2.1 shows schematically the general pattern of solidification and different cooling zones in a continuous casting machine. The details of "primary" and "secondary" cooling and the mode of solidification within the mould are shown in Fig. 2.2. The liquid steel (at a predetermined degree of superheat) is poured in the mould via the tundish at a predetermined or closely controlled rate such that the rate of withdrawl exactly matches with the rate of pouring. This is absolutely essential for smooth operation of machine.

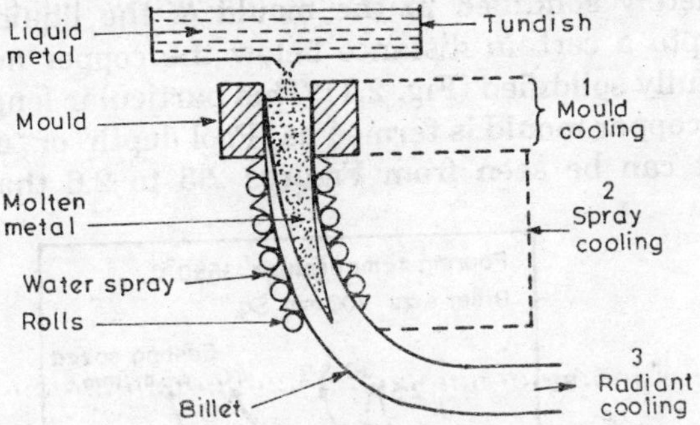

Fig. 2.1. Schematic representation of various cooling zones during continuous casting of billets.

In order to achieve reasonable rates of production, partially solidified strand is withdrawn form the mould. But it has to be ensured that a sufficiently thick solidified skin is formed at the exit point of the mould so that it can withstand the ferrostatic pressure of liquid core (Fig. 2.2). If the solidified skin is too thin it may fracture under the pressure of liquid core and produce cracks or in extreme case, the skin may fracture altogether causing "leakouts" or "breakouts". A proper withdrawal speed is, therefore, required to be maintained for a given section size of the strand and for a

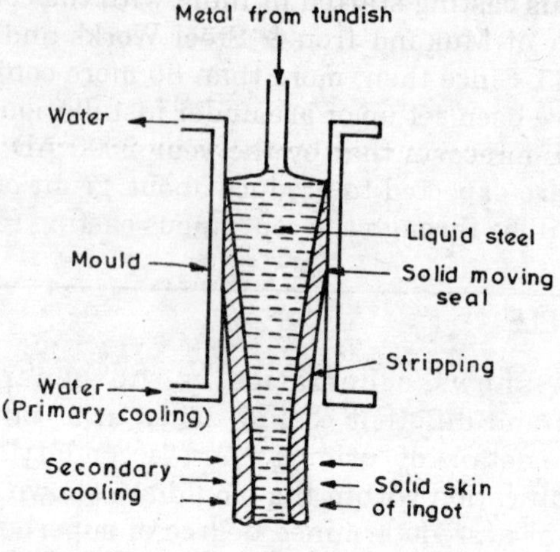

Fig. 2.2. Solidification characteristics of continuously cast ingot.

particular quality of steel and degree of superheat. Figure 2.3 to 2.6 show the effect of casting speed and pouring temperature on the shell thickness and pool depth. This implies that the ingot strand is not completely solidified in the mould & the liquid metal core extends upto a certain distance below the copper mould till the strand is fully solidified (Fig. 2.1). This particular length of strand below the copper mould is termed as "Pool depth" or "metallurgical length". It can be seen from Figures 2.3 to 2.6 that both shell

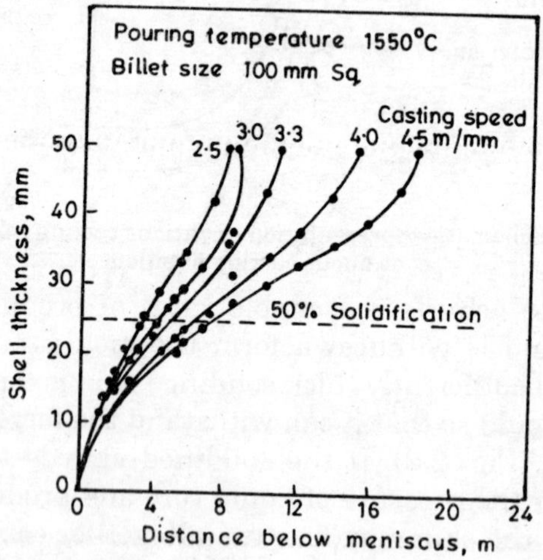

Fig. 2.3. Effect of casting speed on shell formation during billet solidification.

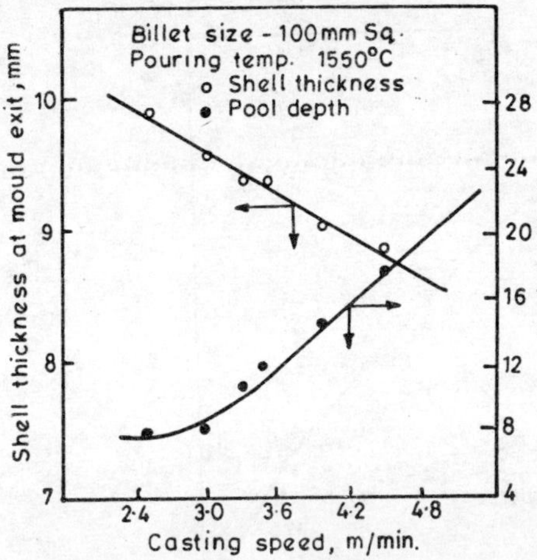

Fig. 2.4. Effect of casting speed on shell thickness and pool depth.

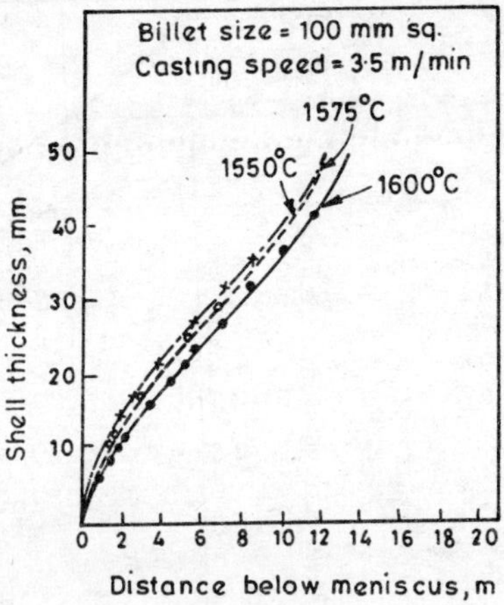

Fig. 2.5. Effect of pouring temperature on shell formation during billet solidification.

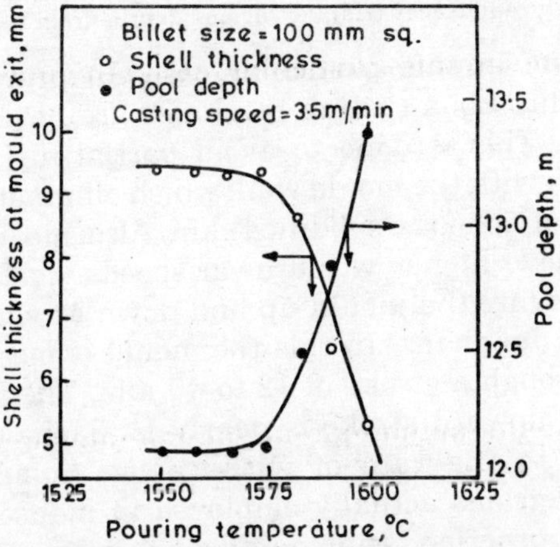

Fig. 2.6. Effect of pouring temperature on shell thickness and pool depth.

thickness and pool depth are greatly influenced by the casting speed (withdrawal speed) for a given size of section cast and the degree of superheat. Sufficiently thick skin of solidified metal is formed in copper mould due to "primary cooling" and therefore, solidification continues upto the centre of cast section due to "secondary cooling". Fig. 2.7 shows solidified shell thickness for a 127 mm square cross-section billet at various depths below the meniscus level. It can be seen that pool depth had extended upto 3.2 m below the

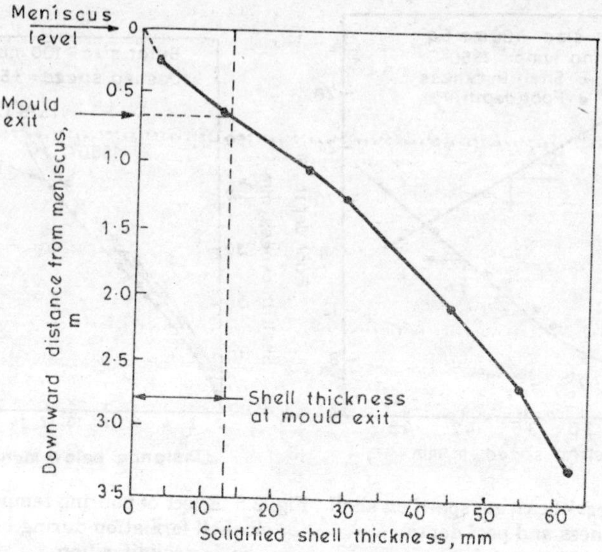

Fig. 2.7. Solidified shell thickness measured from autoradiographs obtained through radiotracer trial, at various depths from the meniscus.

meniscus level in this particular case. In principle, the copper moulds of continuous casting machines are never stationary, but moving type. This is done to avoid partial fusion or welding of solidified skin with the mould wall, which eliminates the possibility of crack formation in the solidified skin. Also, moving mould system helps to achieve higher withdrawal speeds (casting speeds). The concept of moving the mould up and down at various speeds was developed by Junghans. In this the mould is moved up and down variously, through a stroke of 12 to 40 mm. The ratio of speeds of downward to upward stroke is nearly 1 : 3, the downward speed being equal to the rate of withdrawal. Some typical mould oscillation diagrams actually employed in industry are shown in Fig. 2.8. In practice, the moulds are mounted on a special "mould-table", which is attached to an oscillating mechanism, that performs the stripping action of billets. Mould oscillation varies between 0 to 100 cycles/min depending on the type of oscillating mechanism. If the downward speed is even slightly less than that of the withdrawal speed, major transverse cracks are formed. In a new modification, therefore, the speed of downstroke has been increased to little more than the speed of withdrawal. This results in "Negative stripping" of the ingot and is beneficial in the following ways :

1. The initally crystallised skin is further compacted.

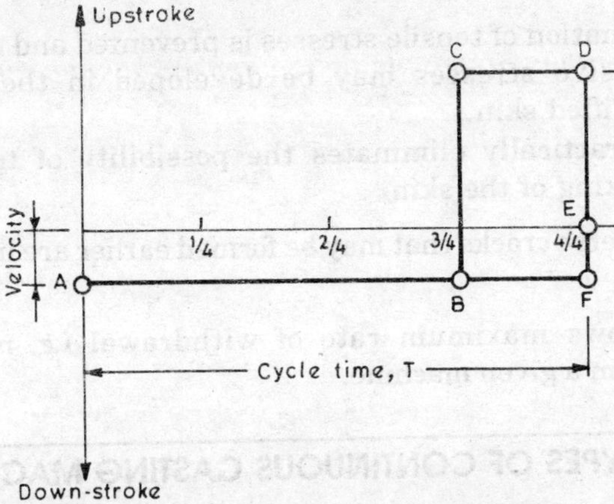

(*a*) Velocity-Time Diagram Junghans - Rossi Cycle Ratio - 3.1.

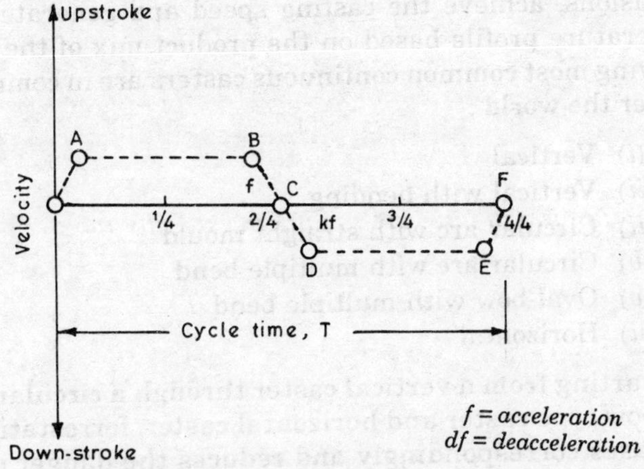

(*b*) Velocity-Time Diagram Savage CCC Cycle.

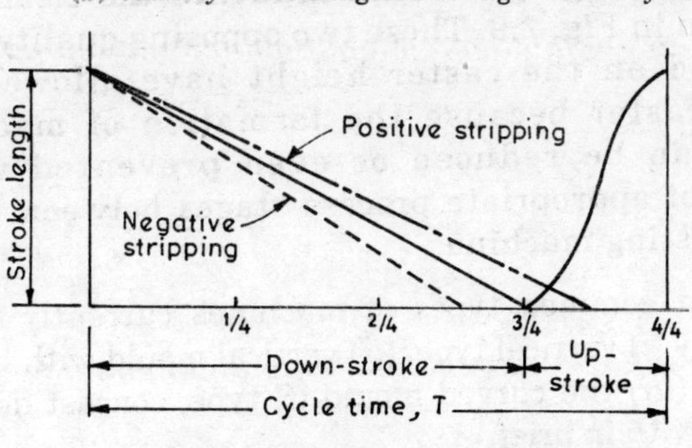

(*c*) Stroke-Length-Time Diagram

Fig. 2.8. Some typical mould oscillation diagrams.

2. Formation of tensile stresses is prevented and even compressive stresses may be developed in the initially solidified skin.
3. It practically eliminates the possibility of transverse cracking of the skin.

(4) Transverse cracks that may be formed earlier are liable to be welded again.

(5) It allows maximum rate of withdrawal *i.e.* maximum production from a given machine.

2.3. BASIC TYPES OF CONTINUOUS CASTING MACHINE

The basic requirements of the caster design are the capabilities to produce the steel grades, required cast dimensions, achieve the casting speed and generate a suitable temperature profile based on the product mix of the plant. The following most common continuous casters are in commercial use all over the world :

(*i*) Vertical
(*ii*) Vertical with bending
(*iii*) Circular arc with straight mould
(*iv*) Circular arc with multiple bend
(*v*) Oval-bow with multiple bend
(*vi*) Horizontal

Starting from a vertical caster through a circular arc to the oval bow type caster and horizontal caster, ferrostatic pressure decreases correspondingly and reduces the danger of internal deformation by bulging and can influence the cleanliness of steel as shown in Fig. 2.9. These two opposing quality features which depend on the caster height have allowed to use "bow-type" caster because the formation of macroscopic inclusions can be reduced or even prevented with the introduction of appropriate process stages between BOF and continuous casting machine.

Three most common types of machines currently in use in practice namely (*i*) vertical type, (*ii*) vertical mould with horizontal discharge and (*iii*) the curved mould (S type, concast design) are described herewith in brief.

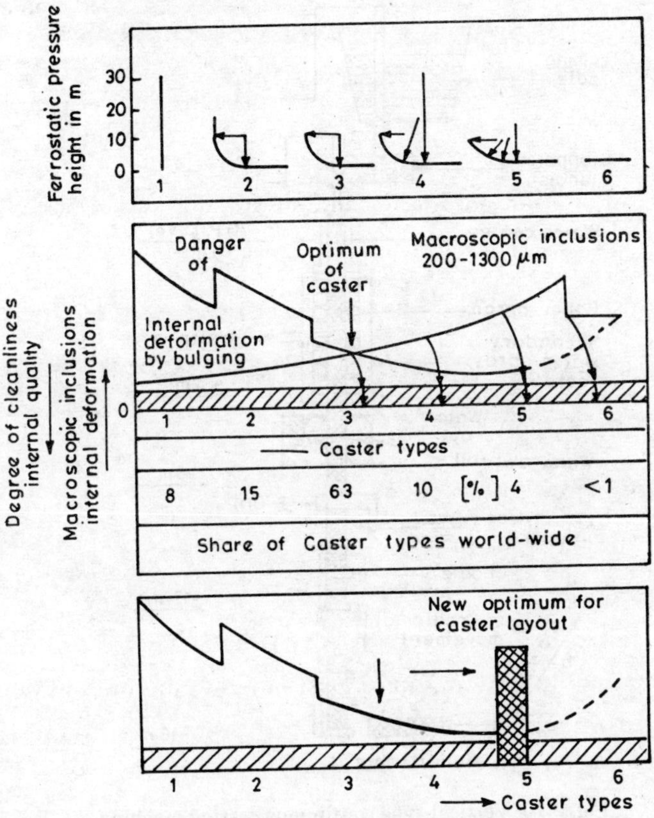

Fig. 2.9. Different types of continuous casters and evolution of optimized caster for clean steel.

2.3.1. Vertical Type (Fig. 2.10)

In this particular case, both the mould and discharge are vertical. Details of machine are shown schematically in Fig. 2.10. The machine was developed mainly for the casting of sections of relatively large dimensions at relatively low speeds. If a large heat is to be cast into relatively small sections, it is essential to have more than one such strand to expedite casting and achieve higher production rates. As many as 8-strand machines are in use. The tundish in this case divides the metal stream in that many equal parts.

The limitation of this design is that the plant is very tall and hence needs either a tall shop or a deep pit to accommodate the machine. The problem is acute if high casting speeds are employed, as a result of which, longer cooling zone is required. The plant is,

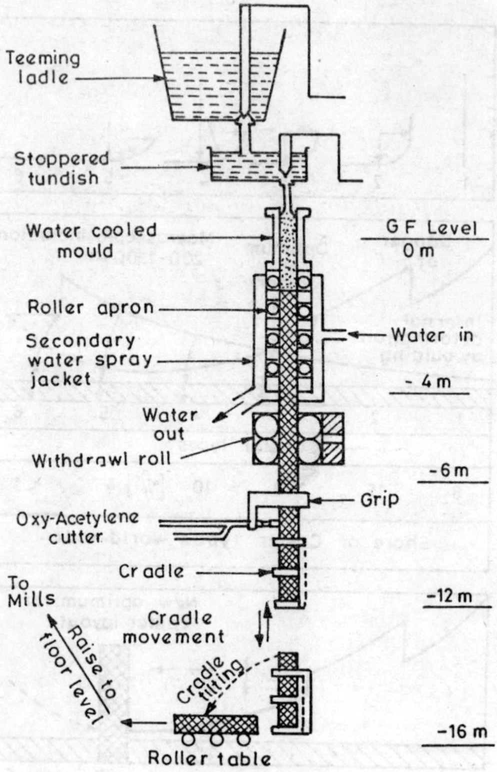

Fig. 2.10. Vertical-type continuous casting machine.

therefore, used for large and medium sections. Slabs can be easily cast by this machine as bending and its associated metallurgical effects are avoided. All steel qualities can be cast and that too at high speeds without fear of damage to the strand by bending and straightening.

2.3.2. Vertical Mould Horizontal Discharge Type (Fig. 2.11)

This design has the advantage of about 30% reduced height enabling the cut-off to be mounted horizontally. Also, dummy bar can be disconnected at the inception of bending and lie in ambush until next required, when it is immediately ready for use, thereby reducing resetting time between casts. Heavy sections being difficult to bend can not be cast by this machine. In the event of breakdown, it is difficult to repair and restart than the vertical machine.

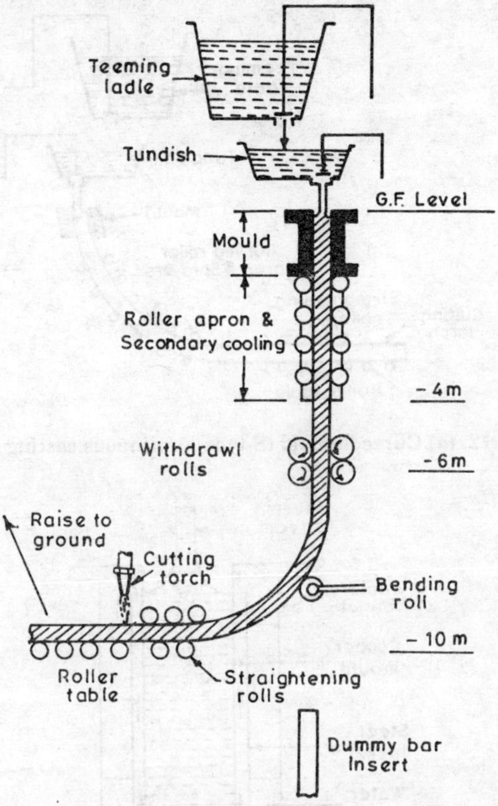

Fig. 2.11. Vertical-mould-bend-discharge-type continuous casting machine.

2.3.3. Curved-mould, curved cooling machines (S type or concast design, Fig. 2.12)

On solidification, the skin and walls of the section take-up not only the cross-sectional shape, but also the curvature of mould and the resulting curved section is continuously withdrawn downward along the curved path and through straightening rolls. The curved mould is reciprocated along the arc of its curvature. Curved-mould machines are being installed for the production of both small and large, and even very large sections. In all cases the section as it passes through the curved spray zone, has to be supported in its curved path by suitably positioned guide rolls. The curvature is designed to reduce the occurrence of sticking and the entire mould is situated at the end of a long arm that imparts a reciprocating action to it.

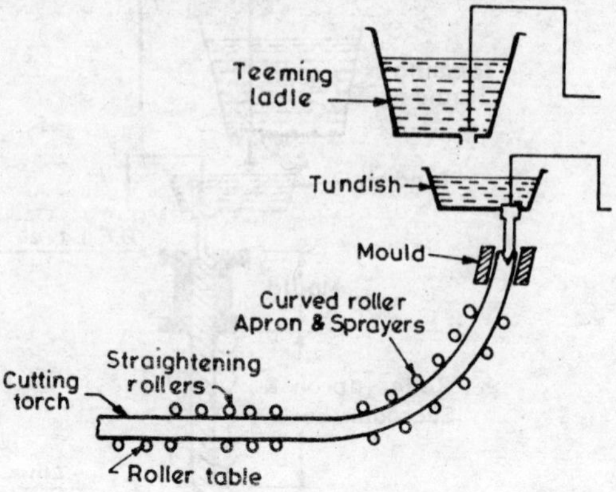

Fig. 2.12. (*a*) Curved-mould (S-type) continuous casting machine.

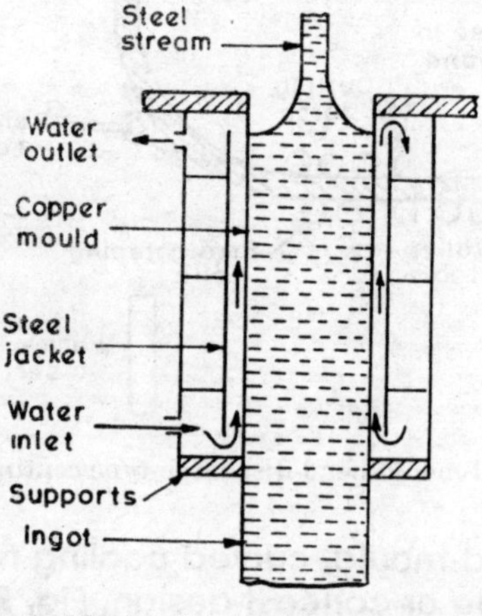

Fig. 2.12. (*b*) Mould of a continuous casting machine.

2.4. PROCESS AND EQUIPMENT DETAILS

2.4.1. Optimum temperature of liquid steel

Casting processes utilizing stoppered tundishes can use lower starting temperature. The temperature loss between the furnace and the casting machine varies with local conditions. To minimise this loss :

(*i*) Ladles are preheated to the highest practicable temperature. (870–1100°C).

(*ii*) Increase the thickness of refractory lining and add special insulating brick layer on the outside.

(*iii*) Lip-pouring ladles have special heating system which is operated during casting. This is fuelled by oil or natural gas through a top burner. Bottom pouring ladles require no heating during casting.

Each casting plant must determine its own temperature losses as related to local conditions.

Casting temperature depends on a number of factors :

(*a*) Temperature losses between tap and start of casting.
(*b*) Temperature losses in the ladle during casting.
(*c*) Nozzle sizes used in the tundish.
(*d*) Viscosity of steel.

It is essential that procedures be such as to keep all temperature losses constant.

Fig. 2.13 shows temperature loss in ladles as a function of heat size for various holding periods. Ladles were preheated to a temperature of 425°C and the heat was tapped at 1650°C. Tapping time was 5 minutes. These heat losses could be minimised by using heavier refractory lining and special insulation.

At Tata Steel, dolomite ladles are employed for the transfer of liquid steel from the LD shop to the C.C. machine. These ladles are adequately preheated to minimize loss of temperature as the liquid steel is being transferred from the LD shop to C.C. machine. Since dolomite bricks are hygroscopic, the ladle must be used only in dry condition. Dolomite is the working linking of these ladles, and it must be perfectly superheated to drive out any traces of moisture present in them. But very long preheating periods can not be used for dolomite linings because of the risk of bond degradation. So, dolomite ladles have to be preheated to a hot-face temperature of 950–1100°C and the mode of heating has to be such that in the first two hours, the face of lining rapidly approaches the peak temperature and then gradually reaches the required temperature during the remaining preheating period. The duration of preheating of a green ladle does not exceed 12 hours in their case and the ladle shell temperature is not allowed to exceed 250°C at the end of pre-heating. Green or freshly lined ladles are fired by Coke oven gas at the rate of 600 Nm^3/hr with 10% excess air and duration of preheating is 12 hrs. maximum. Ladles in circulation have a temperature of about 600°C. at the lining face and outside shell is about 230 °C. These ladles are fired by light diesel oil (LDO) plus Coke Oven gas at flow rates ranging between 250 l/hr to 300 l/hr of LDO along with 400 Nm^3/hr of Coke Oven gas and 10% excess air. The duration of preheating is 75 min., when excess air is not preheated and

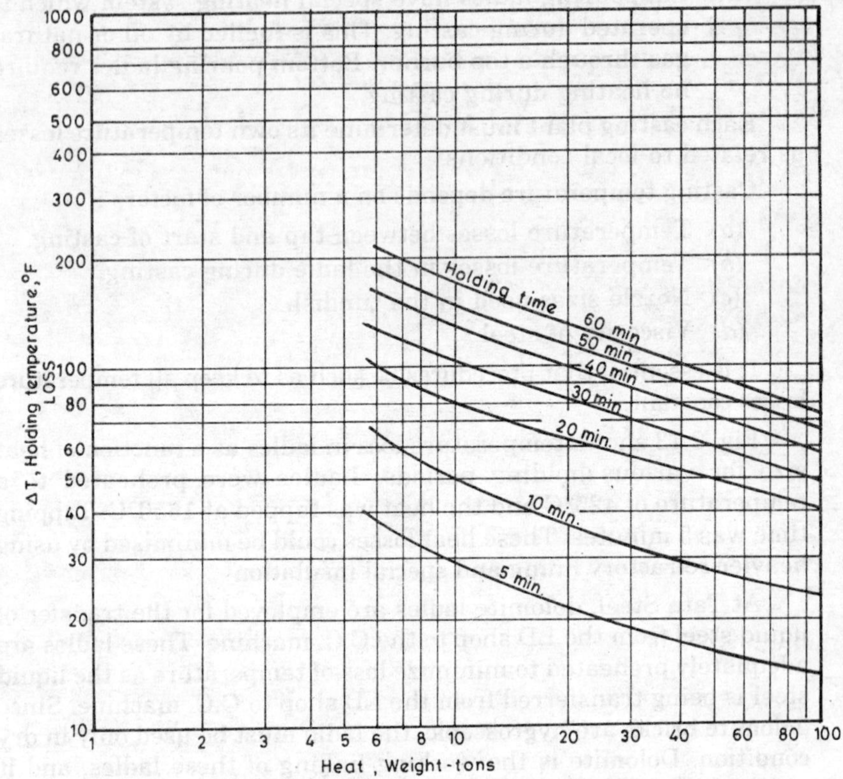

Fig. 2.13. Temperature loss as a function of heat size for various holding times. Steel heats in ladles pre-heated to an average temperature of 800° F. Furnace tapping temperature, 3000° F. Furnace tapping time, 5 minutes.

this period progressively decreases to 50 and 45 minutes, when the excess air is preheated to 200 and 400°C respectively.

2.4.2. Sources of liquid steel for C.C. machines :

Oxygen steel making processes can readily achieve the desired temperature and ensure maximum utilisation of the machine. Open Hearth furnace are used by some companies but require a high standard of ladle pre-heating and hot-metal handling practice.

First continuous casting machines were combined with Electric furnaces (Arc and Induction both). These furnaces give rigid temperature control. But for tonnage steel production, Open Hearth and L.D. are used. Many installations in Europe and USSR have proved that carbon steels can be tapped from these units with good (*a*) temperature control, and the (*b*) consistent quality required by continuous casting. Hence, tonnage steel producers would couple continuous casting machines with L.D. Converters rather than electric furnaces. Furnace capacities must be compatible with continuous casting machine capacities. Excess furnace capacity creates an overload for the casting machine. Higher tonnage is handled by increasing casting speeds, and in most cases this can not be done without increasing the cooling water-volume in the spray cooling zone, but this gives rise to internal defects in the end product - resulting in costly rejections.

Deoxidation and Alloying practice is little more complex than in conventional practice. A high alumina content adversely affects the fluidity of the melt, and with some steels reactions at the nozzle result in premature cessation of the cast. For this reason, deoxidation with Si or other less deleterious elements is practised. If Al must be used, then continuous deoxidation in the mould by feeding Al wire appears to be the only alternative.

It is reported that the type of refractories used for lining the steel transfer ladle has a direct influence on the oxygen levels in the liquid steel. Following oxygen activities in steel melts have been reported at Tata Steel :

(*a*) Dolomite lined ladle - 2 to 6 ppm.

(*b*) Bauxite lined ladle - 2 to 6 ppm.

(*c*) Silica lined ladle - 10 to 20 ppm.

It is evident that oxygen activity is much higher for acid linings than for bauxite or dolomite linings. In TISCO, therefore, dolomite ladles are employed for transfer of liquid steel from LD vessel to the C.C. machines. Also, optimum inert gas purging by nitrogen/argon in steel ladle results in thermal and compositional homogenisation. Besides, purging helps in floating inclusions & scavenging out undesirable dissolved gases (*e.g.* oxygen and hydrogen). Deoxidation of the melt is also carried out with the help of calcium-silicide, based on dissolved oxygen measurements. The basic function of Ca-Si is to deoxidise as well as to control alumina inclusions by forming liquid calcium aluminates, which minimize the possibility of nozzle blockage during continuous casting. Also, the variation of temperature of liquid steel being fed to the caster must be as little as possible - a fluctuation of ±5°C max. can be considered to be satisfactory.

Similarly, liquid steel for the slab caster must have close composition control and low levels of oxygen, nitrogen, hydrogen, sulphur and phosphorus. This requirement becomes essential for (*a*) smooth operation of C.C. machine, (*b*) defect free slab for direct hot- charging and (*c*) defect free finished products. Oxygen above 20 ppm. will lead to higher amounts of entrained alumina, giving rise to clogging of the submerged entry nozzle. This will lead to a reduction of casting speed and ultimately the casting will be interrupted. High amounts of entrained alumina and detached alumina from the clogged nozzle will affect the cleanliness of steel adversely. This will result in the presence of silvers/black line defects in cold rolled flat products.

Both sulphur and phosphorus contents must be kept below 0.02% in the liquid steel. It is because, sulphur gives rise to an increased incidence of cracks in the continuously cast product and tends to segregate to the centre. Phosphorus also gives rise to crack formation due to the presence of a liquid film much below the liquidus temperature. Nitrogen content should also be maintained below 40 ppm. Otherwise it results in the precipitation of AlN (aluminium nitride) which decreases ductility of Al-killed steel in the temperature range of 700–900°C and increases incidence of transverse cracks. The average hydrogen content in steel tapped from LD convertor ranges between 3 to 6 ppm. Hydrogen can be removed from liquid steel either by vacuum application or by using a neutral carrier gas to flush the steel. Hydrogen can be reduced by about 2 to 3 ppm. depending upon the duration of treatment. The hydrogen is sometimes associated with the formation of sub-surface pin-holes in the cast product.

2.4.3. Ladles

Hot metal is supplied to the machine from either a stoppered bottom pouring ladle or a lip pour ladle with a bridge for retaining the slag and a channel to draw metal from a certain depth. Bottom poured ladles can be handicapped by the durability of the stopper rod assembly under the extended casting periods of the process. So, such ladle are not satisfactory if the teeming period exceeds nearly 45 minutes. They are equally ineffective for small capacities since temperature loss during teeming is more. This is partly due to the fact that these ladles cannot be preheated beyond 700°C for fear of overheating the stopper. The bottom poured ladles are therefore used to capacities exceeding 70 tons. A ceramic coating or a protective sleeve assembly around the stopper rod can partly overcome this problem. On the other hand, lip poured ladles can be preheated upto 1400°C so it can be used where teeming continues even upto 2 hours. These are preferred for capacities below 30 tonnes and where a small plant with minimum number of strands is used. Lip poured ladles provide a better control over the flow of metal between ladle and mould, but capital cost of equipment installation and maintenance requirements are generally greater. Proprietory exothermic compounds and insulating materials may replace burners in this case.

Usual firebricks are used to line both the types of these ladles. Normal life of lining is about 10-20 heats. The siphon or the bridge in a lip pour ladle is of high alumina material as this is the weakest part of the ladle. Lip pour ladle is provided with a lid containing facilities for inserting a preheating burner. In TISCO, however, dolomite lined ladles are preferred for transfer of liquid steel from LD vessel to the C.C. machine. (See section 2.4.1 and 2.4.2). These ladles are preheated to a hot-face temperature of 950-1100°C.

2.4.3.1. Ladle Injector/Purging

As mentioned in section 2.4.2, the ladle is first treated by Argon gas before sending the metal to C.C. machine. In special heats CaSi powder is also injected along with Ar gas, whereas in normal commercial grades, only Ar is injected. A thorough purging by Ar gas leads to homogenisation of bath with respect to composition and temperature. CaSi injection helps to achieve very low value of sulphur in the range of 0.001 to 0.003% and in addition helps to control the morphology of oxide inclusions. As mentioned in section 2.4.2, Ca-Si injection converts solid alumina inclusions into liquid calcium aluminates, which minimizes the possibility of nozzle blockage during continuous casting. While purging/injection is going on, a lollypop sample is taken from the ladle and final correction of alloys is done in the ladle. In some plants, Ar injection is carried out through porous plugs fitted in the bottom of the ladle. It is found by some plants that blowing through a porous brick is very effective on the first and second occasions, but subsequently the brick rapidly looses its permeability. For this reason, some plants have changed their practice. Instead now, a perforated rod and a perforated plug (8 mm ϕ) is employed for argon injection. The plug is dipped to some 20 cms from the bottom. The argon injection is carried out at a pressure of 3 kgs/cm^2, taking the precaution of reducing the pressure to 2 kgs/cm^2 during the introduction of the rod into the ladle and untill it is withdrawn after blowing so as to avoid large steel spatterings. The injection time is about 3 mits., the temperature drop before and after the completion of blowing being 15°C. This temperature difference can be as much as 20°C with time of 5-6 minutes. The argon is blown in at an average rate of 300 lit./min.

2.4.3.2. Ladle Stream Shrouding

Fig. 2.14 shows the change of the oxygen content of the molten steel from the time of tapping until it reaches the mould. After ladle argon stirring, the total oxygen content (curve II) attains its minimum value of about 60 ppm. During the pouring from the ladle into the tundish, it again increases by about 15 ppm (curve IIa) and during pouring form the tundish into the mould, it increases by a further 10-15 ppm. (curve IIb). When the casting stream has been oxidised twice, the total oxygen content is about 85 ppm (this is about 40% more than after Ar stirring). The dissolved oxygen content (curve III) behaves in accordance with the temperature curve (curve I). From this illustration, it can be seen that it is just as important to protect the casting stream between the ladle and tundish as it is to protect it between the tundish and mould.

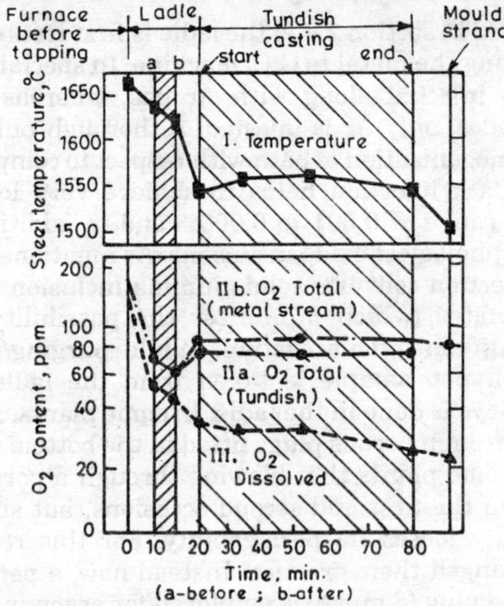

Average values for *C* steels with 0.50% C, 0.25% Si 0.50% Mn.

Fig. 2.14. Temperature curve and oxygen movement during continuous casting.

The following techniques are used to protect the casting streams against oxidation with air (Fig. 2.15*a*) :

(*i*) Pouring with immersion nozzles on the ladle and/or tundish possibly with an additional supply of inert gas (A).

(*ii*) Casting with shrouds on the ladle and/or tundish with the additional use of shielding gas (B).

(*iii*) Casting under a controlled atmosphere using bellows (C).

(*iv*) Casting with a metered addition of shielding gas (D).

(*v*) A combination of these methods.

In order to protect the casting stream between the ladle and tundish, the same measures can be used irrespective of the cross-section of the strand. The best known method of protecting the

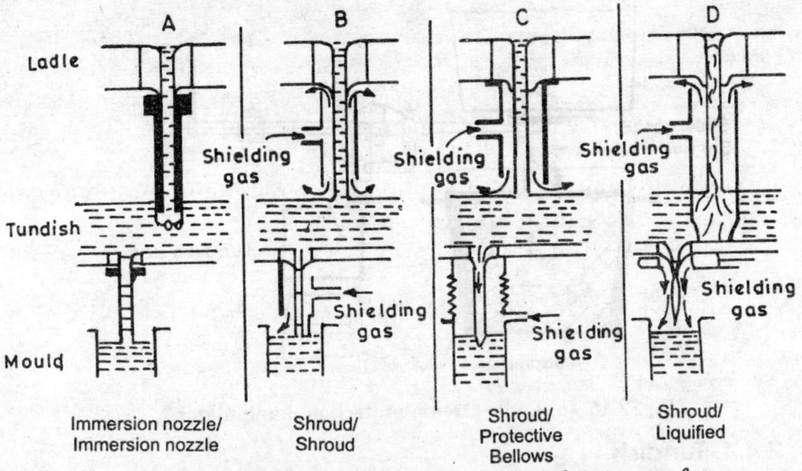

Fig. 2.15. (*a*) Possible methods of shielding the stream of molten metal during continuous casting

stream of molten metal between the tundish and mould is the immersion nozzle. However, in case of strands with cross-sectional dimensions of less than 140 mm square, casting with immersion nozzles can not be applied. For this reason, particularly in case of small cross-sectional dimensions, casting under a controlled atmosphere should gain more and more importance. This process (liquified gas shrouding) offers the following advantages :

(*i*) the stream of molten metal is visible over its entire length, as is the level of steel in the mould and tundish.

(*ii*) if necessary, slag can be fished out without difficulty.

(*iv*) during casting, the nozzle ring remains undamaged even when shielding gas is not used.

(*iv*) it causes no disturbance to the burning open of the nozzle and pulling of the launder.

In another practice, once the ladle nozzle is opened, a ceramic shroud is fitted in the bottom of the collector nozzle. Such a shroud is shown in Fig. 2.15 (*b*). The shroud is made of isostatically pressed graphitised alumina. The tubular shroud is held in a special supporting and manipulating device mounted on the tundish car. Before starting the cast, the shroud is pivoted into position below the collector nozzle of the slide gate ladle and pressed against it by means of a simple counter weight. (Fig. 2.15-*b*). Besides preventing reoxidation of stream, the shroud also reduces the temperature drop between ladle and tundish by 10°C.

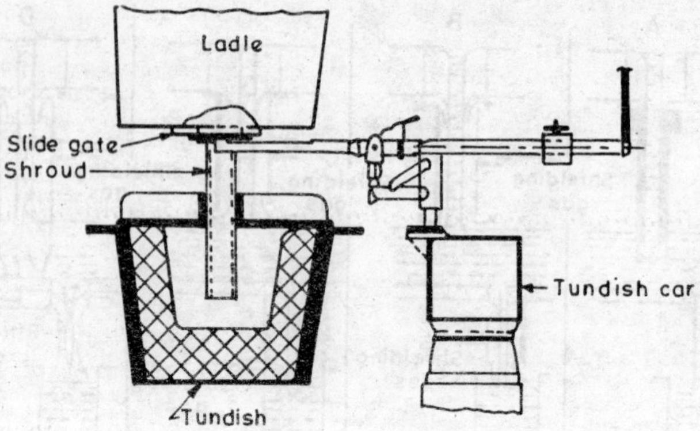

Fig. 2.15. (*b*) Ladle stream protection manipulator.

2.4.4. Tundish

It is basically a refractory lined vessel with a nozzle (or several nozzles) and stopper gear assembly. Liquid metal in general is teemed into a continuous casting mould from a ladle via the tundish. The shape of tundish will be determined by the number of strands and the machine dimensions (by the distances between billet/bloom centres). The liquid metal level in tundish may vary from few inches to about 15/25 inches. A large tundish can be costly if skull formation is allowed to take place, for while removing the skull, the refractory lining can get destroyed. A tundish may last 40-45 heats. The tundish must be preheated to avoid chilling of the first flush of metal. The nozzle should deliver uniform flow of steel right from the beginning. For small billet casting machines, the capacity of tundish may be between 250 to 400 kgs., while for large multiple strand machines (for casting large sized billets and blooms), the capacity may vary between 2 to 3 tonnes.

2.4.4.1. Liquid Steel Flow Control

Stopper operated tundishes are widely used for billet casting and are equipped with automatic level control devices. Sensitive control mechanism must be provided with stopper free tundishes. For tundishes without stoppers, the nozzle size must be established very carefully. The dia must be designed and selected on the required rate of flow to maintain specified casting speed and on the basis of ferro-static head above the orifice. For this reason, highest quality nozzles must be used. An example of Atlas Steel Ltd. is given below to illustrate this fact.

Tundish nozzle data of Atlas Steels Ltd.

1.	Nozzle size (mm) ϕ	23.8	20.6	17.46
2.	Orifice area (mm^2)	444.88	333.29	239.43
3.	Metal supply at 12" (304.8 mm) head (kgs/sec)	4.6308	3.5412	2.4516
4.	Metal supply at 24" (609.6 mm) head (kgs/sec)	6.6284	5.04	3.4958

In modern casters, however, a slide gate system is used in ladle & tundish due to its great reliability & flexibility over the stopper rod system in meeting the stringent requirement of continuous casting. It is possible to change the size of discharge nozzle during casting operation within a short time. The typical requirements of continuous casting are :

1. Precision control of steel flow
2. Continuous regulation
3. Discharging into smaller receptacles, and
4. Intermittent stoppage/reopening

Tundish nozzles are usually made of fireclay (alumina silicates), zirconia or zirconia silicates and they have special shapes.

A very recent development combines the basic advantages of stopper's simple design, low cost, potentially single piece pouring tube, tundish prefilling to any desired level, self drainage when closed-with the superior operational reliability of slide gates. Developed and patented by British Steel Corporation, rotary valves have been in service at Lackenby on the 8 strand bloom caster since November 1985. Some 55,000 tons of various steel grades including low and medium carbon Si killed steels, high carbon rail steels and also various Al fine grained steels have been successfully cast, with rail sequences of upto 14 hours duration.

The main design features of rotary valve are as follows : (A section through rotary valve and schematic of tundish layout is shown in Fig. 2.16 and 2.17 respectively) :

(*i*) Repetitive on-off control; self drainage on closure; ideal for association with tube changer systems.

(*ii*) Control surfaces immersed within the steel bath minimising the risk of air ingress.

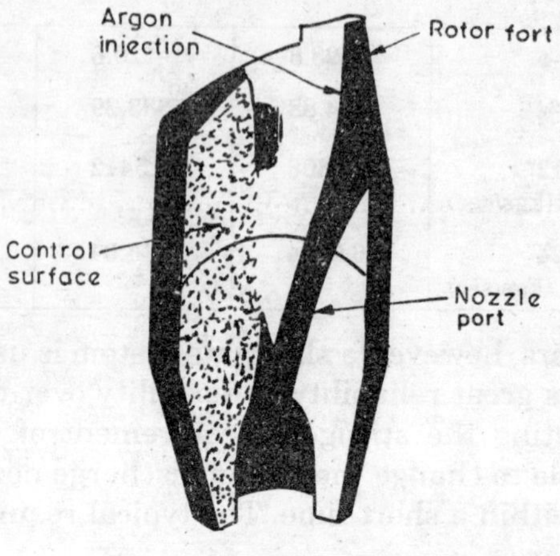

Fig. 2.16. Section through rotary valve.

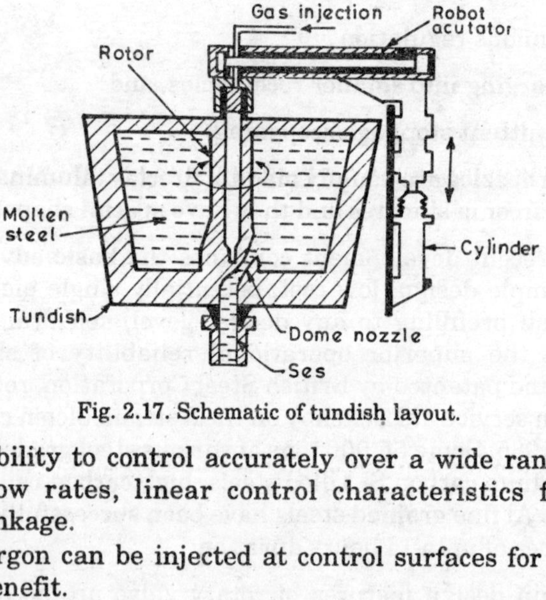

Fig. 2.17. Schematic of tundish layout.

(*iii*) Ability to control accurately over a wide range of steel flow rates, linear control characteristics for M.L.C. linkage.

(*iv*) Argon can be injected at control surfaces for maximum benefit.

(*v*) Slow, controlled mold fill can be achieved, thereby minimising the risk of breakout at strand start-up.

(*vi*) Flexible system allowing easy installation on slab, bloom and billet casting machines.

(*vii*) Refractories like graphitised zirconia, graphitised alumina and graphitised magnesia prepared with carefully controlled friction and corrosion resistance characteristics together with excellent thermal shock resistance are employed.

Like the conventional stopper, but unlike a slide-gate the valve is self draining and can be repeatedly opened and closed without risk of blockage. In practice strands have been shut-off for over one hour whilst the other strands were casting and restarted without oxygen lancing; it is claimed.

2.4.4.2. Tundish Refractories

The refractory lining of tundish consists of three components namely : permanent lining, wear lining, and protective linings. Initially standard dry pressed fireclay (MAXIAL 320) large panels with Al_2O_3 content of 37-40% were tried as tundish runner linings. At the point of impingement of the descending casting stream from the ladle, the standard panel (RESISTAL 057) containing about 57% Al_2O_3 are used. But the use of these high alumina special panels has not been entirely satisfactory. A baffle panel as a fusion MULLIT basis has been developed, which has been found satisfactory. Bricks are mostly used as the refractory material for lining. Coating refractories for the protective linings are mainly installed by trowelling and some are being developed for application by gunning. A new generation of basic insulating gunning mixes was also developed and applied on the permanent lining by spray method. Subsequently, the tundish insulating board was developed and is being widely used in the world. A new tundish lining method has also been developed based on the concept of using preformed precured tiles, applied using a remotely operated industrial robot. Clay (silica-alumina base) was initially used as the aggregate for coating refractories. Since silica was later found to form harmful non-metallic inclusions, the silica based coating refractory was replaced by a high alumina based material. Then, magnesia-based coating refractory was developed and is used widely. High alumina cement was initially used as the binder, but because of its low erosion resistance, slow heating and high cost, it was subsequently replaced by silicates and phosphates.

Cold Tundish Lining System

It consists of prefabricated disposable liner boards made of fibrous insulating refractory materials of low thermal capacity and low thermal conductivity. It does not require heating prior to casting and thus contributes to heat energy saving by about 30,000 kcal/t of cast products. Tundish nozzle, well block and stoppers, however, require heating to about 1100°C to avoid initial chilling of metal in the nozzle area.

Monolithic Tundish Lining

The conventional brick permanent lining is being replace by monolithic lining in several plants whereby a service life of more than 1000 heats has been achieved. The use of high castable technology (using low cement and ultra-fines) provides easy installation. The main advantages are : savings in refractories, reduction in labour cost and increased tundish availability.

Single-Layered/Multiple-Layered Tundish Wear Lining

The system, called DOSSOLITE, allows a lining to be sprayed on the walls of a tundish where each area is protected by a composition specially adopted to withstand the required stress. Multi-layers are also sprayed in sequence on the walls of a tundish. In case of multiple layers, each layer is designed to best achieve the specific role it has to play. It has been reported that hydrogen pick-up is less than 1 ppm in case of hot tundish practice. If the spray system is used according to the cold tundish practice, hydrogen pick-up ranges between 1 and 1.5 ppm. A fully automatic spray system has also been developed to spray the right weight of DOSSOLITE in any type of tundish using an automatic manipulator or robot.

2.4.4.3. Protection of Tundish Stream

Typical kinds of liquid stream protection systems have already been outlined in section 2.4.3.2 in connection with "ladle stream shrouding". The importance of protecting tundish stream need not be over-emphasized.

The reoxidation of liquid stream which flows from the tundish to the copper mould is extremely detrimental. The reoxidation of liquid stream is a normal part of ingot making and the oxides formed during teeming float on the top surface of the metal and do little harm. But in case of continuous casting, the amount of such oxides accumulating on the steel in the mould can become excessive since the teeming process is continuous and therefore these oxides must be skimmed off periodically. Tundish stream shrouding makes the casting process simpler, besides improving the quality of steel cast.

"Submerged pouring" or "immersed teeming" is also considered to be an important and necessary step in the production of clean steels. In this technique the metal stream is shrouded with a refractory tube as it passes from the tundish into the mould, the bottom part of the tube being submerged below the surface of molten steel in the mould. The submerged pouring has the following advantages :

(*a*) protection of steel stream form reoxidation. It is important with steels containing deoxidisers such as Al or where alloying elements are involved.

(*b*) by a suitable choice of tube outlet configuration, the steel stream is directed into the mould such that the flow pattern obtained leads to more even temperature distribution across the width of the mold and hence more favourable conditions for solidification.

(*c*) it prevents splashing of the steel stream and entrainment of the synthetic slag mould cover.

Figs. 2.18 and 2.19 show submerged pouring schematically. In Fig. 2.18 the steel flow from the tundish is controlled by a metering nozzle and the submerged entry tube is separate. This arrangement is common on billet and bloom casters. Fig. 2.19 shows stopper control with the stopper engaging directly with a combined nozzle and sub-entry tube. This arrangement is more common on slab casters. Outlet configurations vary widely, but on slab casters bifurcation angled downwards between 25° and 45° from the horizontal is very common because of the favourable flow pattern produced.

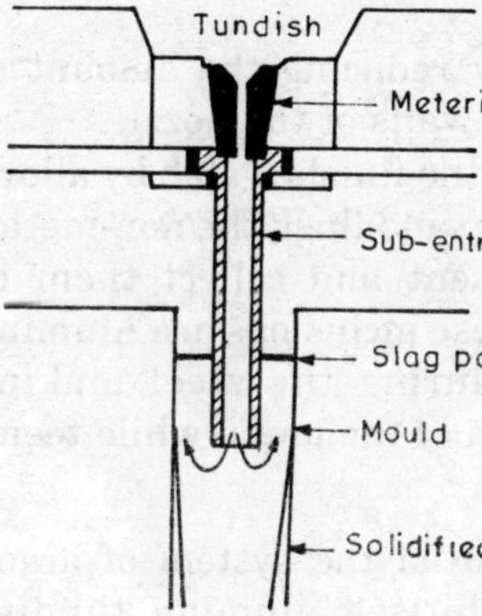

Fig. 2.18. Submerged pouring using a metering nozzle and submerged entry tube.

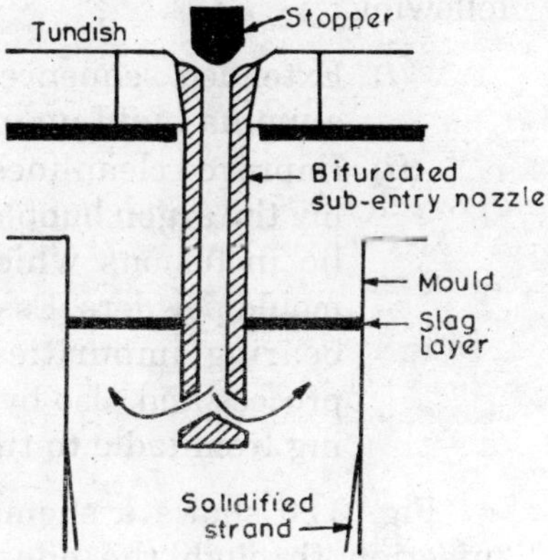

Fig. 2.19. Submerged pouring with stopper control and combined nozzle and submerged entry tube.

Because of severe conditions under which these sub-entry nozzles are used, they must have the following properties :

(*i*) Resistance to thermal shock
(*ii*) Resistance to molten steel
(*iii*) Resistance to synthetic slag
(*iv*) Moderate high temperature strength.

Sintered fused silica which satisfies these requirements is widely used as a standard material for sub-entry nozzles and tubes. However, at manganese content more than 1.2% the wearing rate is rather high and fused silica has been replaced by graphitized high alumina. These materials normally contain 20%-30% flake graphite and 50%-60% alumina and are either carbon, clay or ceramic bonded, formed by isostatic pressing ensuring uniform density and free of forming defects. A special nozzle grade material based on zirconium silicate and zirconium oxide has also been developed for aggressive qualities of steel and sequence casting. Four nozzle grades are presently available according to their zirconium silicate/zirconium oxide ratio. Immersion nozzle with base material alumina graphite and wall in the meniscus region consist of the same material enriched with 10% boron nitride are also used successfully.

In a recent development a technique of injecting argon through the one piece graphitised-alumina stopper into the submerged pouring nozzle has been developed. It provides essentially the following :

(*i*) Extended sequence casting by reducing the amount of alumina build-up in the inner walls of the nozzle.
(*ii*) Improved cleanliness levels in the finished slab by allowing the argon bubbles to pick-up and flush the non-metallic inclusions which are present and collect them in mould powders. Essentially these inclusions are Alumina bearing impurities formed during the steel making process and also by reoxidation of the metal while teeming from ladle to tundish.

Fig. 2.20 shows a shematic drawing of the system of argon injection through the one-piece graphitised alumina tundish stopper. The argon flow rate is 5-10 l/min. depending upon the design of submerged pouring nozzle. For extended sequence casting isostatically pressed graphitized alumina pouring nozzles are becoming more common due to their excellent resistance to erosion

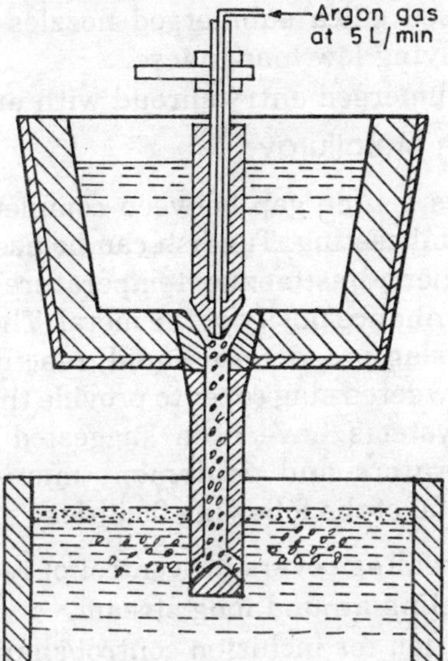

Fig. 2.20. Schematic diagram of argon down the tundish stopper rod.

properties. The basic objective of argon injection is to reduce alumina build-up and to ensure that the maximum amount of alumina is flushed into the mould powder. Tests carried out have shown that by argon injection, the number of sub surface alumina bearing inclusions has been reduced improving the deformability of the finished cold-rolled steel and minimizing a characteristic "black-line" effect of the rolled product. Argon injection together with sub-entry nozzles is becoming more widely used for the low carbon strip and pipe steels, both of which contain high soluble levels of alumina.

The following possibilities are available for reducing alumina build-ups :

(*i*) Additional injection of argon into the immersion nozzle by way of purging slit.

(*ii*) Reduction of porosity of the immersion nozzle or application of sealing glazings to prevent intake of air.

(*iii*) Coating of the tube inner wall with boron nitride, zirconium oxide or SIALON *i.e.* SIALCOR nozzles.

(*iv*) Sitting porous plug on the inner bore of the submerged entry nozzles.

(v) Using fused silica submerged nozzles for casting steel grades having low manganese.

(vi) Use of submerged entry shroud with argon slit.

2.4.4.4. Tundish Metallurgy

There is always a time gap between completion of secondary treatment and actual casting. Tundish can be used as a treatment vessel to make further adjustment of temperature and homogeneity of liquid steel and enhance purity of the metal. The normal practice is to go in for two-slag practice - a liquid, reactive slag to absorb inclusions and a powdered slag cover to provide thermal insulation. Several tundish systems have been suggested or developed by various plant operators and equipment manufacturers which include one or several of the following facilities :

(i) Tundish sealing covers for control of reoxidation.

(ii) Argon sealing around long stream.

(iii) Flux addition for inclusion control/shaping.

(iv) Addition of alloying elements for composition adjustment.

(v) Installation of dams, weirs and slotted baffles to control steel flow pattern and turbulence in tundish.

(vi) Tundish bath heating from top for temperature adjustment.

(vii) Use of ceramic filters for removal of impurities from liquid steel.

(viii) Argon bubbling through porous plugs for homogenisation.

(ix) Inductive/plasma heating of tundish stream for final temperature adjustment just before the liquid steel enters the mould.

(x) Calcium/calcium silicide and aluminium wire feeding.

The three important functions of the slag to be used in the tundish are :

(a) Thermal insulation

(b) Absorption of non-metallic inclusions such as oxides and sulphides.

(c) Prevention of reoxidation and nitrogen pick-up from the surrounding atmosphere.

Fig. 2.21 summarises the various methods available for improving steel cleanliness in the tundish.

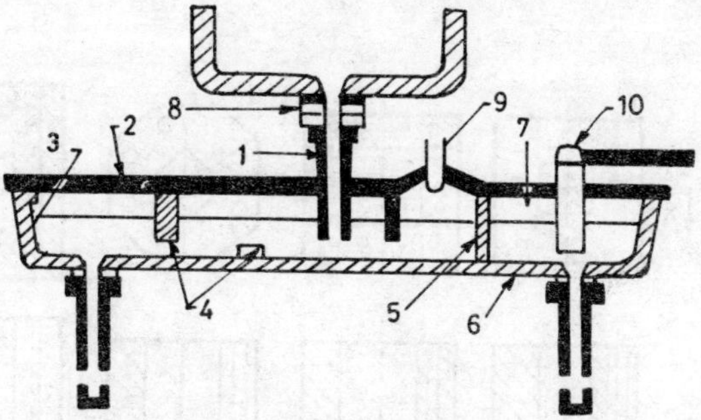

1. Long nozzle + argon
2. Sealing cover
3. Refractory
4. Tundish dam
5. Filter
6. Argon bubbling device
7. Flux
8. Slag detector
9. Heating apparatus
10. Flow control device

Fig. 2.21. Methods for improving steel cleanliness when the tundish is used as a refining vessel.

2.4.5. Moulds

The mould in fact is the heart of continuous casting machine. The mould is made out of extruded tube machined out of a solid block or is a welded plate construction of high conductivity copper. Extruded tube moulds are mostly used for smaller corss-sections of billets upto 100-200 mm square, while block and plate moulds are used for larger square and rectangular sections (blooms and slabs). Prior to casting, the top of the mould is normally protected against adhesion of molten steel splash with either a sheet of asbestos or a refractory coating. With modern stream protection techniques like "submerged pouring" or "immersed teeming", such a precaution may not be necessary. The length of the mould should be such that under the conditions of its cooling, adequate skin must be formed before the billet/bloom emerges out of it. The mould is usually 75 to 140 cm in length. Function of mould is to extract maximum amount of heat during a short time of contact. It is expected to extract 10% of total extractable heat. To avoid transverse cracking of partially solidified strand, the mould is oscillated at the rate of 30-60 oscillations per minute with a negative strip. Tubular or water jacketed moulds are claimed to be cheaper, but solid moulds give safer operation. Different kinds of solid and built-up designs of mould are shown in Fig. 2.22, while Fig. 2.23 shows a typical tubular mould employed for billet casting.

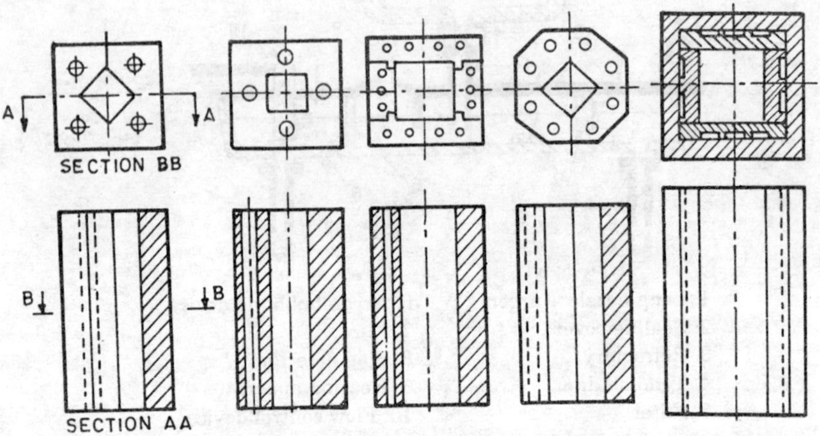

(*a*) Solid copper mould with drilled bores for cooling water Diagonal Arrangement Short-mould ATLAS

(*b*) Solid copper mould with drilled bores for cooling water Parallel Arrangement Short-mould ATLAS

(*c*) Built-up type copper moulds with drilled bores for cooling water TERNI

(*d*) Solid copper mould with drilled bores for cooling water Diagonal Arrangement ATLAS

(*e*) Built-up type copper mould with rectangular water passages

Fig. 2.22. A selection of mould types used in continuous casting of steel billets and blooms.

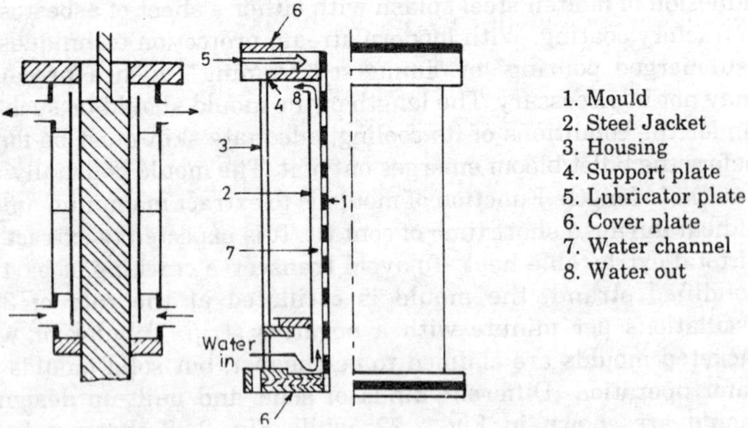

Fig. 2.23. Typical mould design for a billet caster.

The functions of mould are :

(a) To provide the right size and shape.
(b) To provide a solidified skin sufficiently thick to prevent breakout or bulging below the mould.
(c) To provide a good surface to the cast product.

To perform these functions, the mould must have the following characteristics :

(a) Dimensional stability (achieve control of mould distortion).
(b) High heat transfer coefficient
(c) Adequate length
(d) Compensation for shrinkage of the solidifying billet. For this purpose, moulds are tapered downwards to the extent of contraction of steel on solidification.

2.4.5.1. Moulds for billet/slab casting

[A] Mould Wall Thickness :

In normal billet moulds shown in Fig. 2.23, the wall thickness ranges between 6 and 14 mm. It has been observed, however, that thicker walls of the order of 12.7 mm even for casting 100 mm sq. billets gives a better performance. The increase in wall thickness results in a decrease in the oscillation-mark depth and at the same time, thick walled tubes enhance the heat transfer in the meniscus region. This results in reduced incidence of off-corner internal cracks since the bulging resistance is improved. Increased heat extraction could permit higher casting speeds and result in gain in productivity.

[B] Mould Taper

In the mould, a gap forms between the solidifying shell and the mould wall as the shell cools and shrinks. In order to reduce the extent of this gap and to improve the rate of mould heat extraction, moulds used for continuous billet casting are usually tapered. Studies have shown that a single taper of 0.6-0.7%/m does not fully compensate for the shrinkage of the steel shell and different tapers in the same mould are necessary. It is possible to calculate the required mould taper for minimizing gap formation, by combining the profile of the steel billet surface and that of the dynamically distorted hot face of the copper mould. Results of such a study are shown in Fig. 2.24 for a steel with the specifications given below :

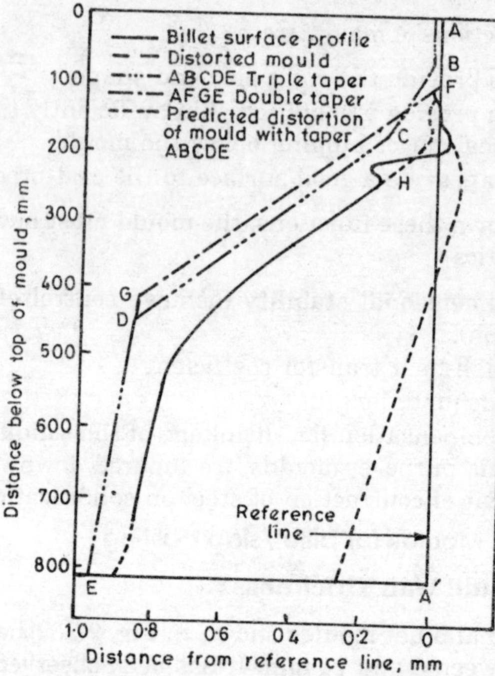

Fig. 2.24. Mould taper prediction based on shrinkage calculation.

Heat details for calculation of mould taper during billet solidification

Chemical composition (%)						
C	Mn	P	S	Cu	Cr	Si
0.36	1.30	0.017	0.028	0.35	0.22	0.190

Billet size (mm sq.)	Casting speed (mm/s)	Ladle temperature (°C)	Tundish temperature (°C)
143.8	38.10	1603	1524

Cooling water conditions		
Pressure (MPa)	Flow (l/s)	Velocity (m/s)
0.448	31.5	9.3

Taper at the meniscus (%/m)	Average taper in the lower part of the mould (%/m)	Maximum distortion (mm)
– 2.02	0.24	0.183

Curve H shows the shape of the distorted mould if a taper ABCDE is introduced. The introduction of additional taper BC helps to minimise the negative taper experienced by the single taper mould in the meniscus region.

The mould taper required to compensate for shrinkage of the shell is influenced by the carbon content of the billets and the cooling water velocity. Higher carbon steel and lower cooling water velocities require greater tapers. A single-taper mould is not the optimum design because the gap is too large in the lower region. This causes bulging of the shell, which can cause off-corner internal cracks. Though single-taper moulds are still in use, there is a clear trend towards using double taper moulds. The apparent absence of bulging in double taper moulds suggests that the rates of heat extraction and resultant shell growth in the upper part of such moulds are significantly greater than in their single-taper counterparts. Thus, a double taper mould is advantageous in that it reduces or eliminates the incipient negative taper, and thereby decreases the depth of oscillation marks.

[C] Example of typical moulds used at Tata Steel

(i) Billet Caster

Mould Size	:	801 mm long
		130 mm × 128 mm (inside) width
		Taper 0.6%/m for 125 mm square billet size.
		0.4%/m for 100 mm sq. billet size.
Material	:	Phosphor-Deoxidised E-copper
		SF-CU DIN 1787
		Cu-99.9%
		P-0.05%
Hardness	:	70-90 HB

(ii) Slab Caster

Mould size	:	900 mm long, 1650 mm wide
Radius of curvature	:	10 m.
Mould oscillation frequency	:	0-250 cycles/min.
Oscillation amplitude	:	0-12 mm
Width	:	Adjustable between the castings.
Slab size	:	800-1500 mm wide 210 mm thick (Fixed thickness) 4.5-11.5 m cut length.

2.4.5.2. Mould Materials

Various copper based alloys (deoxidised Cu, Cr-containing Cu and Ag containing Cu) are being used for different types of mould design in order to ensure optimum performance and long service life due to good thermal conductivity, auto-corrosive and surface finish quality. The main disadvantage with copper alloy is the poor resistance to wear. However, most commonly used material is silver alloyed copper with 0.1% silver content. In order to achieve higher operating life, mould plates are surface plated with chromium to prevent copper penetration into the strand shell and eliminates star cracks.

Age hardening Cu-alloy, namely Elbrodur and SF-Cu are also being used due to better elevated temperature strength and low distortion rate. However, the dimensional stability of Elbrodur is better than that of SF-Cu. Essential characteristics of these alloys are given in the following table :

S. No.		*I*	*II*	*III*
1.	Material	Cu-Cr-Zr Elbrodur	Cu - Ag	SF - Cu
2.	Chemical Composition %	Cu - 99.25 Cr - 0.65 Zr - 0.10	Cu - 99.95 Ag - 0.09 P - 0.006	Cu - 99.9 P - 0.015 - 0.04
3.	Thermal conductivity, W/m.k.	315	370	322
4.	Half softening temperature °C	500	370	350
5.	Elastic Modulus, N/mm^2	128×10^3	125×10^3	120×10^3
6.	Coefficient of thermal expansion, 1/K			
	20 – 100°C	17×10^{-6}	16.7×10^{-6}	16.8×10^{-6}
	20 – 300°C	18×10^{-6}	17.7×10^{-6}	17.7×10^{-6}

In recent years, high strength copper alloys and surface hardened mould have become very popular since they are believed to have greater resistance to permanent deformation. Electro-formed mould which consists of a precipitation hardened copper mould with nickel plating and Cu-Cr-Zr alloy mould, plated which chromium are also used these days. Precipitation hardened copper moulds with about 3 mm thick nickel plating have withstood 500/600 heats. Also, Cu-Cr-Zr alloy plated with Cr was observed to withstand 370 heats without deterioration of mould dimensions.

Use of hot-top mould has also been reported to reduce oscillation marks on the products and in addition, it improves the lubrication of steel mould interface. The method consists of modifying the conventional copper mould by plating the upper surface with an insert, made of material with low heat conductivity, in the meniscus area of the mould. The materials tried for the hot-top include : (*i*) stainless steel inserts, plated by explosion, (*ii*) Electrolytic nickel coating, and (*iii*) Thin refractory inserts (chromium carbide) coated by plasma deposition. Chromium-plated moulds have also been used successfully to avert red-shortness of the strand caused by copper abrasion. Explosion plated mould with coating 6 mm thick, have the advantage of repeated remachining and resultant long service life.

The thermal conductivity of mould is very vital. An increase of 50°C can cause permanent strain in phosphorus deoxidised copper. The influence of increasing temperature on the thermal conductivity and thermal expansion of mould is shown in Fig. 2.25. With the increase in casting speed, the temperature of copper mould increases. At mould temperature above 250°C, the yield strength (0.2% offset Y.S.) of oxygen-free or silver-bearing copper sharply decreases. Also, it has been shown that nickel coated moulds show a marked decrease in the star crack index relative to Cr plated moulds.

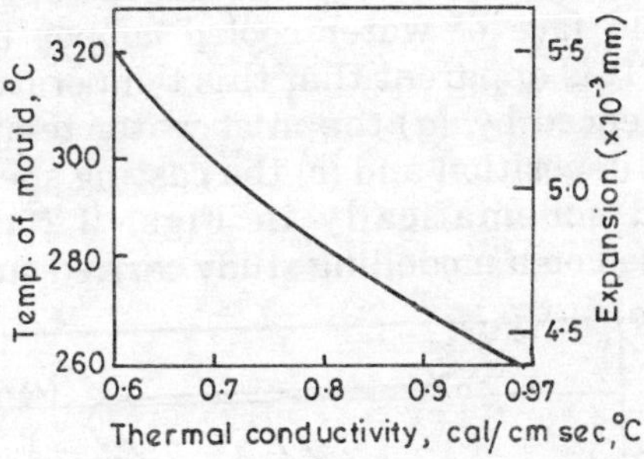

Fig. 2.25. Effect of thermal conductivity on temperature and thermal expansion of the mould.

Mould material employed at Tata Steel, for example, for billet caster conforms to SF-CU grade (Cu-99.9%, P 0.015/0.04%); while for slab caster, the alloy contains about 0.1% Ag (Cu-99.99%, Ag-0.09%, P-0.006%).

2.4.5.3. Heat extraction from the mould and spray cooling zone

Each mould is water cooled (known as primary cooling) in order to accelerate the heat transfer and cause rapid solidification of strand. The heat extraction starts at the meniscus level and as soon as a thin solid skin is formed, a gap starts forming. This gap will continue to widen as the thickness of solidified skin progressively increases downwards in the mould. The gap so formed offers additional resistance for the transfer of heat from liquid steel to the mould. The cooling water maintains the mould temperature in the range of sufficient mechanical strength to withstand the ferrostatic pressure and cooling water pressure. The heat transfer flux determines the shell thickness solidified in the mould which must be certain minimum value for safety's sake. Heat is then most effectively removed from the strand during "Secondary cooling" or "Spray cooling zone". The solidification produced during "Primary cooling" must be such that on discharge from the mould, the billet contains a solid skin of about 5 mm thick, which will prove to be adequate to support ferrostatic pressure. The solidified skin, when discharged from the mould, is about 1100°C externally and 1490°C internally. The total resistances offered in the process of heat transfer form the liquid metal to the cooling water in the mould include : (*i*) thermal conductivity of copper, (*ii*) thermal conductivities of scales/deposits on the mould wall (cold face) and (*iii*) air gap formed due to contracting solidified skin. All these factors put together result in an overall interface heat-transfer coefficient - a gross parameter. This gross heat transfer coefficient would result in the generation of a certain temperature profile on the hot and cold face of water cooled copper mould below the meniscus level. It is apparent that this temperature profile would be greatly influenced by, (*a*) the inlet water temperature, (*b*) the amount of scale deposition and (*c*) the casting speed. These effects are illustrated schematically in Figs. 2.26, 2.27 and 2.28 respectively based on a modelling study carried out at Tata Steel.

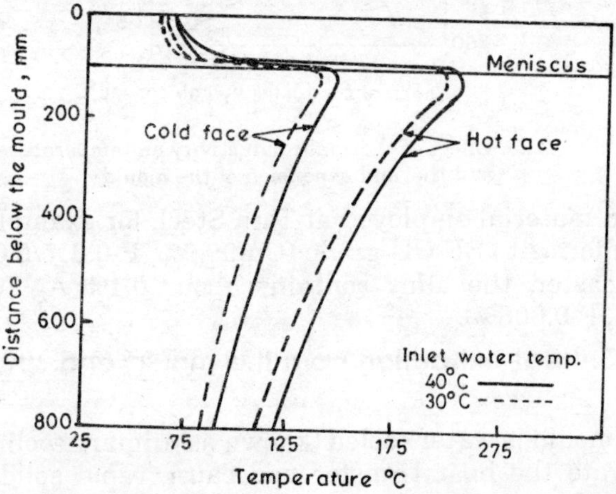

Water channel gap : 5 mm
Mould material : Cu-alloy
Water velocity : 7 m/sec.

Billet size : 125 mm-sq.
Mould thickness : 8 mm
Casting speed : 2 m/sec.
(noscale deposition)

Fig. 2.26. Predicted effect of inlet water temperature on cold-face and hot-face temperature profiles in CC billet moulds.

It can be seen from Fig. 2.26 that an increase of 8 to 10°C occurs all along the length of the mould below the meniscus level, if the inlet cooling water temperature is raised by 10°C (from 30° to 40°C). Also, it can be seen from Fig. 2.27 that deposition of scale on the cold face of the mould has a marked influence on the hot as well

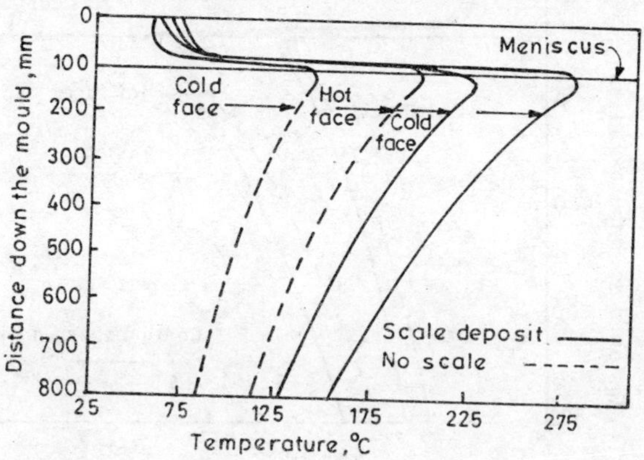

Billet size : 125 mm.sq
Mould thickness : 8 mm
Water velocity : 7m/sec.
Inlet water temp. :35°C
Mould material : Cu-alloy
Water channel gap : 5 mm
Casting speed : 2m/mm
Scale reduced heat transfer co-efficient
0.27248 cal $cm^{-2}sec^{-1}\,°C^{-1}$

Fig. 2.27. Predicted effect of scale deposition on cold-face and hot-face temperature profiles in CC billet mould.

as cold face temperatures all along the mould length. The scale deposition could reduce the extent of heat transfer in actual practice by nearly 50%, thereby increasing the mould temperature by 80-85°C near the meniscus and to 40-45°C at the mould exit. It is because the thermal conductivity of scale (for example, lime deposits containing calcium sulphate) is very poor and is only 0.6 to 2.0 k.cal/m^2 h°C as compared with the thermal conductivity of copper, 320 k.cals/m^2 h°C. Therefore, even a very thin deposit of scale on the cold face of mould wall can seriously deteriorate the extent of heat transfer through the mould. It is basically for this reason that all the cooling water required for primary circuit is properly deionised. This aspect is discussed in a subsequent section. Fig. 2.28 shows that a decrease in casting speed from 2 to 1.5 m/min does not affect the hot face and cold face temperature near the meniscus. But with distance down the mould, a gradual decrease in both hot and cold face temperatures is observed with a decrease in the casting speed. Continuous casting moulds are exposed to average heat transfer rates of $0.8 \times 10^6 - 2.0 \times 10^6$ k. cal / m^2. In the emiscus region, however,

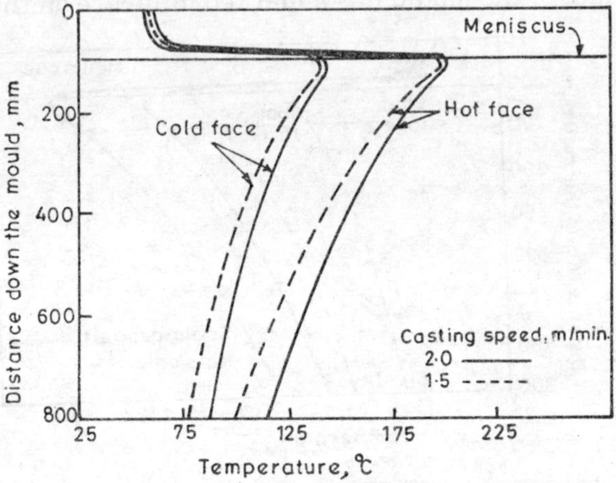

Billet size : 125 mm.sq
Mould thickness : 8 mm
Water velocity : 7m/sec.
Mould material : Cu-alloy
Water channel gap : 5 mm
Inlet water temp. :35°C
(no scale deposition)

Fig. 2.28. Predicted effect of casting speed on cold-face and hot-face temperature profiles in CC billet moulds.

heat flux can reach values upto 40×10^6 k.cals/m^2.h. As mentioned earlier, these heat transfer rates can be considerably reduced if a thin deposit of scale having poor thermal conductivity is formed on the cold face of copper mould.

The fundamental factor with regard to mould heat flux and shell growth is carbon content of the steel grade cast. A single most important process parameter is casting speed. Other factors which also influence mould heat flux are : degree of superheat of melt, mould lubrication and mould taper. Studies have shown that a minimum of heat transfer occurs at around 0.12%C, while a maxima corresponds to about 0.4%C.

The parameter "Cooling Intensity" is usually defined by the amount of water divided by the throughput rate of cast steel (litres of water/kg of steel). Normally 0.8 litres/kg are used for commercial grades of steels. The general trend is towards softer cooling to achieve better uniformity in cooling, lower thermal stresses in the solidified shell and fewer cracks in sensitive grades. For example, in case of Usha Alloy and steel Ltd., Jamshedpur, the water velocity is maintained at 5 m/sec for a cooling slot width of 5 mm. The water velocity may go upto 9 m/sec depending upon the size of section cast. Mould life in their case is between 250 to 410 casts.

Thermal flux in the spray cooling zone

Water sprays used during continuous casting extract 50-65% of the heat during solidification. The water spraying is characterised in terms of the temperature distribution in the billet, water pressure, spray distance, spray angle and spray water flux profile (type of spray nozzle). The billet face temperature varies from around 1100°C at the midface (at the mould exit) to around 600°C at the corners (at the spray zone exit). The gross spray heat extraction parameter, called "spray heat transfer coefficient" in the temperature range of 600°C to 1100°C is most important and is evaluated for different plant conditions. Different types of nozzles, for example, fullcone 'GG' series, a hollow cone with an 80° spray angle etc. are commonly used in practice.

A typical temperature history during spray cooling in the billet caster is shown in Fig. 2.29.

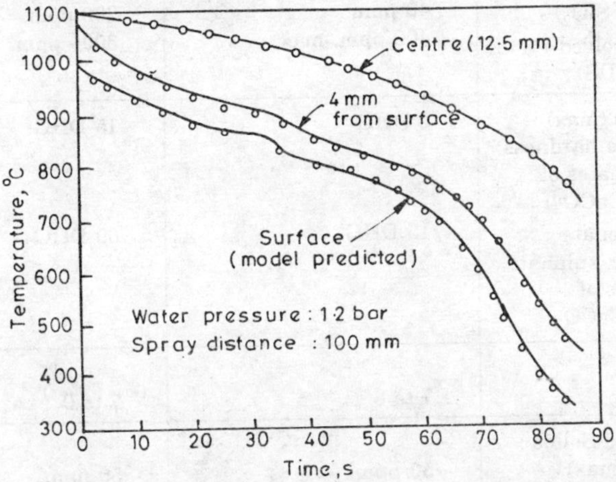

Fig. 2.29. Typical temperature history during spray cooling in the billet caster.

2.4.5.4. Cooling Water for Continuous Casting

The quality of water plays a vital role in the success of continuous casting operation. Poorly treated water may cause corrosion and scaling or carry suspended solids, resulting in erosion or blockages. Because the shape of the mould and mould/shell interaction are governed by the temperature distribution in the mould wall, cooling water characteristics such as velocity, quality etc. are of great importance.

It has been shown that variation of any of the following parameters amounts to varying the resistance to heat flow either in the mould or at the mould/cooling water interface :

(*a*) Cooling water velocity
(*b*) Inlet temperature of cooling water
(*c*) Direction of cooling water flow
(*d*) Water quality and scale formation.

Normally these resistances are much smaller than the resistance of the mould/steel gap and therefore, don't significantly affect the heat flux profile. Over most of the mould length, the combined mould and mould/water resistances constitute less than 20% of the total resistance.

Specifications of water used in Continuous Billet Casting

Item	*Primary Cooling Mould cooling (soft water)*	*Secondary Cooling in open Machine cooling*
A. Salt concentrations :		
— Sulphate (as SO_4)	150 ppm.	600 ppm.
— Chloride (as Cl)	50 ppm. max.	400 ppm.
— Silica (as SiO_2)	40 ppm.	200 ppm.
— Total dissolved Solids (TDS)	400 ppm. max.	3000 ppm.
B. Hardness (max) : — Carbonate hardness (*i.e.* bicarbonates of Ca+Mg, as $CaCO_3$)	2 DHG	15 DHG
— Non-carbonate hardness (*i.e.* sulphate and chlorides of Ca+Mg, as $CaCO_3$)	15 DHG	60 DHG
C. pH Value : — pH range	7 to 8	7 to 9
D. Suspended Solids :		
— Content (max)	50 ppm.	50 ppm.
— Particle size	0.2 mm dia	0.2 mm dia.
E. Water Temperatures :		
— at inlet	40°C max.	40°C
— at outlet	50-55°C	65°C max.
F. Water Pressures :	4 bar (mould) 3.5 bar (at heat exchanger)	7 (at header)
G. Back Pressure :	For assuring over pressure in the mould, in order to prevent critical bubble formation, back pressure must be maintained between 1 to 2 bar.	

2.4.5.5. Automatic Mould Level Control Systems

One of the most important factors affecting the quality of the cast surface is the behaviour of molten steel in the mould. The fluctuations of the level of liquid steel in the mould cause a number of defects like longitudinal cracks and in extreme case may cause breakouts. Accordingly, it is desirable to decrease these fluctuations as much as possible. There are two basic methods by which the level of steel in the mould is controlled :

(*a*) Withdrawal speed control, *i.e.* regulating the strand withdrawal rate to match the feed rate.

(*b*) Stopper rod/slide gate control *i.e.* regulating the metal flow feed rate from the tundish to the mould to match the withdrawal rate.

Out of all the systems developed so far, the following three systems are considered most versatile :

(*i*) Radioactive mould level sensors
(*ii*) Electromagnetic system
(*iii*) Eddy current sensing

Radioactive mould level sensors - operate on the principle of absorption of gamma rays by the molten steel in proportion to its level in the mould. The radiation reaching the detector is converted into an electrical signal, then amplified and used to automatically control the strand withdrawal speed or adjust the stopper rod/slide gate system.

Electromagnetic system - operates on the principle that an emitter coil mounted on one side of the mould produces an A.C. magnetic field, and its effect on the receiver coil mounted on the opposite side of the mould is evaluated. The level of the electrically conducting metal in the mould influences the magnetic field and, thus, by suitably dimensioning and arranging the two coils, an exact level measurement can be obtained.

Eddy current sensing - is based on the principle of eddy current effect. For this purpose, a pencil type molten steel level meter is employed with a sensor head of 30 mm dia. A magnetic field is created by a high frequency current in a measuring coil. This magnetic field is affected by the presence of molten steel, thereby causing a change in the impedance of the measuring coil. Since the magnitude of the eddy current varies with the distance between the measuring coil and the surface of molten steel, it is possible to measure the distance between the coil and molten metal level. The meter is able to measure directly the molten steel level with an accuracy of ± 1 mm in a quick response time of less than 0.1 second. The level measurement is effective to a depth of 100 mm.

In the following table are given the comparative advantage/disadvantages of the three popular systems discussed above:

Item	*Radioisotope system*	*Electromagnetic control (EMLI)*	*Eddy current mould control*
1.	2.	3.	4.
1. Principle	Transmission of gamma rays followed by differential absorption by molten metal and slag.	Electromagnetic flux linkage between two coils mounted on either side of the mould.	Magnetic field created by current in a coil is affected by the presence of molten steel, thereby causing a change in impedance.
2. Measuring range	Up to 200 mm depth from the top end of mould	40 to 160 mm from the top end of mould.	0 to 150 mm from the top end of mould.
3. Accuracy	Measuring ± 5 mm controlling ± 10 mm	Measuring ± 5 mm controlling ± 7 mm	Measuring ± 1 mm controlling ± 3 mm
4. Response time	1 second	0.3 second	0.1 second
5. Position of mounting	Side wall of the mould.	Firmly set on the top of the mould.	From top of the mould independent of the mould.
6. Effect of mould powder	Yes, slightly	No	No
7. Size	25 mm dia source and 200 mm long detector.	Transmitter 50 × 50 × 200 mm Receiver 50 × 50 × 200 mm	60 mm dia 150 mm length.
8. Application	All moulds	Slabs only.	Slab, bloom, billet
9. Safety	Radiation hazard	Water leakage.	Good, only air cooling
10. Obstruction to the operators	Nil	Nil	Care to be taken not to hit the sensor.
11. Ruggedness	Rugged	Rugged	Fragile
12. Cost	Rs. 200,000 to 300,000 for one set.	Higher	Highest
13. Maintenance	Negligible because of less parts, but requires regular checking.	Negligible because of modular characteristic.	Regular checking.
14. Mould change	Recalibration required.	Recalibration required.	No recalibration required.

2.4.5.6. Mould powders and their characteristics

Mould powders are used in the continuous casting process to form a glassy-viscous "lubricating film" between the solidifying steel strand and the cooled continuous casting mould and to optimise the heat transfer between the solidifying steel and the mould. The main functions of mould powders are :

(*a*) To protect the liquid steel at the meniscus from reoxidation.

(*b*) The provide thermal insulation for the liquid steel, thereby preventing premature solidification or skulling.

(*c*) To provide lubrication and facilitate strand withdrawal from the mould.

(*d*) To absorb and dissolve high melting point scum, such as deoxidation products, and thereby prevent its entrapment in the solidifying shell.

(*e*) To control the heat transfer from the solidifying metal to the mould, through the solidifying shell.

(*f*) To control interfacial tension at the meniscus, therefore reducing the severity of oscillation marks.

A schematic arrangement of the phases in the meniscus region is shown in Fig. 2.30.

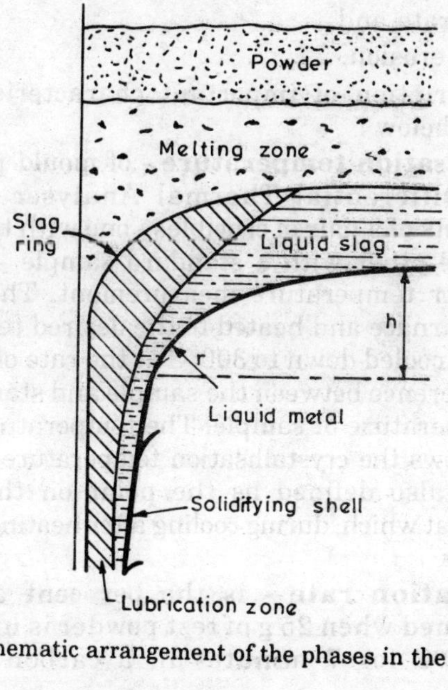

Fig. 2.30. Schematic arrangement of the phases in the meniscus region.

Continuous casting mould powders are synthetic slags which can be broadly divided into three categories : (*a*) Fly-ash based, (*b*) Synthetic, (*c*) Prefused, fritted or granulated. Casting mould Powders are essentially a mixture of SiO_2, CaO, Al_2O_3, Na_2O and CaF_2 with minor additions of MgO, MnO, FeO and K_2O. Categories (*a*) and (*b*) are simple blends of the powder raw materials in a number of permutations, while category (*c*) are blends that have been melted (at least partially) and sized after mixing. Presently, granulated mould powders of different chemical compositions are being used increasingly because they show less tendency of segregation and display a more consistent melting behaviour. Mould powders now used in India are primarily cement based with graphite, cryolite, flourspar, silica flours as the major constituents.

The following tests are usually conducted to get an idea of the general properties of mould powders :

(1) Chemical Analysis
(2) Moisture content
(3) Viscosity
(4) Crystallisation temperature
(5) Vitrification rate.

For critical applications, however, following tests are also required :

(1) Alumina dissolution kinetics (ADK)
(2) Fusion rate and
(3) Shroud erosion.

A brief description of important characteristics of mould powders is given below :

(*a*) **Crystallisation temperature** - of mould powders can be measured by Differential Thermal Analyser (DTA), which essentially consists of a pair of crucibles - one with the sample to be analysed and the other with a standard sample - with attached thermocouples for temperature measurement. The crucibles are placed inside a furnace and heated to the desired temperature, say 1400°C and then cooled down to 300°C at the rate of 25°C/min. The temperature difference between the sample and standard is plotted against the temperature of sample. The temperature at which the peak appears shows the crystallisation temperature. Crystallisation temperature is also defined as the point on the viscosity vs. temperature plot at which, during cooling after heating, an abrupt rise in viscosity occurs.

(*b*) **Vitrification rate** - is the per cent amount of the molten slag formed when 25 g of test powder is unidirectionally heated at 1400°C for 7 minutes in a carbon crucible with internal dimensions of 30 mm dia and 40 mm height. The vitrification rate serves as an indicator of the powder's melting characteristics.

(*c*) **Alumina dissolution kinetics (ADK)** - The ADK test measures how rapidly a mould slag dissolves a standard size piece of pure alumina. For this test, a 25 mm long alumina tube is attached to a carbon rod and immersed in the hot, molten mould slag. The time taken to dissolve the alumina tube is taken as the ADK time. Always a short ADK time is desirable. This test basically measures the ability of a mould powder to absorb alumina. This test is important in practice because alumina is major impurity in aluminium-killed steels and the mould powder is required to absorb significant amounts of alumina from steel as quickly as possible.

(*d*) **Fusion rate** - or the melting rate of mould powder is an important functional property. It is controlled by addition of carbon powder which retards the fusion and agglomeration of molten slag. It is thus, possible to regulate the feeding rate of molten slag to the meniscus region and the thermal insulation properties of the powder layer.

(*e*) **Shroud Erosion Index** - The duration of casting of a heat is often limited by the erosion of shroud (extended nozzle shroud) by the mould slag. Therefore, it is essential to know the erosion rate of mould slag towards the shroud. Shroud erosion Index is generally defined as :

$$\text{SEI} = \frac{\text{\% loss of perimeter in the test mould powder}}{\text{\% loss of perimeter in the standard mould powder}}$$

A typical example of mould powder recommended for applications in billet casting at Tata Steel are given in the following table :

Recommended mould powders for continuous casting at various carbon equivalent levels and casting speeds

Carbon equivalent	0.18-1.35	0.08-0.18	<0.08
Casting speed (max.), m/min	1.27	1.77	2.29
Chemical composition range, %			
SiO_2	36.5-38.5	35.0-37.0	33.0-35.0
CaO	30.0-32.0	32.0-34.0	28.0-30.0
CaO/SiO_2	0.78-0.87	0.86-0.97	0.80-0.90
Al_2O_3	6.50-7.50	6.50-7.50	5.50-6.50

MgO	1.00-2.00	1.00-2.00	1.50-2.50
Na_2O	7.50-8.50	6.50-7.50	—
K_2O	0.10-0.40	0.10-0.40	—
F	6.50-8.50	7.50-9.50	6.00-8.00
Fe_2O_3	0.50-1.00	0.50-1.00	0-1.00
Li_2O	0.50-1.00	0.75-1.25	—
C free	3.50-4.50	3.50-4.50	3.50-4.50
C total	4.50-5.50	4.50-5.50	4.50-5.50
H_2O at 700°C	0.75 max.	0.75 max.	0.50 max.
R_2O	8.10-9.90	7.35-9.15	12.0-14.0
Fusion temperature range °C			
Initial deformation	1040-1060	1060-1090	1110-1130
Softening	1070-1090	1080-1100	1130-1150
Hemispherical	1090-1110	1090-1120	1150-1170
Fluidity	1100-1130	1115-1135	1160-1180
Vitrification rate, %	75.0-80.0	75.0-80.0	75.0-80.0
Crystallisation temperature, °C	1060-1090	1060-1080	1160-1180
Viscosity at 1300°C, poise	2.80-3.20	1.80-2.20	1.30-1.70
Thermal conductivity,			
Theoretical, K	1.3	1.0	1.0
Integral, BTU/sq. ft.	3350	2900	2900
Bulk density	47.50-52.0	48.0-52.0	50.0
Particle size, % (-200 mesh)	75	75	75

2.4.5.7. Mould Oscillation

The mould is mounted on a special "mould table" and oscillated by electric drive on a curve corresponding to the casting radius. Mould oscillation performs the stripping action of billets. Steel will "adhere" to the copper mould by friction and form a weak bond at the interface, and stripping action will break this contact. This shearing of bond leaves behind what is known as "stripping marks" on the surface of cast product (billets, blooms or slabs). The solidified skin must be sufficiently thick to withstand the force of stripping. The operation is performed by the upstroke motion of the mould, usually within a very short span of time. A variety of oscillation cycles are employed in practice; a few typical ones are shown on Fig. 2.8. The downstroke prevents constant friction of the solidified shell with the mould wall, after which, the upstroke strips the skin from the mould surface in a more effective manner than the sliding and dragging that occurred previously. The velocity of upstroke is usually three times that downstroke velocity. Effort is always made to minimise the depth of oscillation marks.

The stroke length can be adjusted between 12 to 40 mm. The frequency can be chosen between 100-150 cycles/min. In a new modification, the speed of downstroke has been increased by 30-40% than the average withdrawal or casting speed. This results in "Negative Stripping" of the ingot and is beneficial in the following ways :

(*a*) Initially crystallised skin of the ingot is further compacted.

(*b*) Formation of tensile stresses is prevented and even compressive stresses may be developed in the initially solidified skin.

(*c*) It practically eliminates the possibility of transverse cracking of skin.

(*d*) Transverse cracks that may be formed earlier, are liable to be welded again.

(*e*) It permits maximum rate of withdrawal.

The relationship between stroke, frequency, casting speed and negative strip can be derived from the following expression :

$$\text{Strip } \% = \frac{V_c - V}{V_c} \times 100$$

where, V = Mould speed, m/min

V_v = Casting speed, m/min

when the velocity of downstroke (V) is higher than the casting speed (V_c), the value of % strip obtained is negative. This is schematically shown in Fig. 2.8. (Section 2.2)

2.4.5.8. Mould Lubrication

The purpose of mould lubrication is to achieve the following :

(*a*) decrease the friction between the mould wall and the billet surface.

(*b*) to avoid the sticking tendency of solidifying billet to the mould.

Common lubricants employed are rapeseed oil, mineral oil, spent transformer oil and paraffin wax. About 1 ml of lubricant per min. per inch of mould parameter is a common average. This practice, however, has now been largely replaced by mould powders, which perform several other functions in addition to normal lubrication of mould (for details refer to previous section 2.4.5.6).

2.4.5.9. Multistage Moulds

The billet energing from the mould jacket enters the "multistage mould". The objectives of installing MS mould are as under :

(*i*) to attain higher casting speeds in commercial grades.
(*ii*) to reduce breakout rates.
(*iii*) to reduce rhomboidity in billets.
(*iv*) to produce sounder billets of special steels, where casting speeds are not increased.

The multistage (MS) mould is bolted to the bottom plate of the water jacket of the primary mould. These are four cooling plates, which during casting, are gently pressed to the billet by springs. The cooling plates are made of copper. The heat extracted from the billet through the cooling plate is removed by direct spray cooling. Water is being sprayed onto the plates through full cone spray nozzles placed in the extension of the diagonal of the billet. Due to the fact that the billet corners are not covered by the plates, the corners are directly spray cooled. Thus, the multistage (MS) moulds enable the caster to operate at casting speeds higherto not possible due to danger of frequent breakouts. Recently, a combination grid support system (comb grid) has been shown to provide improved support relative to previous containment system even at high casting speed.

2.4.5.10. Mould breakout prediction

Breakouts can be predicted by visualising the temperature irregularities with the help of thermocouples installed at two different levels in the mould and linking them to reduce the casting speed or even stop the casting. Several foreign machines have facilities for mould friction measurement in order to predict any breakout and to take suitable preventive measures.

2.5. MODERN TECHNOLOGICAL DEVELOPMENTS

2.5.1. Caster Design

Fig. 2.9 (section 2.3) shows the different types of casters available today and their market share. In this figure, the influence of the type of caster on steel cleanliness and internal quality are also depicted. It can be seen that bow type casters are the most popular and accounts for 62% of the casters built. The casting radii of these casters varies between 8 and 15 m. The increasing use of bow type casters with multiple straightening points have permitted the utilization of higher casting speeds without incurring the disadvantage of deformation at a single straightening point and formation of internal cracks. It also satisfied the ever increasing demand for higher internal quality.

2.5.2. Electro-Magnetic Stirrers (EMS)

Application of EMS in mould area reduces central segregation while maintaining higher casting temperature and optimum casting speed, reduces central porosity and improves strand surface. Stirring of metal in mould also helps in uniform shell growth in the mould, which prevents breakout during casting.

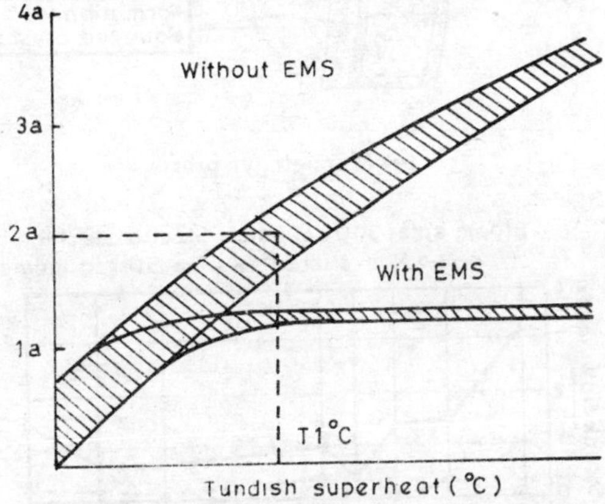

Fig. 2.31. Influence of EMS and degree of superheat on the extent of columnar zone.

The rotation produced in liquid steel helps to eliminate the residual superheat due to improved heat exchange between the solid and liquid. This reduces the thermal gradient ahead of the solidification front and promotes the formation of equiaxed crystals. A wider equiaxed zone helps in a more uniform distribution of solute atoms such as carbon, manganese, sulfur and phosphorus, and reduces the degree of axial segregation and porosity at the middle thickness of the slab. Fig. 2.31 shows the influence of EMS and degree of superheat on the extent of columnar zone. It can be seen that application of EMS arrests the columnar growth mode and promotes equiaxed crystallization and a more uniform distribution of solute in the cast structure. The trend has been found for low, medium and high carbon steels.

Applying EMS at the secondary cooling stage for billets, blooms and slabs have resulted in a significant improvement in the internal soundness of the products. The columnar growth is minimised and the ratio of equiaxed crystals within a cast strand is increased, which reduces the centre segregation and porosity. The choice of location of EMS and employment of independent or combined EMS method will depend upon the type of cast product and quality of steel to be cast. Fig. 2.32 shows the principle of EMS

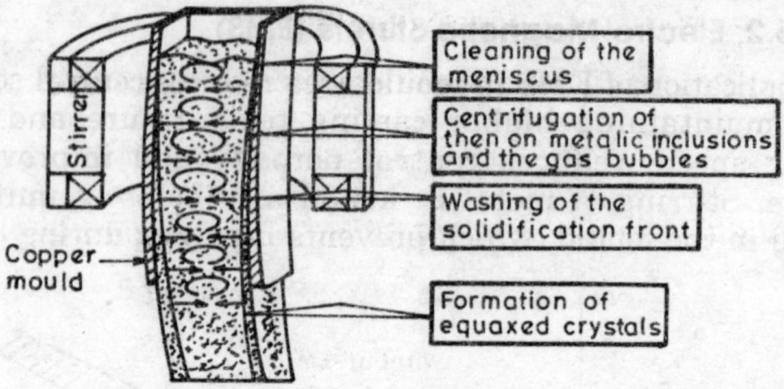

(*a*) Magnetogyr process

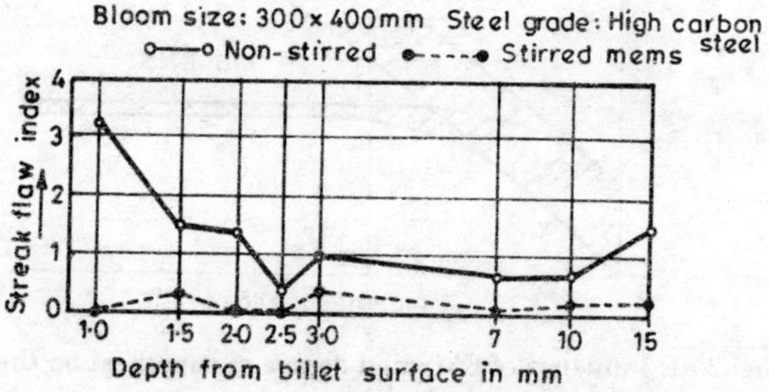

(*b*) Sub-surface cleanliness of aluminium-killed steel.
Fig. 2.32. Principle of electro-magnetic stirring.

and cleanliness ratings in an Aluminium-killed steel. EMS also leads to reduction of bulging in slabs. It is well known that when the support rolls are too far apart, the slabs may bulge along its central axis under the effect of the ferrostatic pressure, leading to positive or negative segregation. Roller stirrers have been designed to take care of the "bulging" phenomenon in slabs. EMS systems are today used for the continuous casting of slabs, blooms and billets, especially in Japan. It is particularly used for the continuous casting of high carbon steel blooms, which are susceptible to severe axial segregation.

2.5.3. Forced Acceleration of Solidification Technology (FAST)

The CSM and Terni Special Steel Co. have developed a new continuous casting technology known as FAST to improve surface and internal quality of slabs as an alternative route to EMS to counteract columnar solidification. The new technology involves feeding a cored wire into the continuous casting mould through the tundish stopper rod and the submerged entry nozzle. The solid reagents within the metal sheath act as cooling and/or alloying element for liquid steel in order to control the alloy content as well as the amount of superheat of the liquid metal and hence the primary solidification structure. The application of FAST on slab continuous casting can ensure the modification of primary solidification structure upto 100% equiaxed zone.

2.5.4. Sequence Casting and Mould Change Device

The trend today is to cast long sequences, sometimes as many as 100 heats in a sequence. This increases the productivity of caster, improves the yield and allows the casting operation to take place with minimum fluctuations in operational variables. Adjustment/changing of the mould size during sequential operation has been made possible and put to practice by some advanced steel plants in the world.

2.5.5. Atomised Mist Cooling System

Atomised air mist secondary cooling has been developed in replacement of full-spray water cooling system. The system mixes water and air in a chamber and sprays the mixture over the strand.

The size of water droplets produced by spray cooling devices in secondary cooling regime controls the cooling efficiency of the system. This is because the size of the droplets determines the total effective surface area of water in contact with the steel to be cooled. A fine atomised spray with 150-250 μm droplets provides a higher rate of heat transfer than usual 700-900 μm droplets from a hydraulic spray.

The spray pattern is important also. Recently developed caster jet air/water mist nozzles not only provide the desired atomising spray, but are designed to also produce a spray pattern 2-3 times thicker than a conventional flat spray nozzle. This is the most effective system to obtain the desired degree of solidification of the strand and results in uniform cooling of the strand surfaces, better slab surface, use of only one nozzle type for all casting sizes, less demand with regard to water quality and impurities in water due to bigger nozzle diameter. It helps in reducing both surface and internal cracks, improve surface and edge quality due to soft cooling characteristics.

2.5.6. Dry Casting

The formation of surface defects is promoted by heat stresses caused by uneven strand cooling and surface temperatures, which are unfavourable for the ductility of material. These temperature variations can be reduced and surface defects largely eliminated by employing dry casting. In dry casting, almost all heat from the strand is transferred to the surrounding strand support structure through the specially designed sleeved rollers. Dry casting by virtue of its high strand temperatures and avoidance of critical temperature level is ideally suited for direct rolling/hot charging. Dry casting is being practiced successfully on low head Oval-bow caster of Mannesmann-AG.

2.5.7. Soft Reduction Process

The soft reduction technique was first introduced by Nippon Steel's Kimitsu Works in 1988 to minimise centre line segregation in slabs. This technique has been reported to provide an effective means of preventing the movement of molten steel (due to bulging) and decreasing segregation in a continuous cast slab. However, there is a limitation to the prevention of molten steel movement during the soft reduction process. This has led to the subsequent development of the controlled plane reduction (CPR) process at Nippon Steel. Fig. 2.33 shows a comparison of the soft roll reduction

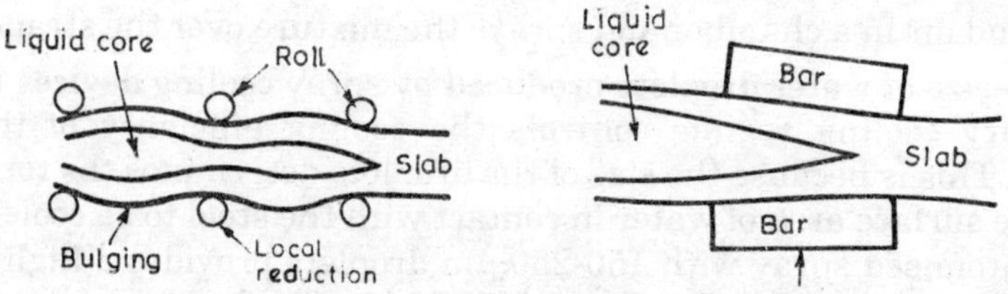

(*a*) Roll reduction (*b*) Controlled plane reduction

Fig. 2.33. Comparison of soft roll reduction and controlled plane reduction processes.

and controlled plane reduction process as they softly reduce a slab. The advantages of the CPR process are as follows :

(*i*) **Continuity of slab reduction :** The CPR process continuously reduces a slab and compensates for its solidification shrinkage by flat bars with a constant taper provided in the casting direction. Thus excessive localised reduction of the slab and concurrent slab bulging, which is harmful for centreline segregation, can be basically ignored.

(*ii*) **Effect of crater end position :** When the crater end position is located between the rolls, the soft reduction process cannot fully compensate for the shrinkage of the slab at the crater end. The segregation level during soft reduction process varies with the position of the crater end. In the CPR process, the crater end is located within the reduction zone, and therefore is less sensitive to the crater end position.

Fig. 2.34 shows the effect of controlled plane reduction on centreline segregation. As the slab reduction taper increases, the average centreline segregation index decreases. The segregation index is also a function of the reduction taper difference (δ) between the two bar blocks. In order to completely alleviate segregation by the CPR process, it is important to apply an optimum reduction taper to the slab and minimise δ between the two bar blocks.

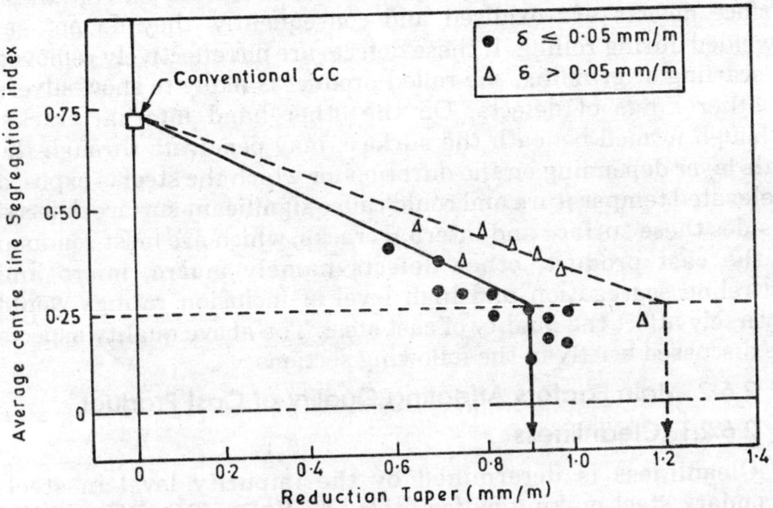

Fig. 2.34. Effect of controlled plane reduction on centre line segregation.

2.5.8. Computerised Process Control

For smooth and efficient performance of a continuous caster, complete automation and control of the various operational parameters is essential. Suitable algorithms have been developed to control the water flow rates in the different zones of the secondary cooling regime as a function of the casting speed. Sensors have been developed and are presently commercially available which helps in obtaining feedbacks from different operational facilities *e.g.* bath weight in ladle, slide gate operation during steel pouring, degassification of metal, metal level in mould, viscosity of slag in tundish, prediction of metal break-out and shape control of semis.

2.6. METALLURGICAL QUALITY AND DEFECTS

2.6.1. Introduction

The metallurgical quality of cast product (billets, blooms and slabs) is of importance to quality conscious customers. The quality of cast product can be directly attributed to the following technological parameters :

(*i*) Cleanliness (*ii*) chemical homogeneity, (*iii*) freedom from segregation, porosity and cracks (internal unsoundness), and (*iv*) conformance to desired shape. The formation of cracks is the most common problem encountered during continuous casting of steels. These cracks may appear almost anywhere – at the surface or in the interior of slabs, blooms or billets. Due to contact with air, these surface defects are oxidised and consequently, they do not get rewelded during rolling. If these defects are not effectively removed by scarfing or grinding, the rolled product is likely to show silvers or other kinds of defects. On the other hand internal cracks, although formed beneath the surface, may penetrate through the scale layer depending on the duration for which the steel is exposed to elevated temperature and could cause significant surface defects. Besides these surface and internal cracks, which are most common in the cast product, other defects namely macro, micro and centreline segregation and high level of inclusion ratings would adversely affect the quality of cast steel. The above quality aspects are discussed briefly in the following sections.

2.6.2. Main Factors Affecting Quality of Cast Product

2.6.2.1. Cleanliness

Cleanliness is determined by the impurity level in steel. Secondary steel making by facilities like VAD, VOD, RH and DH are used today to reduce levels of impurities, particularly sulfur, phosphorus, nitrogen & hydrogen in steels. In future it is planned that total impurity content in special grades of steel like off-shore steel and sour gas resistant pipeline steel etc., shall be restricted below 45 ppm. The objective of continuous casting is not only to maintain the cleanliness level of the liquid steel as delivered to the caster but also to promote further inclusion removal. The following factors are controlled during continuous casting to prevent reoxidation of steel and also promote removal of inclusions simultaneously :

(1) Minimisation of oxygen transfer from the air or refractories.

(2) Pickup of exogenous inclusions by ladle, tundish and casting powders.

(3) Control of fluid flow in the tundish and mould to maximise inclusion flow out.

(4) Adoption of optimum mould powder.

Exogenous inclusions (slag patches) are observed on plate surfaces in the form of patches, concentrated occasionally in a particular plate and absent in others. They have a typical egg-shaped appearance, varying in size from heat to heat and also from plate to plate for the same heat. Fig. 2.35 shows a photograph of exogenous inclusions observed in IS : 2002 plates. A typical inclusion measures as much as around 50 mm along the major axis and 30 mm along the minor axis. Most of such inclusions are found to contain SiO_2, Al_2O_3, CaO, MgO and K_2O along with a high concentration of carbon.

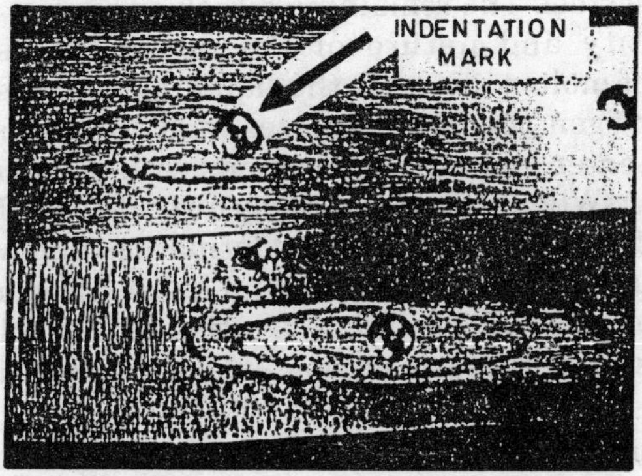

Fig. 2.35. Photograph showing typical exogenous inclusions on the surface of IS : 2002 steel plates.

Exogenous inclusions can arise on plate surfaces due to the following factors :

(1) Slag carry-over from ladle to tundish.
(2) Abnormal casting condition leading to drop of metal level in the mould.
(3) Erosion of tundish refractory lining.
(4) Reoxidation during casting leading to increase in non-metallic oxides.
(5) Casting powder.

Recent investigations have suggested that the casting powder is the main source of exogenous inclusions. The casting powder melts, spreads and covers the metal surface in the form of a scum. If the scum has a high melting point, low fluidity and high interfacial tension with molten steel, it has a tendency to be drawn in and get trapped on the slab surface, especially when the metal level fluctuates in the mould. Other investigations have shown that low ΔT (lower than 20°C) and speed fluctuations are the primary factors responsible for the incidence of exogenous inclusions.

Another investigation involving low carbon Al-killed steel showed that high slag patch density is related with ΔT less than 10°C.

Oxygen can be present in steel in two forms : (*a*) dissolved form and (*b*) combined form in the form of oxides of elements, which have high affinity for oxygen. As a result of deoxidation, the dissolved oxygen is extensively converted into a stable combined form. Also, the product of deoxidation must be precipitated quickly and extensively form the molten steel. The combined oxygen remaining in the solid steel (indigenous oxide inclusions) indicates the degree of purity of the steel. The remaining dissolved oxygen is dependent on the quantity and nature of the deoxidant used and the temperature of molten steel. As the liquid metal cools further, the equilibrium oxygen content decreases, and the excess oxygen so liberated can enter into further reactions with the deoxidants (like Mn, Si, Al etc.). When this excess oxygen reacts with Mn and Si, it results in the formation of slag (Scum formation) and when it reacts with C, it results in the formation of pin holes. These reactions are known as "secondary deoxidation reactions". The following table shows the result of these reactions, when the steel is deoxidised with Mn/Si.

Oxygen equilibrium with Mn/Si deoxidation as a function of the temperature (Hilty and Crafts)

	Oxygen in equilibrium, %		*Δ O, %*	*Slag ($MnO + SiO_2$) corresponding to ΔO, kg/t steel*
	1650°C (ladle)	*1525°C (mould)*		
Molten steel with				
0.10% Si+0.40% Mn	0.0118	0.0060	0.0058	0.25
0.25% Si+0.50% Mn	0.0090	0.0045	0.0045	0.20

It must be emphasized that similar "secondary deoxidation reactions" can also take place as a result of renewed oxidation of the molten steel (secondary oxidation). Therefore, during the production of quality steel with high degree of purity, it must be ensured that until it solidifies, the deoxidised steel is protected against further reactions with oxygen from slag, refractory linings and, in particular from the atmospheric air. A number of techniques are in practice these days to protect the casting stream against atmospheric oxidation (see section 2.4.3.2 and 2.4.4.3). In order to avoid oxidation of the stream of molten metal the atmospheric oxygen content close to the casting stream must be reduced to less than 0.8%.

An important prerequisite for the avoidance of secondary deoxidation reactions is that the difference in temperature between the deoxidised steel in the ladle and in the mould is kept as low as possible. This also makes a contribution towards reducing other casting defects *e.g.* stress cracks, segregations and core porosity. But whenever there are long casting times due to low caster capacity, a high steel temperature in the ladle must be used. In such cases, the use of cold tundishes with good insulation properties or the use of high temperature tundishes can bring an improvement.

It can be seen from Fig. 2.36 that even after vacuum treatment, the oxygen contents lie in a scatter band which is, to some extent, still above the equilibrium with Mn and Si. Although the oxygen contents can be reduced advantageously and extensively via a gaseous deoxidation product during vacuum treatment, they are to some extent still so high that, the danger of secondary deoxidation reactions is not entirely ruled out. In such cases, it is necessary to reduce the dissolved oxygen content further by the additional use of stronger deoxidants. The additional use of Al or Ca in addition to the normal Mn/Si deoxidation therefore becomes necessary.

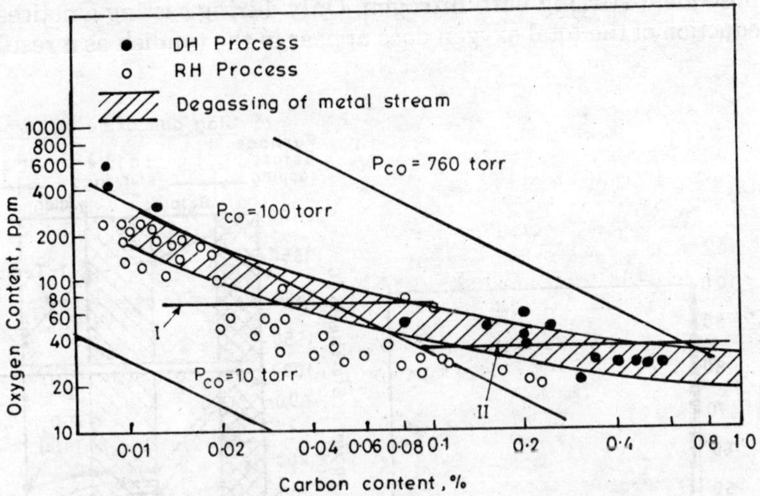

Theoretical oxygen content (Hilty and Crafts[4]) for Mn/Si deoxidation; I, steel with 0.10% Si. 0.40% Mn, 0.06% C, T liquid 1525°C; II, steel with 0.25% Si, 0.50% Mn, 0.50% C, T liquid 1490°C

Fig. 2.36. Dependence of the oxygen content on the carbon content of the steel with vacuum treatment.

When brittle inclusions are not prejudicial to subsequent application of steel, the dissolved oxygen contents can be reduced to very low quantities using metered additions of Al. But the residual Al content of the steel in the tundish must not exceed about 0.007% or better still 0.005%, or otherwise, the pouring nozzles would become chocked or blocked by deposits of alumina (Al_2O_3).

Referring to Fig. 2.37, it can be seen that the amount of dissolved oxygen is greatly dependent on the Al contents. With Al concentrations of more than 0.005%, in comparison with Mn/Si deoxidation, negligibly low amounts of oxygen in solution are found. This Al content is within the order of magnitude, which can be considered permissible when casting strands with small cross-sectional dimensions.

The Al can be added in metered larger quantities before or during deoxidation with Mn and Si on in small quantities (less than 0.2 kgs/t) after Mn/Si deoxidation. In both cases, contact with the slag should be avoided. Gas stirring is to be recommended.

When producing low carbon steel grades the idea of determining the oxygen content of the steel before tapping and adding Al in suitable quantities to the ladle has proved satisfactory. These results are shown in Fig. 2.38. Here steel was already initially deoxidised with Al in the furnace. The total oxygen content (curve II) is reduced greatly as a result of deoxidation in ladle but only slightly during the subsequent stirring with nitrogen. Only during casting a noticeable reduction of the total oxygen does appear in the tundish as a result of

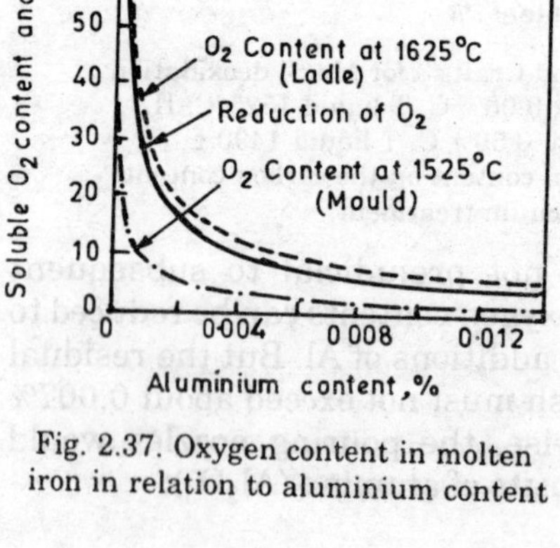

Fig. 2.37. Oxygen content in molten iron in relation to aluminium content

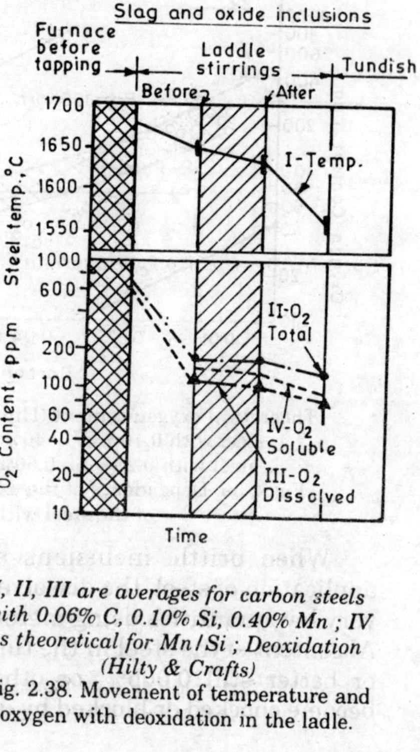

I, II, III are averages for carbon steels with 0.06% C, 0.10% Si, 0.40% Mn ; IV is theoretical for Mn/Si ; Deoxidation (Hilty & Crafts)

Fig. 2.38. Movement of temperature and oxygen with deoxidation in the ladle.

favourable flotation conditions for the products of deoxidation containing alumina. The dissolved oxygen (curve III) reduces in accordance with the temperature drop (curve I).

Brittle alumina inclusions are considered undesirable for many applications. For example, in the case of wire rod, they may lead to breakages when drawing to small diameters. So, in such cases, deoxidants other than Al have to be used.

Calcium and Magnesium which are alkaline-earth metals have a very great affinity for oxygen and sulphur, with calcium being the strongest available deoxidant and desulphuriser. Calcium is insolubles in molten iron and is present in vapour form at molten steel temperatures. In contrast to Al, it causes no problems during continuous casting. But as it is evaporated prematurely, it is necessary to increase the efficiency of calcium by means of special alloys and/or special addition methods.

The addition of calcium to the ladle in the form of Ca, Si, Mn (1 kg/t molten steel) brought a clear improvement over pure Mn/Si deoxidation when producing steels with higher carbon contents. In order to obtain further improvement of the efficiency of calcium, standard commercial calcium alloys are injected into the ladle by means of an inert carrier gas via an immersion lance. When doing so, the lance should be immersed until it is a little above the bottom of the ladle. In addition to good desulphurisation, the oxygen content is also reduced and, as a result, the degree of purity is improved. In this case, deoxidation was aimed at the production of heavy plate with isotropic properties. Steels with other applications would have to be treated in a different manner. By Ca-Si injection in the ladle, the sulphur content of steel can be reduced to a value of 0.001 to 0.003% and also the amount and form of oxides can be controlled.

2.6.2.2. Chemical Homogeneity

One of the pre-requisites for obtaining uniform mechanical properties & micro-structure is chemical homogeneity of the steel. The liquid steel is homogenised by rinsing with argon or nitrogen at a high pressure for 3 to 10 minutes. This not only helps the alloy elements to redistribute uniformly into the metal bath, but also aids in attaining a uniform temperature and float-out of inclusions to the surface of the bath.

2.6.2.3. Segregation and Porosity (Internal Unsoundness)

Micro segregation and porosity is a result of the solidification process and occurs in the axial zone of the slab or billet. The origin of microsegregation lies in the distribution of elements like carbon, manganese, sulphur and phosphorus between solid and liquid during the freezing of steel, as well as the solidification structure and convection process near the bottom of the liquid pool. The fundamental issue of micro segregation is therefor, concerned with what occurs at the advancing solidification front. The nature of segregation depends on three factors :

(*i*) the rate of advance of the front

(*ii*) the temperature gradient at the front

(*iii*) the degree of mixing which occurs in the liquid steel ahead of the front.

These factors influence the physics of crystal growth of the solidification front and generate a structure which may be equiaxed trapping small volumes of rejected solute and leading to micro segregation when low temperature gradient exist ahead of the solidification front (*i.e.* low super. heat and poor mixing). At the opposite extreme, when high temperature gradient exist just ahead of the front, a columnar structure develops. This allows solute to be continuously rejected and macrosegregation results. So, the solidification process needs to be controlled to minimise segregation. It is aimed to promote preferential growth of equiaxed crystals (as opposed to growth of columnar crystals) in the core of the cast product. As mentioned earlier, the equiaxed zone can be propagated by controlling the factors like (*i*) degree of superheat (*ii*) steel composition (*iii*) section size and (*iv*) induced fluid flow ahead of the solidification front.

Fig. 2.39 shows the effect of superheat on the thickness of columnar and equiaxed zones in the as-cast slabs. It can be seen that a higher degree of superheat promotes columnar growth and, in order to obtain a reasonably wide equiaxed zone, it is necessary to impose a limit on the degree of superheat. It is fairly an established practice today to permit a maximum degree of superheat of 20°C for ordinary grades of steel and 15°C for steels required for critical applications. Carbon and phosphorus contents in steel have an influence on the size of equiaxed zone. Medium carbon range, from about 0.17 to 0.38%C, favours an equiaxed structure. Similarly, phosphorus in the range of 0.02-0.025% causes the columnar zone to shrink.

Section size of the blooms and slabs have a moderate influence on the solidification structure ; a larger section size favours the growth of an equiaxed structure. Similarly, induced fluid flow, such as by electromagnetic stirrers, increases the relative width of the

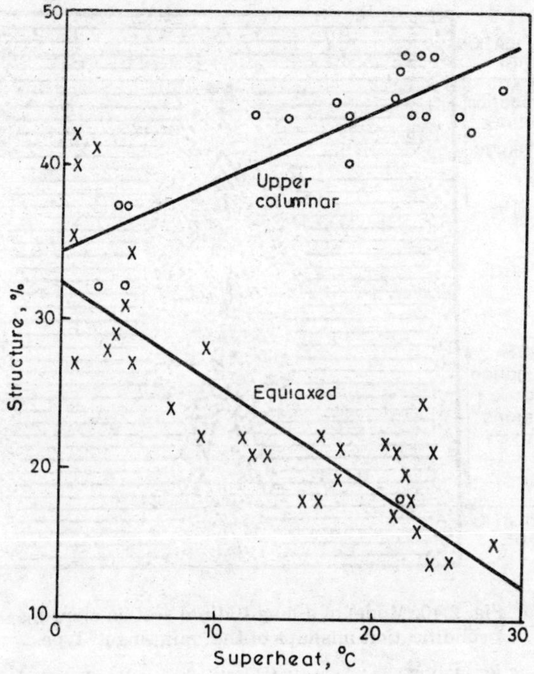

Fig. 2.39. Effect of superheat on width of equiaxed and columnar zones.

equiaxed zone. This is particularly effective at moderate degrees of superheat.

Secondary cooling and casting speed can have an appreciable effect indirectly on central unsoundness by determining the resistance of the solidified shell to bulging forces exerted by the ferrostatic pressure of the liquid core. When bulging occurs, the shell is rolled down by the subsequent support rolls causing a continual flexing of the shell. This can interfere with the crystal structure at the solidification front and allows excess solute to be rejected, promoting macrosegregation. The degree to which "bulging" effects occur depends on the distance below the meniscus, shell thickness, temperature and elastic properties and machine design. A good correlation has already been established between the segregation indices and bulging.

To limit the columnar growth and prevent the formation of "solidification bridges" causing segregations of the "Mini ingot" type, the steel temperature on the tundish should be very close to that of the steel liquidus. The casting temperature from the furnace should thus be regulated with great accuracy. Also to prevent the

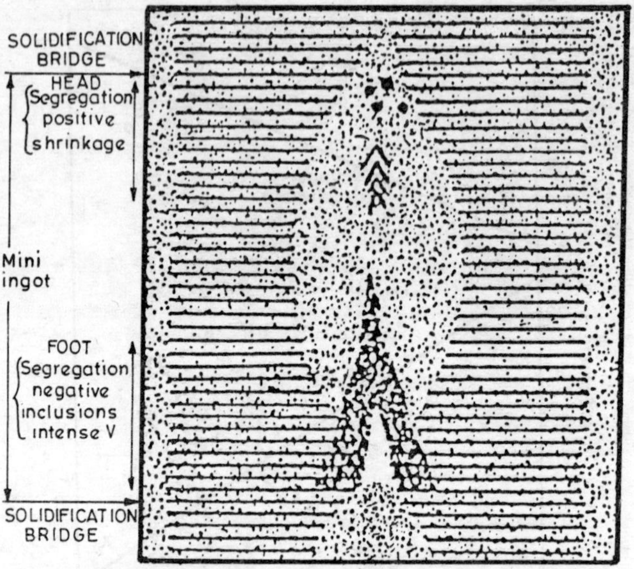

Fig. 2.40. Model of a longitudinal section showing solidification mishaps of the "mini-ingot" type.

formation of "Solidification bridges", the withdrawal rates of the machine must be carefully established. The technique of electro-magnetic stirring enables the cast structure to be controlled definitely; irrespective of the temperature, the section-size and the grade of steel produced. A schematic view of the "solidification bridge" causing segregation of the "Mini-Ingot" type is shown in Fig. 2.40. Electromagnetic stirring arrests systematically the growth of the columnar zone and promotes the formation of fine and regular exiaxed zone. Also, this process prevents the formation of solidification mishaps and "Mini Ingots" in the axis of continuously cast products like billets, blooms and slabs. Thus electromagnetic stirring reduces the intensity and frequency of the principal defects which affect the internal soundness of the products : segregation, porosity and accumulations of inclusions. It has been shown that the stirred products have a very high degree of chemical homogeneity, a better distribution of inclusions and more regular service properties.

Studies on 32 mm thick plate have revealed the presence of non-metallic inclusions and impurities segregated in a chain-like fashion in the axial zone. The inclusions were embedded inside voids and porosities, formed during the solidification process of the cast slab. These inclusions were found to contain C, O, Si, Fe and also Ca, S and Al in addition to the above elements. Further studies on 36 mm thick plates indicated that C and Mn have a strong tendency to segregate at the mid thickness of the plate, while S and P do not show any appreciable tendency to segregate. The segregation of C and Mn not only leads to a higher pearlite content at the core of the plate, but also results in the localised formation of a hard phase, namely Bainite. These bainite pockets due to its poor ductility and impact toughness properties, are preferential sites for crack nucleation. Cracks nucleate within these pockets and propagates through the band leading to rejection of the plate.

The concept of thicker slab and higher reduction ratio was put to practice for the production of internally sound thick slabs at Bhilai Steel Plant. One of the casters was converted to 320 × 1500 mm^2 section from the existing 250 × 1500 mm^2 section. A higher starting slab thickness ensured a higher reduction ratio which in turn resulted in :

(*i*) Lessening of segregation on account of higher diffusion rates at plate rolling temperature

(*ii*) Welding of internal defects/discontinuities within the slab/plate and

(*iii*) Thinning out and lower incidence of surface defects in plates.

An all round improvement in the quality of slabs and plates was achieved. Rejections on account of internal defects came down to practically nil. Mechanical properties also showed distinct improvement with almost a two-fold increase in the impact toughness at – 20°C. SEM examination of the mid section of the plate showed only a diffused centreline with no evidence of any major internal defects such as centreline cracks, voids and porosities. While conventionally processed 36 mm thick plate did show evidence of centreline segregation and axial cracks. An examination of the concentration of C, S, P and Mn along the centreline does show that C, S and P do not show much tendency for segregation while Mn does so.

2.6.2.4. Formation of different kinds of Cracks in C.C. billets and Slabs

Fig. 2.41 is a schematic representation of the different types of cracks observed in continuously cast slabs and billets. Cracks are basically of two types : surface and internal cracks. Surface cracks are a serious problem as they oxidise during rolling giving rise to oxide-rich seams in the rolled product. Internal cracks also constitute a quality problem, particularly when they do not close

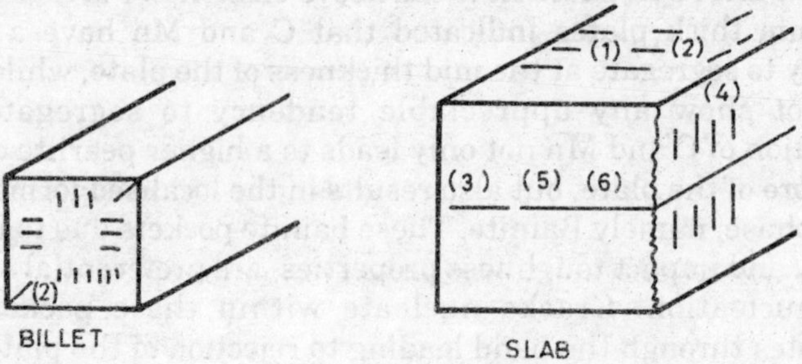

1. Midway cracks
2. Rhomboidity/diagonal

1. Transverse surface cracks
2. Longitudinal mid-face cracks
3. Triple-point cracks
4. Midway cracks (radial streaks)
5. Centre-line cracks
6. Centre segregation

Fig. 2.41. Schematic representation of different types of cracks observed in concast slabs and billets.

during rolling, giving rise to voids in the plate product. The mechanism of formation of each type of crack and the influence of continuous casting and steel making parameters on cracking have been investigated thoroughly in recent times. For example, it is now established that high sulphur contents or excessive superheat adversely influence the formation of internal cracks. It is because the high superheat of melt favours the formation of columnar structure having easy crack paths while a high sulphur content reduces the hot ductility of steel in the temperature range where most cracks form. Such studies have helped a great deal in achieving zero defect cast product.

(1) Longitudinal Cracks

Longitudinal cracks generally originate in the mould and are predominantly associated with the following factors :

(*a*) A low ductility trough 30-70°C below the solidus temperature.

(*b*) Strains generated due to peritectic transformation, especially for the carbon range 0.08-0.2%.

(*c*) In sufficient slag film thickness, or high fluidity of slag film, leading to large frictional stresses on the solidifying shell.

(*d*) Metal level fluctuations in mould leading to non-uniform growth of solidifying shell, localised regions of weakness and increased thermal stress.

Propagation of these longitudinal cracks already nucleated in the mould or formation of new ones are further activated by excessive non-uniform cooling in the secondary cooling zone immediately below the mould. Blockage of spray water nozzles caused by carry over of scale or corrosion products along with the water stream can be responsible for causing non-uniform cooling of strand. This problem can be minimised by the use of stainless steel pipes for water transportation in secondary cooling regime.

Longitudinal corner cracks in C.C. billets are similarly associated with retarded cooling at the corners. Also, most important operational factor responsible for this defect is the reversal of mould taper due to distortion. This may lead to comparatively slow cooling and, hence, the liquid film at the grain boundary gets retained for a longer period and makes the corner comparatively weak and prone to cracking. When this defect occurs under correct casting conditions, it is usual to change the mould immediately and this always eliminates the defect. The defective mould is then reconditioned by cleaning the water cooling channels and surface grinding or machining.

Experiments on square section cast billets (75 mm to 170 mm size) of Si-Mn spring steel (0.55-0.65% C, 1.5-1.8% Si, 0.7-1.0% Mn) have revealed that for moulds of the same dimension, if the radius of quadrant of the corner was large, a long longitudinal crack of several mm in depth occurred at the top of the quadrant of the cast billet. On the other hand, if the radius of quadrant of the corner was small, the longitudinal cracking occurred at the end of the quadrant of the cast billet. If the radius of quadrant of the corner is correct, there is no tendency to longitudinal cracking, (Fig. 2.42). Experiments on 130 mm square billets of the same Si-Mn spring steels indicated that longitudinal surface cracks are eliminated when primary cooling is even on all the four surfaces. It was also found that longitudinal cracks of several mm in depth occur, when the casting temperature is too high. The safe casting temperature for Si-Mn spring steel was found to be 1490°C. When the casting speed of 115 mm square billets is increased from 1200 to 1600 mm/min, the total amount of cooling will decrease from 32.7 to 27.9 Kcals/kg of steel. Consequently, even though the decrease in cooling is small, the increase of casting speed substantially increases the delay of solidification during the early stages of solidification, or causes a rise in the average temperature of the outer solidified shell. Thus, the increase of casting speed as well as casting temperature will increase the strand's susceptibility to longitudinal surface cracking. Generally, in continuous casting, the rate of descent of the billet is comparatively high. So, the ratio of the static pressure of molten steel (in the partially solidified strand) to the thickness of

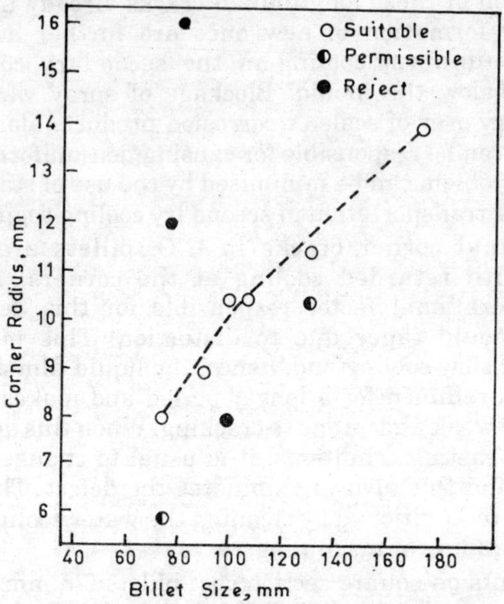

Fig. 2.42. Effect of corner radius on longitudinal surface cracks of square billet.

the solidified shell increases. For this reason longitudinal surface cracks are more liable to occur on the surface of billets.

Mid-face longitudinal cracks in slabs could be like fine cracks to long deep ones, usually in the middle half of the broad face. It is generally accepted that mid-face longitudinal cracking of C.C. slabs is initiated in the mould and that the severity of cracking can be influenced by the secondary cooling and slab support system immediately below the mould. A carbon level of around 0.12% is particularly found sensitive to surface defects, which are predominantly longitudinal cracks. At higher and lower carbon levels, the quality improves. These results apply to the three thicknesses of slabs (180, 230 and 305 mm).

The effect of Mn/S ratio on hot shortness and susceptibility to cracking is fairly well established. Increasing Mn/S ratio, typically in the range of 20 to 60, is found to have some positive influence on surface defects. But this influence declines as the slab thickness increases and casting speed progressively falls. The correlation between increasing Mn/S ratio and better surface quality of high Mn/S ratios (25-55) suggests that low ductility fracture resulting from the formation and propagation of austenite grain boundary micro cracks linking together below 1204°C, might be an important factor in the generation of longitudinal cracks.

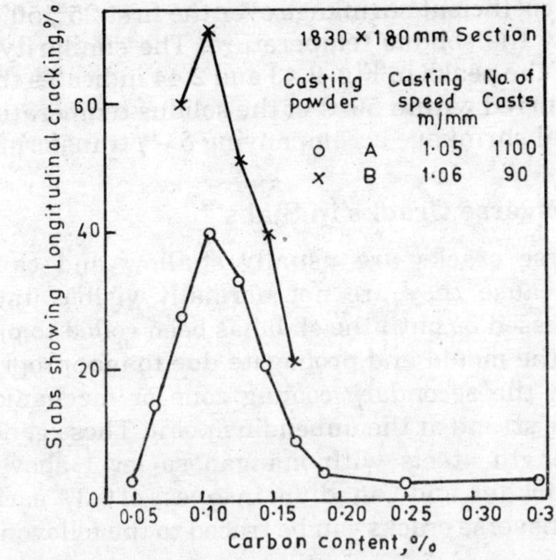

Fig. 2.43. Influence of carbon on incidence of longitudinal cracks in slabs.

Fig. 2.43 shows the influence of carbon content on the incidence of longitudinal cracks in slabs. It can be seen that a carbon level of 0.11 to 0.15 leads to a significantly higher crack incidence level compared to steels containing lower (<0.11%) and higher (>0.15%) carbon levels. Fig. 2.44 illustrates the effect of carbon content on the

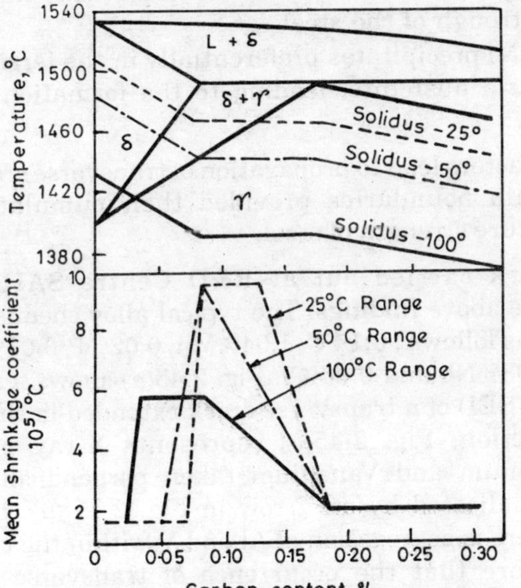

Fig. 2.44. Effect of carbon content on shrinkage after solidification.

mean linear coefficient shrinkage over the first 25°, 50° and 100° of cooling below the solidus temperature. The similarity in position and shape of the peaks in Fig. 2.43 and 2.44 indicates that cracking probably occurred within 50°C of the solidus temperature owing to the additional shrinkage accompanying $\delta - \gamma$ transformation in the steel shell.

(2) Transverse Cracks in Slabs

Transverse cracks are usually shallow and causes severe problems because they are not normally visible until the slab surface is dressed or until the slab has been rolled to plates. These originate at the mould and propagate due to inappropriate cooling conditions in the secondary cooling zone or mechanical stresses applied to the strand at the unbending point. These generally occur in high strength steels with manganese level above 1.0% and containing Niobium and Vanadium in excess of 0.3% each. The basic reason of transverse cracks can be traced to the following causes :

(*a*) The hot ductility of microalloyed steel is significantly lower than that of plain C manganese steel.

(*b*) Addition of Nb and V not only reduces the hot-ductility, but also increases the temperature range of the low ductility trough.

(*c*) Al, V(N) precipitation take place in the temperature range of 700-900°C, which coincides with the low ductility trough of the steel.

(*d*) Al, V(N) precipitates preferentially in the ferrite as compared to austenite, leading to the formation of brittle ferrite.

The above factors lead to propagation of transverse cracks along the ferrite grain boundaries provided their cumulative effect exceeds the facture strength of steel.

Recent work carried out at R&D Centre SAIL, Ranchi corroborates the above findings. The typical alloy chemistry of the steel used was as follows : 0.1%C, 1.04%Mn, 0.02%P, 0.01%S, 0.26% Si, 0.02%Al, 0.03%Nb and 0.06%V. Fig. 2.45(*a*) shows a secondary electron image (SEI) of a transverse crack extended in the through thickness direction. Fig. 2.45(*b*) represents X-ray profiles of sulphur, aluminium and Vanadium taken perpendicular to the crack path as indicated by an arrow in Fig. 2.45 (*a*). It is quite evident that there is segregation of Al and V within the crack. It is inferred therefore that the occurrence of transverse cracks is possibly associated with Al, V(N) precipitates.

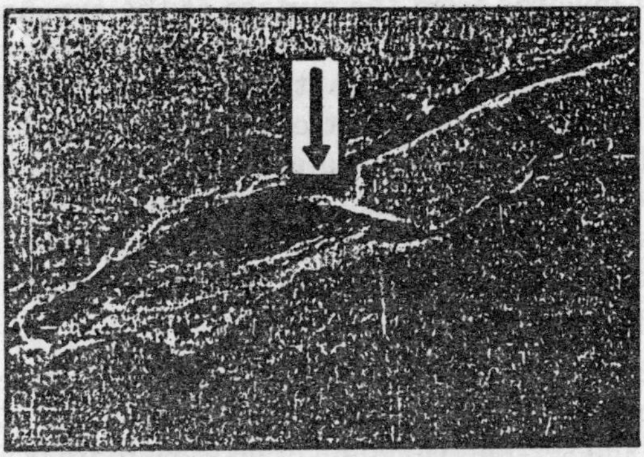

Fig. 2.45. (*a*) Scanning electron image of a transverse crack

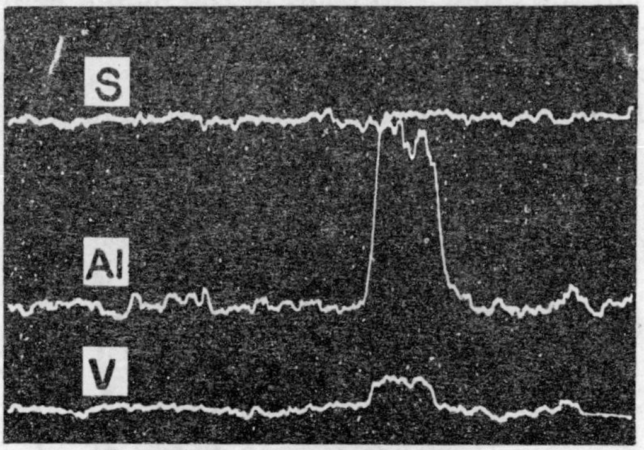

Fig. 2.45. (*b*) line profiles of S, Al and V.

To minimise the incidence of transverse cracks, the following technological measures were successfully implemented at Bhilai Steel Plant for sensitive grades of steels :

(*a*) Microalloying with Nb only (0.03% max.).

(*b*) Al restricted to 0.05% max.

(*c*) Softer cooling regime.

(*d*) Temperature at unbending point in excess of 900°C.

(3) Transverse Midway Cracks in Billets

These types of cracks can be detected in sulphur prints and macroetched samples of transverse sections. Basically, this type of defect appears as dark lines running normal to a given face in a region roughly midway between the surface and the centreline. There are various factors which contribute to mid-way crack formation in billets. High casting temperature is one of the significant factors, which favours an enlarged columnar zone, extending inwards from the surface at the expense of the central equiaxed zone. Midway cracks are formed much more easily between the dendrites in the columnar zone, which run perpendicular to the tensile stress, compared to the equiaxed zone. Other related factors associated with the formation of midway cracks are excessive secondary cooling and high concentrations of sulphur and phosphorus, which have a deleterious effect on the strength and ductility of steels in the high temperature range.

(4) Diagonal Cracks in Billets

Fig. 2.46 shows a typical diagonal crack which is associated with rhomboidity and usually runs between the obtuse corners of the rhomboid section. Diagonal cracks originate from the distortion of the billet which arises when two adjacent faces are cooled more rapidly than the other two faces in the mould or in the secondary cooling zone. Such cracks initially form in the high temperature zone of low ductility but may grow outwards, essentially towards the corners, depending upon the magnitude of strain.

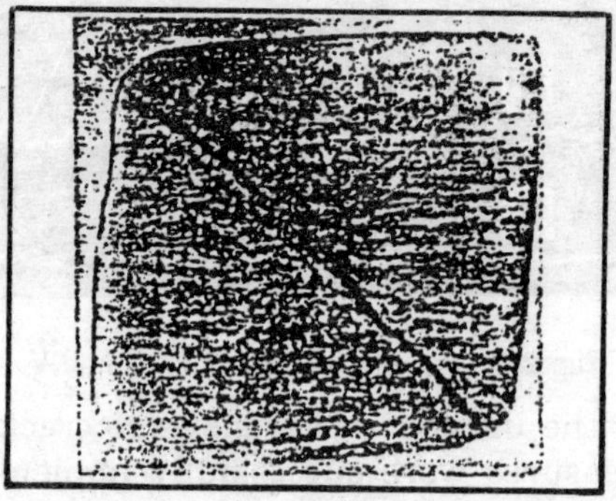

Fig. 2.46. Macrograph showing a diagonal crack in a billet.

(5) Off-squareness/Off-corner Internal Cracks in Billets

Off-squareness is basically related to the cooling characteristics in the mould and if severe off-squareness occurs, it must involve the sprays as well. Off-corner cracks are caused by the retardation of shell growth in the off-centre locations, due to non-uniform heat extraction as well as deep oscillation marks (across the billet face). This results in the generation of large strains in the regions of low ductility, which lead to crack formation. Off-corner cracks are formed most easily adjacent to the billet corners, due to the locally weakened shell, and predominantly appear at sites that have deepest oscillation marks. Fig. 2.47 shows macrograph of a billet showing an off-centre corner crack.

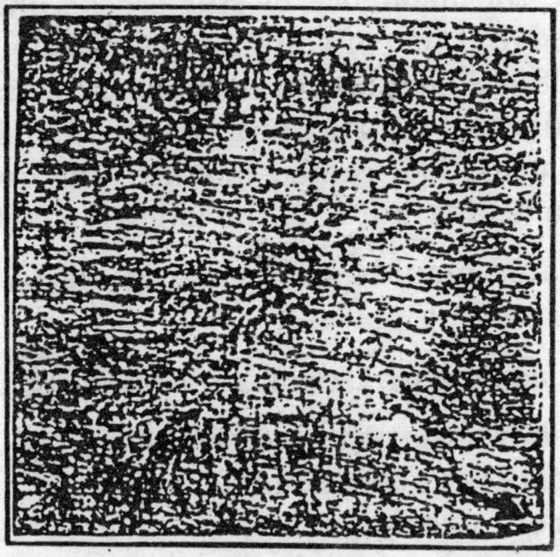

Fig. 2.47. Macrograph of a billet showing an off-centre corner crack.

(6) Star Cracks in Slabs

Star cracks are usually shallow and are observed only after scarfing of the slabs. The source of this defect is related to diffusion of copper from the mould surface to the surface of the slab. Star cracks usually arise due to sticking of the slab with the mould surface. This primarily arises due to the non-uniform thickness of the slag film formed by the mould powder. By adjusting the viscosity and melting characteristics of the mould powder, a step wise reduction in the frequency of sticking and star crack incidence is obtained.

A direct relationship was established between basicity and star cracking. It can be seen from Fig. 2.48 that as the mould powder

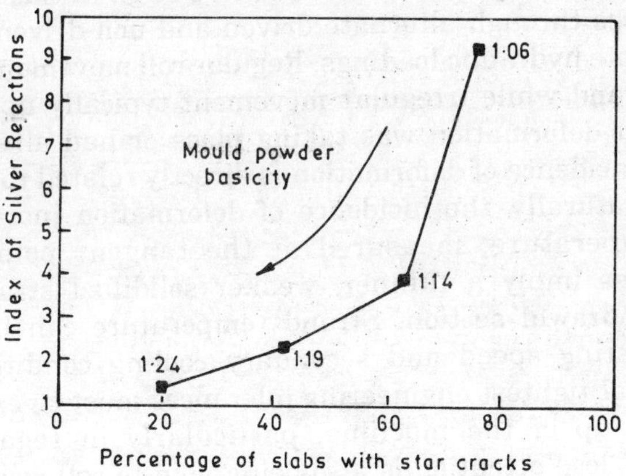

Fig. 2.48. Variation of star crack frequency and silver rejections with mould powder basicity.

basicity decreases, the incidence of star cracks and silver defect increases. Experimental use of Chrome and Nickle plated moulds was initiated in 1986. Since then, the incidence of star cracks have decreased drastically and is today practically nil in Japanese Steel mills, where they use 100% coated moulds. The decrease in star crack incidence can be attributed to the following factors :

(*i*) use of coated moulds and hence no diffusion of copper and (*ii*) improved casting powder.

(7) Internal Centreline Cracking in Billets and Blooms

The evidence shows that the problem of centreline cracking is strongly influenced by the operational factors such as casting speed and extraction roll pressure but is largely unaffected by metallurgical factors such as sulphur content and superheat. It is, therefore, thought that this defect is not associated with chemical macrosegregation but is more associated with the machine design factors and operating conditions. Operational factors such as the casting speed, machine hydraulics and secondary cooling are considered vital.

Casting Speed

Trials have shown that centreline cracking is caused by some form of alternate rolling and bulging of the strand near the point of total solidification which coincides with the first pair of extraction or pinch rolls in the curved withdrawal section (12.32 m below the meniscus for the 1830 × 180 mm slab section cast at 1.05 m/min. speed with secondary cooling water of 0.74 l/kg). In this region, the strand passes through alternate driven and non-driven roll pairs with alternate hydraulic loadings. Regular roll movement indicates a stable strand while irregular movement typically upto 3.5 mm showed that deformation was taking place or had already taken place. The incidence of deformation is directly related to centreline cracking. Naturally, the incidence of deformation increases with strand temperature, measured at the tangent point. Higher temperatures imply a thinner, weaker solidified strand in the curved withdrawal section. Strand temperature can be affected both by casting speed and secondary cooling conditions. It is apparent that tightest engineering tolerances must be exercised in the setting up of the machine, particularly in regard to roll alignment. Close attention is also being paid to roll wear and roll bending. In the above instance, a limit of 1.0 m/min was put or the casting speed. With this casting speed, total strand solidification would occur before the withdrawal section.

Secondary Cooling

In the range of 0.83 and 0.95 l/kg of cooling has shown that centreline cracking incidence can be reduced by increased secondary cooling water in the lower spray zones. Increasing the secondary cooling intensity increases the resistance of the solidified shell to bulging by slightly thickening the shell and reducing the length of the liquid core. The strand is also cooler and therefore has increased strength. Apart from the observed reduction in centreline cracking, increased secondary cooling measured reduction in :

(*i*) Withdrawal power by 30%

(*ii*) Ferrostatic loading on the segments by 5%

(*iii*) Strand temperature at the tangent point, from 935°C to 895°C.

In the case of square section cast billets, the pinch rolls are in contact with the two opposite faces, and operate at a pressure high enough to support the total weight of the cast billet. A stress caused by the pinch rolls thus arises in the interior of the billet. This stress acts as tension in a direction parallel to the pinch roll axis and acts as compression in the perpendicular direction. It is certain that internal cracks do not arise until the cast billet passes the pinch rolls. Also, internal cracks develop in the direction perpendicular to the pinch-roll axis, but never develop in other directions. Internal cracks are also found to be associated with alloy elements or the enrichment of impurities. The thermal and mechanical stresses which cause internal cracks of the cast billets, both originate owing to the cooling process used in continuous casting. Such internal cracks of the cast billet are therefore peculiar to continuous castings.

2.6.2.5. Influence of Compositional Factors

(1) Effect of Carbon

It is now established that steels with carbon content ranging from 0.08 to 0.2% undergo a peritectic reaction during cooling, leading to enhanced shrinkage as shown in Fig. 2.44. This gives rise to discontinuities in the mould heat flux, which makes such steels very prone to cracking. This aspect has already been emphasized with respect to the mechanism of longitudinal crack formation in section 2.6.2.4. (Item 1).

(2) Effect of Mn/S Ratio

When Mn/S ratio is less than 20, it influences the crack formation arising out of the formation of FeS. Liquid droplets of FeS precipitate at the austenite grain boundaries and result in low ductility in the intermediate temperature zone near the solidus and below 1200°C. Manganese is beneficial as it preferentially combines with sulphur to form less harmful MnS which precipitates in the matrix rather than predominantly at the grain boundaries. Fig. 2.49 shows the nature of changes from iron sulphide to manganese sulphide with increasing Mn/S ratios. The solidification temperature of the sulphide inclusions is related to the manganese level as shown in Fig. 2.50. A Mn/S ratio greater than 20 minimises the tendency for cracking and above 60, sulphur is tied in a stable phase and does not give rise to embrittlement. However, a high Mn level (>1.6%) is reported to lower ductility possibly due to matrix hardening. The effect of Mn/S ratio on hot shortness and susceptibility to cracking (longitudinal cracks) has also been discussed in section 2.6.2.4 (Item-1).

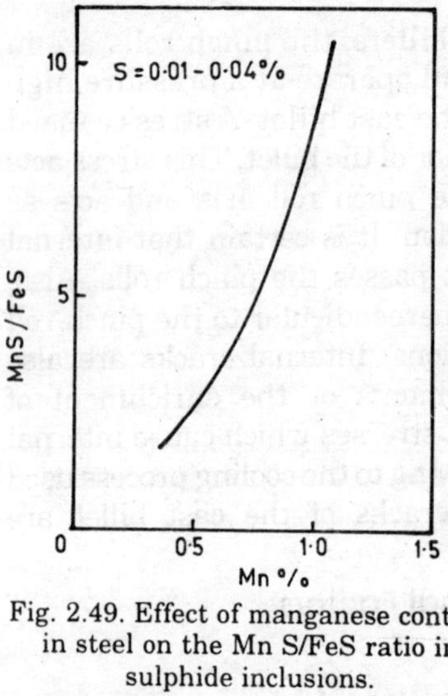

Fig. 2.49. Effect of manganese content in steel on the Mn S/FeS ratio in sulphide inclusions.

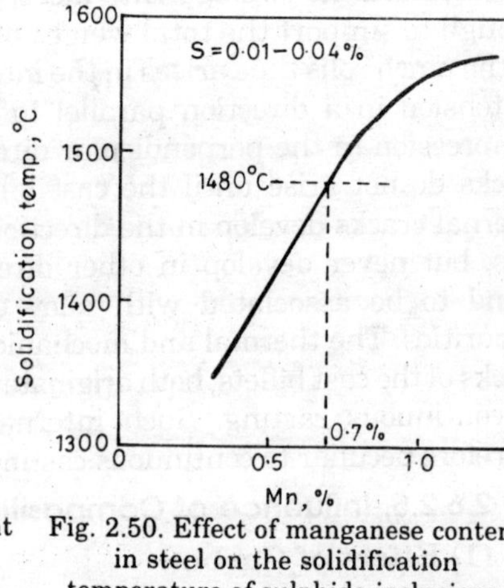

Fig. 2.50. Effect of manganese content in steel on the solidification temperature of sulphide inclusions.

(3) Effect of Phosphorus

Phosphorus segregates in the interdendritic regions giving rise to the presence of liquid films which do not freeze until temperatures well below the solidus are reached. Phosphorus concentrations of 0.2 to 0.5% have been observed at the grain boundaries at a matrix level of only 0.02%. A phosphorus concentration of about 0.18% has been associated with internal crack formation when the matrix has a concentration of 0.02%. The influence of phosphorus on fracture strength and ductility of steel can be seen in Figs. 2.51 and 2.52 respectively. Steels with a phosphorus content of 0.039% have drastically reduced strength and ductility levels, compared to steels containing 0.017% phosphorus in the high temperature as well as low temperature

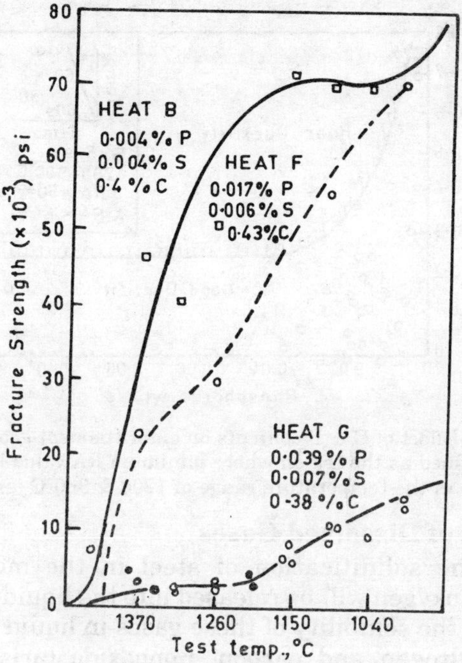

Fig. 2.51. Effect of phosphorus on fracture strength of SAE 4340 steel.

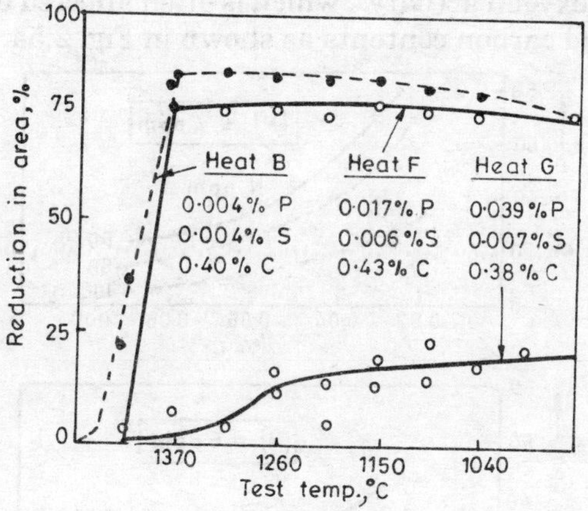

Fig. 2.52. Effect of phosphorus on the hot ductility of SAE 4340 steel.

zones. In a recent investigation, shown in Fig. 2.53, it has been established that embrittlement due to phosphorus is observed for carbon steels having a carbon content higher than 0.25% and it becomes more detrimental as the carbon content increases.

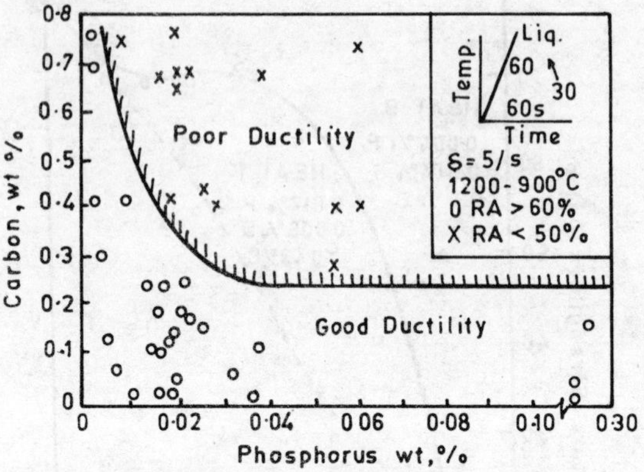

Fig. 2.53. Effect of C & P contents on embrittlement-good ductility region is defined as the region where minimum RA value is higher than 60% in the temperature range of 1200 & 900°C (ε = 5/s)

(4) Effect of Dissolved Gases

During the solidification of steel in the mould, nitrogen, hydrogen and oxygen will be released into the liquid steel due to the differences in the solubility of these gases in liquid and solid steel. Nitrogen, hydrogen and carbon monooxide (arising out of the reaction with oxygen) can cause pin-holes near the surface of billets. The oxygen activity below which pin holes do not occur, is termed as the "critical oxygen activity", which is often affected by the nitrogen, hydrogen and carbon contents as shown in Fig. 2.54. The formation

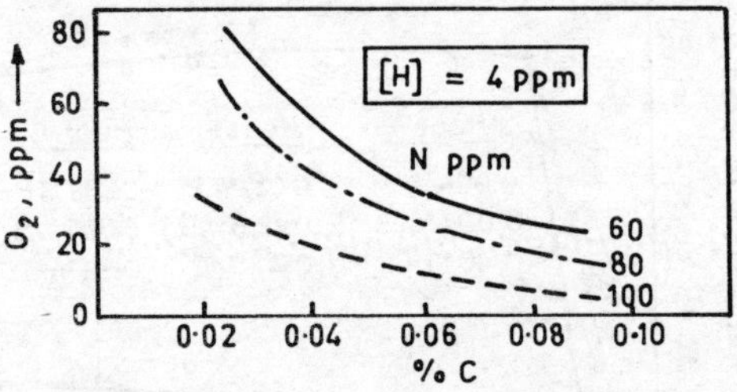

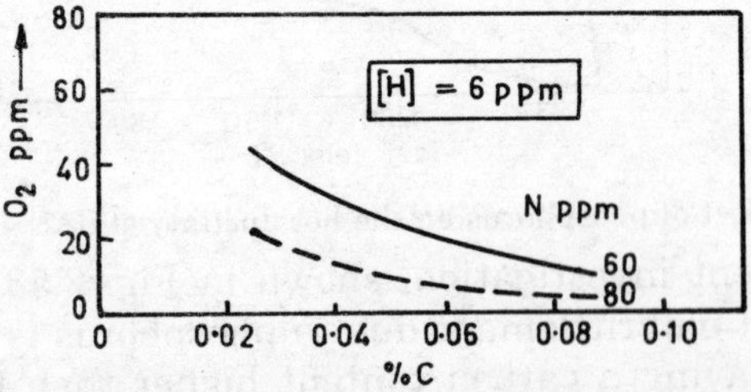

Fig. 2.54. Effect of hydrogen and nitrogen on the critical oxygen activity for pin hole formation.

of pin holes/blow-holes is associated with this difference in the solubility of various gases in liquid/solid steel.

2.7. TECHNOLOGICAL ADVANCES

2.7.1. Near Net Shape Casting (NNS)

The near-net-shape casting units are classified into categories based on the thickness of slab/strip :

(*i*) The thin slab caster producing 20-70 mm thick and 1000 - 2000 mm wide slabs which could be directly fed without any conditioning into the finishing strand of the hot-strip mill.

(*ii*) The strip caster producing 5-20 mm thick strips which may need some limited hot-rolling due to metallurgical reasons.

(*iii*) The strip caster producing strips less than 5 mm in thickness which directly go to cold rolling mills.

(*iv*) Foil caster produces strips in the thickness range of 0.02 to 0.5 mm which can directly be used as equivalent to the cold rolled sheets.

The main advantage of NNS casters are : lower capital investment and operating cost, reduced production cost, greater flexibility of production, improved quality and more sophisticated products.

Fig. 2.55 shows the shortening of hot-strip production routes in various cases and following table shows a comparison of energy consumption between three carbon steel sheet producing processes :

Table : Comparison of energy required to produce steel strip/sheet.

Process	*KwH/t produced*	*KwH/t shipped*	*Material loss %*
Ingot Casting	1760	2394	27
Continuous Casting	1442	1993	25
Near-Net-Shape Casting	721	750	4

The continuous casting technology has developed to such an extent that continuous cast slabs are hot charged to a reheating furnace, without cooling for intermediate inspection. This requires a high internal and surface quality of the slabs. Nippon steel is today hot charging more than 50% of its slabs directly to the reheating furnace, thereby resulting in utilisation of the sensible heat and reduction in energy consumption. Futuristic plans envisage direct rolling of the semis with marginal heating for the corners and edges of the cross-section, resulting in further savings in energy. For direct rolling, the continuous casting operation

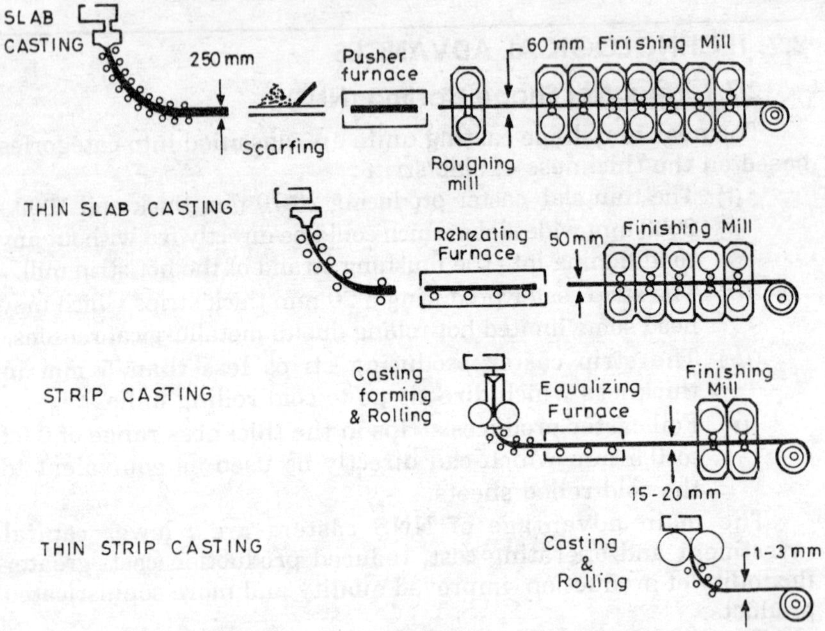

Fig. 2.55. Schematic diagram showing developments in continuous casting technology to shorten the process route.

should lead to hot slabs with uniform skin temperature of around 1200°C. The hot slabs must be transferred to the rolling mill without delay for processing into plates/strips. The facilities for production of desired quality of hot slabs for direct rolling are : (*i*) combined air and water cooling, (*ii*) minimum roll intervals to prevent segregation, (*iii*) edge insulation, (*iv*) insulation of roller tables, (*v*) hot inspection, (*vi*) a furnace for edge heating and for adjustment of the rolling start temperature and (*vii*) a high degree of automation.

2.7.1.1. Thin Slab Casting

The world wide status of thin slab casting is shown in the table on next page.

Various types of NNS caster designs for thin slab production can be classified into following broad categories :

(*i*) Stationary mould caster

(*ii*) Moving mould caster -

(*a*) Inclined twin belt thin caster

(*b*) Vertical and horizontal twin belt casters.

Worldwide Trend in Thin Slab Casting

Type	System		Typical size cast (mm)		Ladle (t)	Company	Country
	Angle	Scale	Width	Thickness			
Oscillating Mould	Vertical	Production	1220	75	30	Danieli/Fenlung	Tai-Wa
		Pilot plant	1350	50	100	SMZ/Nucor	U.S.A.
			1400	30-70		Danieli	Italy
			1200	50	16	SMS	F.R.G.
			1000	28		Danieli	Italy
			1200	60	220	Mannesmann	F.R.G.
	Horizontal		400	20	3	Krupp Stahl	F.R.G.
Twin belt	Horizontal	Hot model	100	54	0.2	Kawasaki	Japan
	Inclined	Pilot plant	180	70	40	Krupp Ind.	F.R.G.
	(Hazerlett)		430	25	30	Bethlehem/USX	U.S.A.
			1300	38	32	Nucor	U.S.A.
			600	40	50	Sumitomo I	Japan
			1320	30-50	50	Sumitomo II	Japan
	Vertical	Light pilot	1200	30	5	Kawasaki	Japan
			600	50	7	NSC/MHI	Japan
			430	25	1	Bethlehem/USX	U.S.A.
Block moulds	Horizontal	Light pilot	500	30-75	4	BSC	U.K.
		(single)				NKK/IHI	Japan
		(twin)				Kobe Steel	Japan
Twin roll	Vertical	Hot model	200	35	0.4	Hitachi Zosen	Japan

Stationary Mould Caster

Nucor has been operating such plant based on SMS design for the slab thickness of 50 mm and upto 1600 mm width at a casting speed of 4-6 m/min. The unique feature of this caster is the funnel-shaped stationary mould with a specially contoured submerged nozzle to prevent steel reoxidation as shown in Fig. 2.56. Recently MDH has also developed a modern caster for producing 40-70 mm thick and 1200 mm wide slabs at a casting speed of upto 6 m/min. with latest features such as continuously changeable width of slab during casting, dry casting techniques and in-line rolling of strand with liquid and/or solidified core. Voest Alpine and Danieli are also operating the super low head thin slab caster.

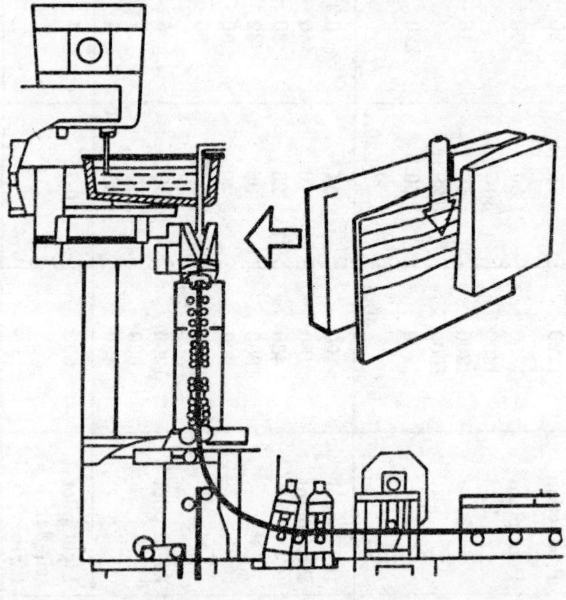

Fig. 2.56. SMS Schloemann-Siemag stationary mould thin slab caster.

Moving Mould Caster

The difficulties associated with direct casting of strips arise from the combined requirement of high solidification rate and high order of shape control. High solidification rate necessitates the use of moving mould which minimizes the interaction between mould wall and cast product when casting thin sections at high speed. For heat flow reasons, the travelling mould belts or blocks are provided with a suitable coating.

(*a*) **Inclined Twin Belt Thin Caster**

Moving mould concept permits the production of thin slabs (20-70 mm) at a reasonable high speed (8-10 m/min). A twin belt caster was developed by Hazelett, which is practically a horizontal arrangement with slight inclination, thus permitting the use of long mould and obtain minimum strand bending without the risk of crack formation. Such a twin belt caster which is being used in Sumitomo Industries is shown in Fig. 2.57.

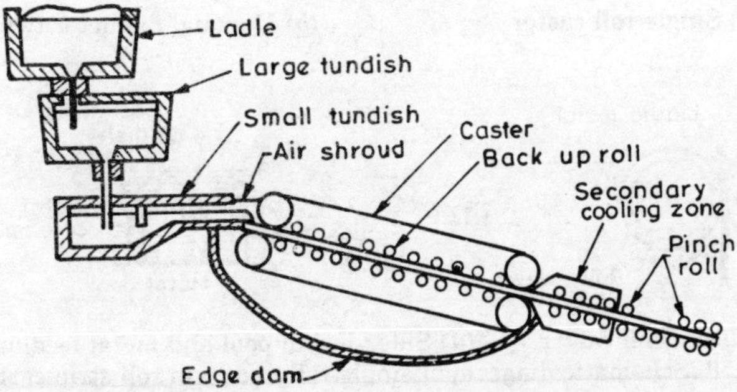

Fig. 2.57. Hazelett twin-belt thin slab casting system.

(*b*) **Vertical and Horizontal Twin Belt Caster**

A vertical configuration of a twin belt caster has been successfully tried by Kawasaki Steel Corporation to cast 30 mm thin slab at a casting speed of 10 m/min. In this caster a V-shaped mould with conventional submerged entry nozzle is employed. A horizontal twin belt thin slab caster is being used by Kawasaki to continuously cast 30 mm thin, 100 mm wide slab with maximum withdrawal speed of 12.5 m/min. It forms the wide faces of the mould while the narrow faces of the mould remain stationary. A similar type of the caster is the Alusuisse Caster II which employs travelling chill blocks. A pilot plant based on the moving mould principle has also been developed by British Steel Corporation to cast thin slabs of size 75 × 500 mm at a casting speed of 13.7 m/min. An insulating substrate is fed into the mould before the teeming point to produce sound slabs.

2.7.1.2. Thin Strip Casting

The thin strip casters are principally different than thin slab caster in respect to the method of metal feeding, containment of the molten pool in contact with the rolls and number of solidification fronts. A typical caster for thin strip casting is schematically shown in Fig. 2.58. This new technology will eliminate the need for slab or ingot casting and hot strip mill rolling presently practicised in steel mills. The table on the next page shows current world wide status of thin strip casting.

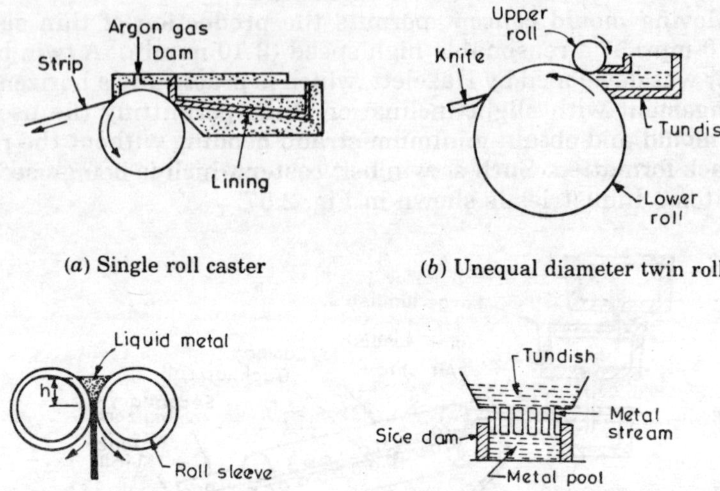

(*a*) Single roll caster (*b*) Unequal diameter twin roll

(c) Twin roll caster (*d*) Side view of pool and metal feeding

Fig. 2.58. Schematic diagram of single roll and twin roll strip casters.

Single Roll Caster

The machine uses a basic principal of the Melt Drag process. It employs a single horizontal casting roll at the top of which the liquid steel is held by a tundish. The strip formation begins in the liquid pools at the tundish exit as metal begins solidifying immediately upon contact with the roll and continues to thicken by the roll drag until it emerges from the liquid pool. The strip is continuously carried away from the pool by the roll and coiled up after rolling.

Twin Roll Caster

In this method the liquid metal is poured into the nip of the moving rolls. The molten pool is contained in the side by a pair of side plates pressed into the moving rolls. Both the rolls participate in the heat extraction process as there are two solidification fronts which converge at the plane joining the axis of rolls. Other key feature of this method is the dynamic roll gap control to ensure complete solidification just at the kissing point of the rolls. Bethlehem Steel Corporation has developed a pilot scale twin roll caster, which can cast the strip in the thickness range of 0.5-2 mm at a speed varying from 19-24 m/min. respectively. Recent laboratory trials at Kobe Steel has led to successful production of stainless steel strip (1.5 mm thick and 300 mm wide) cast between twin rolls (400 mm dia) at speeds ranging from 12 to 36 m/min.

Worldwide Trend in Thin Strip Casting

Type	System: Angle	System: Scale	Typical size cast (mm): Width	Typical size cast (mm): Thickness	Ladle (t)	Company	Country
Twin roll	Vertical roll dia	Light pilot	700	5	3	CSM II	Italy
	> 1000		850	2-10	6	IRSID II	France
	roll dia > 500		140	5	3	CSM I	Italy
	> 1000		200	5	0.3	ISID I	France
	roll dia > 500		300	5		Thyssen Stahl	F.R.G.
Belt + (Roll)	Horizontal	Hot model	152	5	2	LTV	U.S.A.
			200	5-10	0.15	T.U. Clausthal	F.R.G.
Spray deposition	Vertical	Hot Model	760	12-18		Mannesmann	F.R.G.
Twin roll	Vertical roll dia	Light pilot	400	3	4.00	BSC II	U.K.
	< 1000	Hot model	200	3	0.10	IHI	Japan
	> 500		500	0.4		Kawasaki	Japan
			400	3.5	0.25	NKK	Japan
	roll dia < 500		76	2	0.03	BSC I	U.K.
			50	2	0.02	MPI	F.R.G.
			300	1.5	0.12	Hitachi/Nisshin	Japan
			200	1.7	0.10	NSC	Japan
			305	2	0.30	Bethlehem & al	U.S.A.
	Inclined		300	1.5	0.10	Kobe	Japan
	Down		300	1.2		Nippon Yakin	Japan
			100	1.5	0.02	Waseda Univ.	Japan
	Unequal		600	1.5	3.00	NMI/Krupp Stahl	F.R.G.
			300	2	0.30	NMI	Japan
	Inclined up		600	1.5		NMI	Japan
Single roll	Inclined up	Light pilot	250	1.5	5	VOEST-Alpine	Austria
			560	1	3	Allegheny-Ludlum	U.S.A.
		Hot model	76	0.5	0.50	Armco-Westing house	U.S.A.
			100	1	0.09	Ribbon-Tech.	U.S.A.
				1.25		National Batelle	U.S.A.

Spray Deposit Casting

For casting high quality NNS products, a new shaping process known as continuous metal spray deposition has been developed recently. The basic steps involved in the process are :

(*i*) Atomisation of the liquid metal stream with an inert gas,

(*ii*) Extraction of latent heat of solidification of the liquid droplets during movement to a substrate, and

(*iii*) Compaction of solid particles by impact or fusion of liquid droplets at a substrate.

Flat products with a length of 2 m, maximum width 1 m and thickness 8 mm and 20 mm have been cast employing an experimental spray deposition unit of Mannesman Research Institute. Properties of this isotropic material are reported to be good and comparable with those of conventional Cast Steel. Another variation of this process is the spray rolling process in which an atomised spray is directed evenly over the surface of a water cooled drum. Deposition occurs at a relatively high rate, 50-100 kg/min. with individual droplets solidifying completely. Small amount of porosity in the cast product is eliminated by subsequent hot or cold rolling.

2.7.2. Horizontal Continuous Casting (HCC)

HCC has been termed as a next generation caster offering the direct rolling of cast steel. The HCC process is characterised by the direct connection of the tundish and the mould and the absence of bending throughout the entire facilities linearly arranged on the ground level. The main features are :

1. Molten steel is not subjected to secondary oxidation resulting in an excellent cleanliness of cast product.
2. Absence of casting flux gives more intense heat extraction by the mould but at the same time missing mould lubrication may increase mould wear and may enhance sticking problem.
3. More suitable to Steel grades having a high crack sensitivity.
4. It is possible to cast round and square sections for tubular products.
5. Easy automation and maintenance, low investment and operating cost etc.
6. Easy installation of electromagnetic stirring.
7. Excellent metallurgical properties and surface finish.

The main components of HCC mould are : a copper section for initial solidification, a breakring at which solidification is initiated and a secondary cooling section. Different types of continuous casters which have been developed are shown in Fig. 2.59.

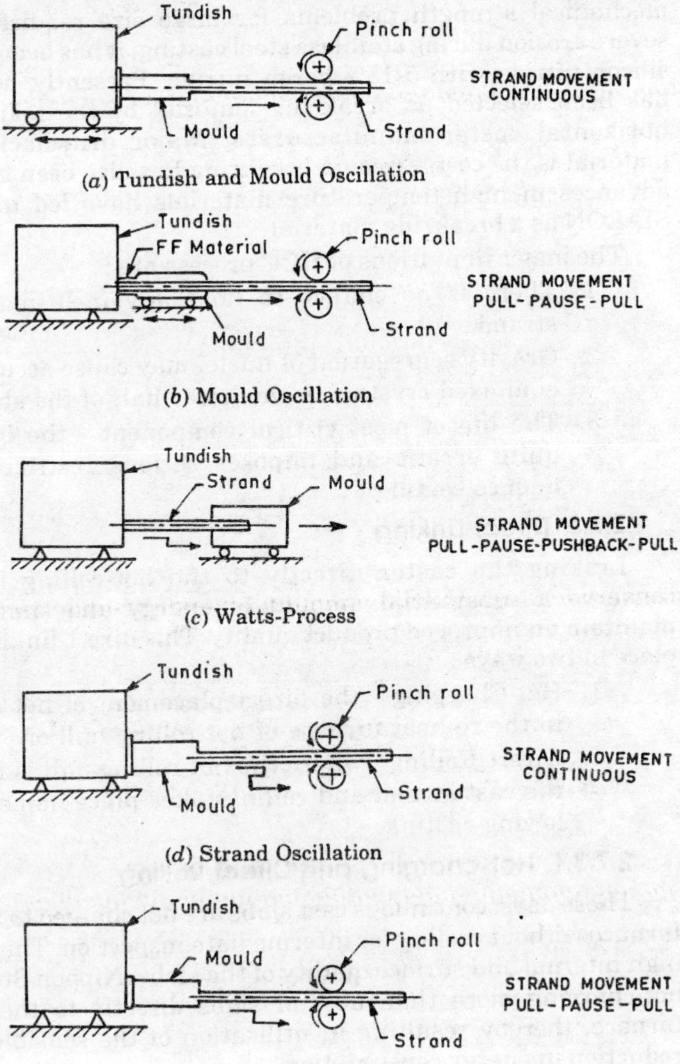

(*a*) Tundish and Mould Oscillation

(*b*) Mould Oscillation

(*c*) Watts-Process

(*d*) Strand Oscillation

(*e*) Strand Oscillation

Fig. 2.59. Development of tundish-mould arrangement in horizontal continuous casting

In horizontal continuous casting method, the direct connection of the mould to tundish in a horizontal plane eliminates the formation of meniscus and an artificial meniscus is created by a ceramic breakring component. Early horizontal casters utilised silicon nitride as a breakring material. Due to limitations of being susceptible to surface crack formation, machinability problem, mechanical strength problems for large size requirements, and severe erosion during stainless steel casting, it has been replaced by silicon nitride with 5-15% boron nitride. Presently, boron nitride has been selected as breakring material by the majority of the horizontal caster manufacturers. Major drawback with this material is the cost. Zirconia breakring has also been tried. Recent advances in high temperature materials have led to the use of SIALON as a breakring material.

The major limitations of HCC process are :

1. There is no chance to float any inclusions from the strand,
2. Gravity segregation of nuclei may cause accumulation of equi-axed crystals in the lower half of the strand and
3. The life of most critical component - the breakring is quite erratic and imposes serious restrictions on sequence casting.

2.7.3. Direct Linking

Linking the caster directly to the hot-rolling facility can conserve a substantial amount of energy and simultaneously maintain an improved product quality. This direct-linking can take place in two ways :

1. Hot-Charging - the direct placement of hot cast-strand in the re-heat furnace of hot-rolling mill or
2. Direct-Rolling - where the hot-rolling mill is in line with the cast strand and rolling takes place immediately following casting.

2.7.3.1. Hot-charging and Direct Rolling

These days, continuous cast slabs are hot charged to a reheating furnace without cooling for intermediate inspection. This requires a high internal and surface quality of the slabs. Nippon Steel is today hot-charging more than 50% of slabs directly to the reheating furnace, thereby resulting in utilisation of the sensible heat and reduction in energy consumption.

Future plans include direct rolling of the cast-product with marginal heating of the corners and edges of the cross-section, resulting in further savings of energy. To produce a strand with the requisite heat content for direct rolling, the continuous casting operation should lead to hot-slabs with uniform skin temperature of around 1200°C. The hot slabs must be transferred to the rolling mill without delay for processing into plates/strips. The facilities for production of desired quality of hot slabs for direct rolling are :

(*i*) Combined air and water cooling
(*ii*) Edge insulation and insulation of roller tables
(*iii*) hot inspection
(*iv*) A furnace for edge heating and for adjustment of the rolling start temperature.
(*v*) A high degree of automation, which includes quality evaluation by computer.

The main limitations of the process are : lower capacity of caster compared to that of the rolling mills and costly coordinated operation of casting and rolling. When the caster is out of operation, the rolling mill has to stop. When the rolling mill is under shut-down, the caster has to slow down but can produce products, which have to be stored and rolled at a later date.

2.8. INDIAN SCENARIO

Continuous Casting Status in Mini/Midi Steel Plants

At present out of about 99 CCMs installed in the mini and midi sectors, only two are for slabs, 43 for bollms/billets, and 54 for billets. The all new upcoming mini-steel plants are equipped with continuous caster and it is expected that by the end of century the mini-steel plants would be producing around 7 Mt of crude steel per annum, out of which over 90% would be through concast route only.

Continuous Casting Status in Integrated Steel Plants

In the integrated steel sector, there are two billet casters, seven bloom casters, four slab casters and one slab cum bloom caster. Presently, none of the continuous casting plant has neither round nor near-net-shape caster. However, combination casting of slab and bloom in the same machine has been introduced in one of the plants as per the recent trends in the world.

The corporate planning of the Steel Authority of India Ltd. (SAIL) envisages that by 2000 AD, the integrated steel plants would be producing around 17 mt of crude steel of which around 98% will be through concast route. TISCO may also completely change over to CC route during its modernization programme by the end of the century. VSP which has commissioned bloom casters recently will continuously cast all its liquid steel. The expected crude steel production through the CC route and its estimated share in the projected total crude steel production in the integrated steel plants by the end of year 2000 AD is shown in a Table on next page.

Developments in Bhilai Steel Plant

After the installation of continuous casting machines in BSP, a number of modern features such as slide gate system in ladle, aluminium wire feeding during rinsing, cold tundish practice, calcium wire injection in tundish, ladle to tundish and tundish to mould shrouding and monoblock stoppers in tundish have been adopted for improving product quality and smooth operation. Beside these measures, several automation schemes, namely, tundish weighing and metal level control, mould level measurement and control, water flow in secondary cooling, etc. have been introduced. With these measures BSP has achieved sequence casting of 29 heats for slab caster and 31 heats for the bloom caster. Several other features, viz., ladle weight measurement, length optimization and automatic gas cutting, strand withdrawal force measurement at mould exit, strand temperature measurement, slab receiving roller table automation, roll gap checker and automatic marking machine are being planned to be implimented.

ANTICIPATED CC RATIO IN INTEGRATED STEEL PLANTS

	Crude Steel Production Mt	Continuous Cast Steel, Mt	Continuous Cast Ratio, %
1989–90			
SAIL–plants	11.23	1.50	18.1
TISCO	2.40	0.35	
VSP	0.75	0.75	
	14.38	2.60	
1994–95			
SAIL	13.95	9.88	71.0
TISCO	3.50	1.70	
VSP	2.82	2.82	
	20.27	14.40	
1999–2000			
SAIL	17.07	16.69	98.39
TISCO	3.50	3.50	
VSP	2.82	2.82	
	23.39	23.01	

Development of Special Steel at BSP

As a result of successfully establishing the continuous casting technology, following types of special steel slabs and blooms with improved internal and surface quality are being continuously cast regularly.

(*i*) Structural and ship building quality steels,
(*ii*) High tensile weldable quality steel,
(*iii*) Micro-alloyed steel,
(*iv*) Boiler quality and pressure vessel steels,
(*v*) Deep drawing quality LPG cylinder grade steels,
(*vi*) Estra deep drawing steel for auto body sheets, and
(*vii*) Medium and high carbon wear resistant steels.

Stainless Steel Production through CC in ASP

ASP has a bow-type curved mould slab cum bloom caster. The caster is equipped with some of the modern features which include ladle turret for sequence casting, ladle weighing and slide gate (Flocon type), ladle to tundish shrouding, tundish to mould SEN, automatic mould level controller, automatic start of casting powder and different casting powders for different grades, and length optimization to improve productivity and product quality. EMS technology is being planned to be added for both slabs and blooms. The grades which are continuously cast are 304/304L, 316/316L, 310, 321, 409, 430, 3Cr12, duplex and low nickle stainless steel.

Continuous Casting in Maharashtra Elektrosmelt Ltd. (MEL)

MEL is having a two-strand billet caster with an installed capacity of 0.7 Mt of mild steel per year. Following measures have been adopted to facilitate continuous casting of stainless steel (AISI-304) and special steels (20 MnCr5, CK-45) of improved quality. Nitrogen/argon purging in ladle, aluminium wire feeding in the mould, nitrogen shrouding of ladle stream and slide gate, and tundish stopper mechanism and SEN in tundish is being employed.

Continuous Casting at VSP

VSP has commissioned six four-strand casters with an installed capacity of 2.8 Mt of continuous casting blooms. The modern features of the machine are ladle turret and weighing, slide gate system, monoblock stopper, submerged entry nozzle and cold tundish practice, process control computer with dynamic model and sublance, and automatic process control of continuous casting machines.

THREE

Secondary Steel Making

3.1. GENERAL

For many applications, the surface quality, internal quality, microcleanliness and mechanical properties of steels, produced by conventional processes like open hearth, L.D., Kaldo, BOP, Q-BOP and electric steel making, are entirely satisfactory. However, for certain applications, especially in alloy steels, the quality, performance and mechanical property requirements are so stringent that they can not be met by conventional steel making processes, for example, the steels used for rotors, special forgings, aircraft industry components, plates for ships and submarines, special fabricated components, cryogenics and electrical steels. It may be worthwhile to mention here that besides other processing conditions, such as deoxidation practices, the manner of transferring molten steel from one container to other, the reaction of molten steel with slag covering and refractory linings and the solidification conditions in a mould, affect the quality of steel.

As the specifications are becoming more stringent, the conventional primary steel making processes are finding it increasingly difficult to meet these requirements. Even if the existing processes are able to meet these requirements they often do so at the expense of loss in production or underutilization of the steel making unit. The shortage and the high cost of energy as well as the need to conserve it, demand that remelting or reheating of steel be kept to the unavoidable minimum and large number of processes have been developed to treat liquid steel outside the primary steel making vessel.

The secondary steel making processes have, therefore, been developed in producing steels conforming to stringent specifications.

Secondary steel making processes are adopted, primarily to meet stringent requirements, in the following areas :

(*i*) Achieving extremely low values of Sulphur, Oxygen, Nitrogen, Carbon, Phosphorus and Hydrogen.

(*ii*) Inclusion control to extremely low levels and control of morphology of inclusions.

(*iii*) Temperature control particularly in case of continuous casting and specially for sequence casting which may also include heating.

(*iv*) Alloying and analysis control for optimum utilisation of ferroalloys to achieve desired properties consistently.

(*v*) Improving the fluidity of liquid steel for easy teeming.

(*vi*) Increasing production rates by decreasing furnace refining times.

The basic principle behind secondary steel making is to shift the major part of the lengthy refining, deoxidation, desulphurization and alloying operations outside the primary steel making unit. Therefore, when compared with the conventional treatment of steel in melting furnaces, one achieves, besides metallurgical advantages, a higher productivity of the entire plant, lower direct cost per ton of treated steel and better operational and environmental conditions. The secondary steel making-refining units along with their accessories are relatively cheap and by installing one of these, in the steel making plant, the primary steel making unit can be more fully utilized for primary refining, thereby, increasing considerably, the overall production rate along with the improved quality of the end product.

The secondary steel making-refining processes can be performed in ladles, special vessels or in the moulds. There are number of secondary steel making refining processes development, over the last four decades, for producing quality steels. However, the basic processes employed may be classified into three major categories, namely :

(*i*) Vacuum degassing process.

(*ii*) Inert gas purging process.

(*iii*) Injection treatment and Plunging Process.

Processes incorporating typical combinations of the primary processes have been developed over the years and are given below :

1. Vacuum arc remelting
2. Vacuum arc degassing
3. Vacuum carbon deoxidation

4. Vacuum oxygen decarburization
5. Argon oxygen decarburization
6. TN process
7. Ladle furnace
8. Electroslag refining process.

PART - 1

3.2. VACUUM DEGASSING PROCESSES

Hydrogen is the cause of hair-line cracking and flake formation. Degassing of teeming jet was adopted as early as in the 1950s for the first time to reduce the hydrogen percentage in industrial heats to approximately < 2 ppm *i.e.*; the limit beyond which there is a risk for flake formation, using a vacuum chamber with pressure upto 1-3 mbar.

3.2.1. Principle

Vacuum treatment has advantage of protection against oxidation, nitrogen and hydrogen pickup. Other important feature of vacuum treatment is that carbon may be used as powerful deoxidant with deoxidation product which is not left behind in the metal as inclusions. Freedom from inclusion and dissolved gases are the primary aims of vacuum treatment.

The solution of gases like hydrogen and nitrogen depends primarily on the partial pressure of these gases in contact with the metallic phase. Solution equilibria of these gases may be expressed as :

$$1/2\, H_2 \leftrightarrow \underline{H} \qquad \text{...(1)}$$

$$1/2\, N_2 \leftrightarrow \underline{N} \qquad \text{...(2)}$$

The equilibrium constant for the above solution equilibria may be expressed as :

$$K_H = \frac{[H]}{pH_2^{1/2}} \qquad \text{...(3)}$$

$$K_N = \frac{[H]}{pN_2^{1/2}} \qquad \text{...(4)}$$

This amount of hydrogen and nitrogen dissolved may be given as

$$[Wt\%\ H] = K_H\ pH_2^{1/2} \qquad \text{...(5)}$$

$$[Wt\%\ N] = K_N\ pN_2^{1/2} \qquad \text{...(6)}$$

The above expressions were first derived by Sievert and the solubility of gases is determined by well known Sievert's law, mathematically given as expression (5) and (6).

To lower down the amount of these gases dissolved at any stipulated temperature, one may bring down the partial pressure of these gases in atmosphere by creating vacuum of desired level or purging with some inert gas.

Carbon could act as a strong deoxidiser under vacuum. The reaction between carbon and oxygen dissolved in liquid melt (steel) may be represented by considering the Henrian activities of carbon and oxygen,

$$[C]_{1wt\%} + [O]_{1wt\%} = \{CO\} \qquad ...(7)$$

with the equilibrium constant :

$$K = \frac{p_{CO}}{[h_c]\,[h_o]} \qquad ...(8)$$

the activity product is given as :

$$[h_c]\,[h_o] = \frac{p_{CO}}{K} \qquad ...(9)$$

which may be expressed in terms of Wt% C, Wt% O, f_c and f_o as,

$$[f_c.\ Wt\%\ C]\ [f_o.\ Wt\%\ O] = \frac{p_{CO}}{K} \qquad ...(10)$$

where, h_c and h_o are Henrian activities and f_c and f_o are Henrian activity coefficients of carbon and oxygen w.r.t. 1 Wt% standard state.

At any chosen partial pressure of carbon monoxide the carbon and oxygen are inversely related to each-other. Since the value of K increases with decrease in temperature, carbon would react with oxygen to form CO with decrease in temperature. This reaction would be favoured with decrease in partial pressure of CO. Partial pressure of CO may be decreased by applying vacuum or by purging with an inert gas. In a vacuum chamber the pressure over the melt becomes essentially the CO pressure in the system. Reducing the pressure by a factor of 10, 100 or 1000, that is from 1 atm to 0.1 atm, 0.01 atm or 0.001 atm will tend to cause a corresponding lowering of oxygen and carbon level.

Combination of gas purging and vacuum accelerate the process of degassing since a very small amount of purging gas causes violent stirring of the melt which results in effective exposure of melt to the vacuum.

3.2.2. Stream Degassing

Molten steel is exposed to vacuum in the form of a stream of metal flowing from one vessel to another. Sudden exposure to vacuum results in such a rapid rate of degassing that the stream of steel explodes into tiny droplets. The large increase in the surface area of steel in the form of falling droplets leads to very efficient degassing and surface reactions. The following processes are used in commercial practice wherein degassing is accomplished by pouring steel from one vessel to another :

(*i*) Ladle to mould degassing
(*ii*) Ladle to ladle degassing
(*iii*) Tap degassing
(*iv*) Gero vacuum casting process
(*v*) Therm-I-Vac process

Ladle to ladle, ladle to mould and tap degassing processes are shown in Figs. 3.1, 3.2 and 3.3 respectively. These processes although hasten the degassing but involve loss of temperature which requires higher superheat of the steel. Higher superheat has its inherent drawbacks.

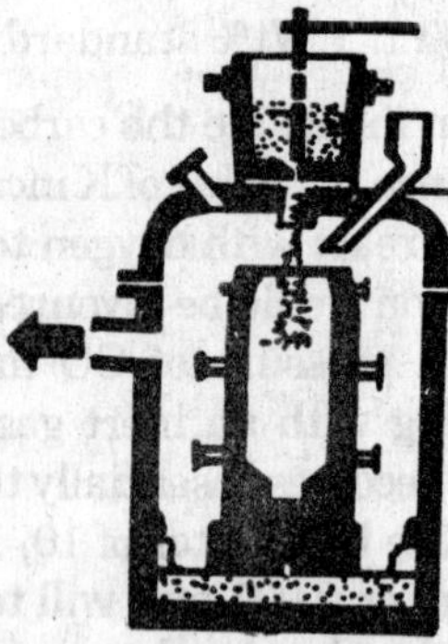

Fig. 3.1. Schematic representation of vacuum ingot teeming (vt)

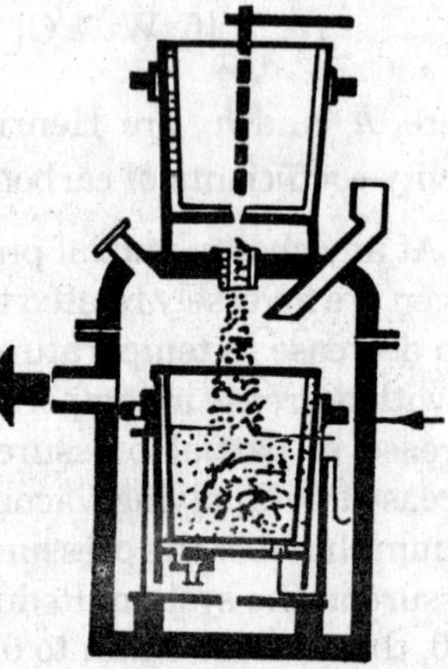

Fig. 3.2. Schematic representation of ladle-to ladle stream degassing.

Effect of reduced pressure on hydrogen content of the steel melt in different degassing processes is shown in Fig. 3.4.

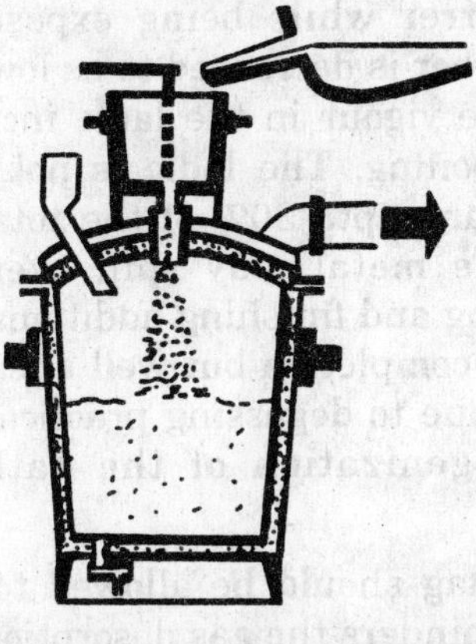

Fig. 3.3. Schematic representation of tap degassing (TD).

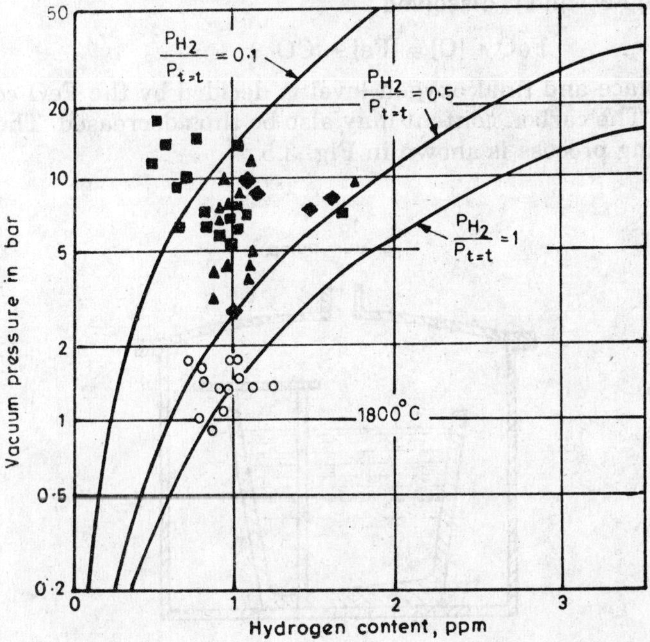

■ Tap Degassing (unkilled) ▲ Tap Degassing (Killed)

◆ Ladle to Ladle ○ Vaccum Ingot Teeming

Fig. 3.4. Final hydrogen content vs. vacuum pressure in stream degassing.

3.2.3. Ladle Degassing

Liquid steel is held in a ladle which is put inside a vacuum chamber. Steel may be stirred by bubbling an inert gas or by an electro-magnetic stirrer while being exposed to vacuum. The pressure in the chamber is decreased to as low as 1-10 mm Hg. As the pressure falls the vigour in the ladle increases and the bath appears as if it is boiling. The ladle is not filled completely. A considerable free board upto 30% of the total volume should be provided or else the metal may spill over the edges during degassing. Deoxidising and finishing additions are made when the degassing is nearing completion but well ahead of the point when stirring of the bath due to degassing practically stops. Stirring is essential for homogenization of the bath composition and temperature.

A minimum of slag should be allowed to flow into the ladle during tapping as it hinders the gas desorption process. The (FeO) content of slag keeps on supplying oxygen to the bath. The reaction between FeO and C dissolved :

$$FeO + [C] = [Fe] + \{CO\} \quad \text{...(11)}$$

takes place and final oxygen level is decided by the FeO content of slag. The carbon content may also be thus decreased. The ladle degassing process is shown in Fig. 3.5.

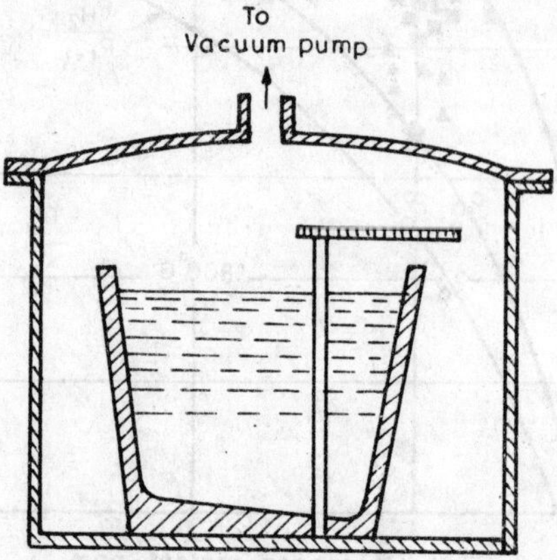

Fig. 3.5. Schematic diagram showing ladle degassing.

Mould Degassing : This is similar to the ladle degassing. It consists of teeming moulds in air and then subjected to vacuum during solidification. It does not need extra-superheat in the metal at the beginning. This process is not as effective in degassing as compared to other processes.

3.2.4. Recirculation Degassing

In these processes molten steel is exposed to vacuum in the form of either a circulating stream or a small portion temporarily lifted from the rest of the metal pool into the evacuated chamber. After this exposure the partial quantity is returned to ladle and further amount is drawn from the ladle. This process is repeated continuously to degas, decarburize and clean the melt. These processes were developed in mid fifties in Germany and are known as Ruhrstahl-Heracus or for short R-H process [circulating stream] and Dortmund-Hoder or for short D-H process [small portion lifted from the ladle].

3.2.4.1. R.H. Process

R.H. process, developed in 1958 at the Hattingten Thyssen Henrichs-Hulttee works is now used in almost 100 plants for variety of applications. The salient features of the degassing chamber are shown in the Fig. 3.6. The chamber is a cylindrical steel shell with two legs, called snorkels and openings at the top side are provided for exhaust, alloy additions, observations and control.

It is lined from inside with fire bricks in the upper portion and high alumina bricks in the lower portion which directly come in contact with steel. The snorkels are also lined from inside with high alumina refractory. A refractory tube is attached to each snorkel since it is to be dipped in liquid steel for degassing. These are made of commercial grade pure alumina. One snorkel works as inlet and

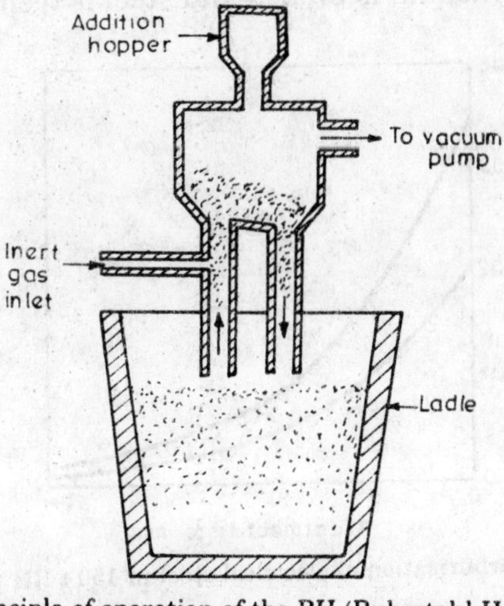

Fig. 3.6. Principle of operation of the RH (Ruhrstahl Heracus process).

the other as outlet. Lifter gas like argon is introduced in the inlet snorkel at the point where alumina tubular tip is attached to it.

Before use of the vacuum chamber it is usually heated in the temperature range of 900-1500°C by firing fuel up in the snorkels or by electric heating rods. After the snorkels are dipped to an appropriate depth in the melt kept in the ladle the chamber is evacuated and argon is let in the inlet snorkel. The steel thus rises in the vacuum chamber. The gas bubbles expand and burst in the chamber exposing larger surface area of the melt to the vacuum. The net result of this is that degassing takes place very efficiently. The melt thus degassed becomes more dense and flow back to the ladle under the force of gravity through the other snorkel. The degassed steel is slightly cooler and dense than the melt in the ladle and hence it forces the lighter undegassed steel upwards thereby ensuring adequate mixing and homogeneity of the bath.

The rate of circulation is controlled by controlling the vacuum level and the argon gas flow. The vacuum level reaches to about 0.6-0.3 torr. The average rate of circulation is nearly 12t/min and nearly 20 mts are required to treat 100t of steel to bring down the gas contents to the usual desired levels.

At the end of the degassing alloy additions may be made depending upon the superheat available in steel. Circulation ensures homogeneity of the bath composition. The melt cools down by about 25-50°C depending upon the ladle size. The amount of argon required is 0.015-0.075 m^3/ton of steel treated. At the end of the treatment the vacuum is broken and steel is teemed in normal manner in air.

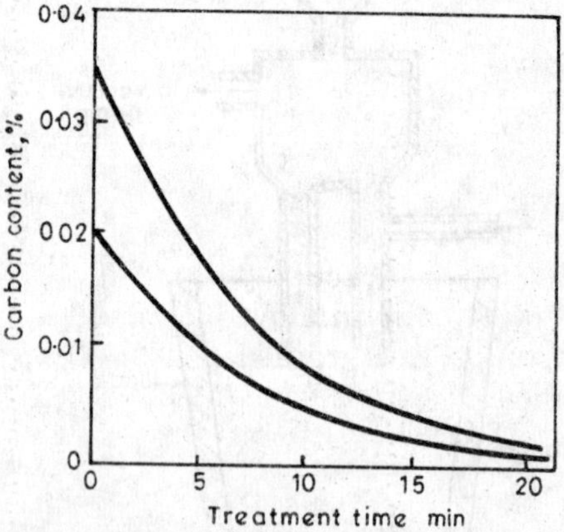

Fig. 3.7. Decarburisation of electrical steel in 150 t RH unit.

The effect on carbon, hydrogen and loss of temperature during R-H process is shown in Figs. 3.7, 3.8 and 3.9. Further developments in this process are RHO and RHOB processes.

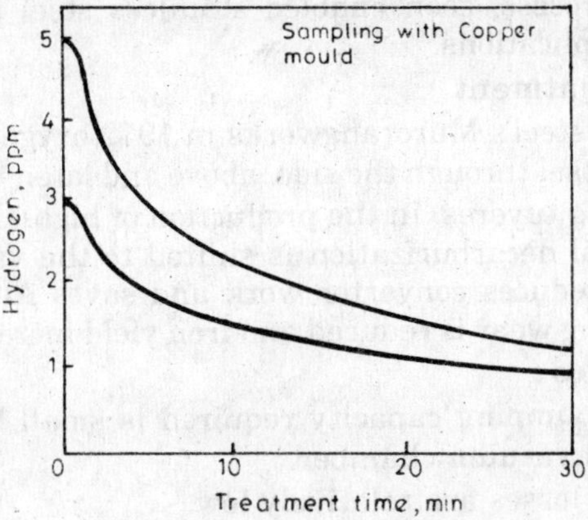

Fig. 3.8. Hydrogen content, 150 t RH unit.

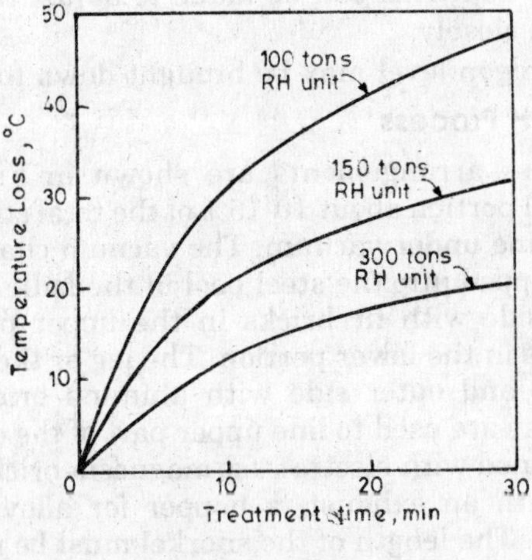

Fig. 3.9. Temperature loss in RH units depending on treatment time (without alloy additions).

This process may be effectively used for production of high quality steels such as plain unalloyed or alloyed structural steels, HSLA grades, Al-killed fine grained steels, deep drawing grades, Al-free vacuum deoxidised steels etc.

Further Developments

RHO Treatment

The stainless steel and other chromium bearing steels may be decarburized under vacuum without any significant loss of chromium. Under vacuum, chromium oxidation starts only at lower carbon levels. Oxygen lancing on to the bath surface in an RH vessel was first carried out at Thyssen Henrichshutte works. This method called RHO process, soon enabled stainless steel production for heavy plate applications.

RHOB Treatment

At Nippon steel's Murorangworks in 1972 oxygen was injected into the RH vessel through the side, above and later, below the steel bath level using tuyeres. In the production of high-alloy-chromium steels, the final decarburization is shifted to the vacuum facility. This method reduces converter work and saves Al as deoxidant. Converter lining wear is reduced and iron yield increased.

Advantages :

(*i*) The pumping capacity required is small because of the small vacuum chamber.

(*ii*) Heat losses are relatively low.

(*iii*) It can produce steels with less than 0.02% C.

(*iv*) Alloy additions can be made to adjust the specifications more closely.

(*v*) Hydrogen level may be brought down to 1.5 ppm.

3.2.4.2. D-H Process

The process arrangements are shown in Fig. 3.10. In this process a small portion about 10-15% of the total steel in the ladle is treated at a time under vacuum. The vacuum chamber has a long leg which is dipped into the steel pool in the ladle. The chamber is lined from inside with firebricks in the upper portion and high alumina bricks in the lower portion. The leg or the snorkel is lined on both inner and outer side with alumina bricks. Now-a-days magnesia bricks are used to line upper part of the chamber and the lower part is lined with electro-cast magnesia bricks. The chamber is provided with an exhaust, a hopper for alloy addition and a heating device. The length of the snorkel must be more than 1.4 m because atmospheric pressure would force steel to rise to such a height when the chamber is under vacuum.

A preheated chamber is lowered in the ladle, so that the tip or the snorkel is well below the metal surface. (more than 60 to 70 cms. deep). The evacuated chamber is moved up and down to make the steel enter the chamber and exposed to the vacuum and then drain down to the ladle. The chamber is moved through 50-60 cm with a cycle time of about 20 sec. Since 10-15% of steel is exposed at a time some 7 to 15 cycles are required to expose the entire steel once. Adequate degassing is possible in about 20-30 cycles and additional 5-10 cycles may be required, depending upon the alloy addition, to

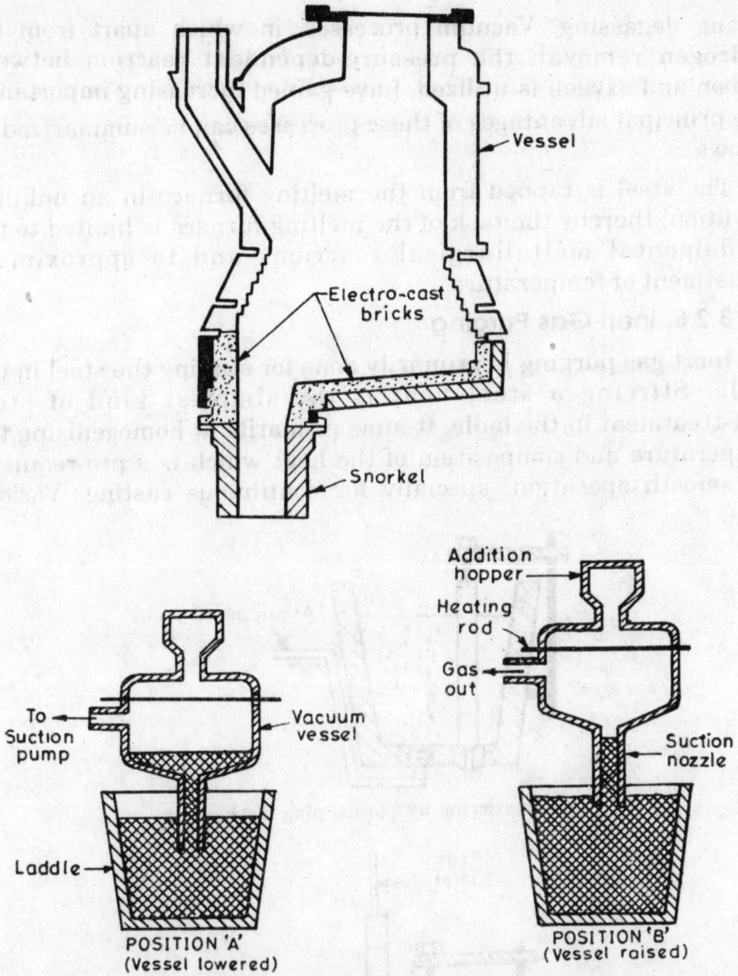

Fig. 3.10. Profile of D-H vessel and principle of operation.

ensure homogenization. Therefore nearly 15-20 min are required for degassing a ladle of almost any size. A slag layer is deliberately kept on the steel bath in the ladle to minimise heat losses. Slag, however, must not enter the snorkel. The snorkel lining lasts for about 90-100 heats, the lower part of the chamber for nearly 400 heat and the upper part lasts for some 800-1000 heats.

The D-H process possesses most of the advantages of R-H process and in addition it is possible to work with much less superheat because heating device like graphite resistor rod is provided in this process to compensate part of the heat losses during degassing. Vacuum processes, in which apart from the hydrogen removal, the pressure-dependent reaction between carbon and oxygen is utilized, have gained increasing importance. The principal advantages of these processes can be summarized as follows.

The steel is tapped from the melting furnace in an unkilled condition; thereby the task of the melting furnace is limited to the fundamental meltallurgical reactions and to approximate adjustment of temperature.

3.2.5. Inert Gas Purging

Inert gas purging is primarily done for stirring the steel in the ladle. Stirring a steel heat is the simplest kind of steel post-treatment in the ladle. It aims primarily at homogenising the temperature and composition of the heat which is a pre-requisite for smooth operation, specially for continuous casting. Various

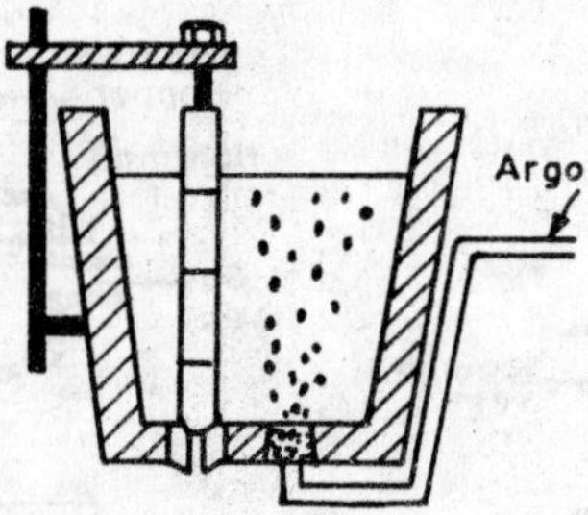

Fig. 3.11. Argon stirring by porous plug method.

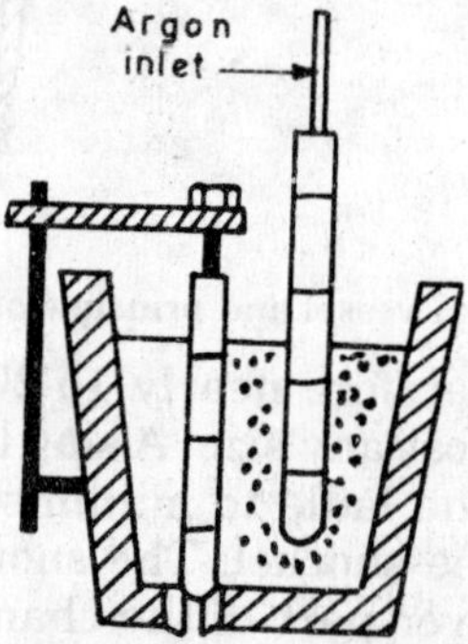

Fig. 3.12. Argon stirring lance method.

stirring systems are in use today. Wide spread use has been reported of one or several porous bricks in the ladle bottom or wall, and lances through which inert gases like argon or nitrogen are introduced into the heat as shown in Figs. 3.11 and 3.12.

Heats of steel are argon stirred in ladle also by lowering a refractory protected lance into a ladle of steel to within 0.03 m above the bottom and blowing argon through the lance shown in Fig 3.12. For argon stirring practices precautions must be taken to minimise the carryover of the oxidising slag into the ladle during tapping. If it happens, the intimate mixing of slag and the steel during argon purging can deleteriously affect steel quality and chemical composition of the steel. Usually the flow rate of argon is regulated to such an extent that new heat-radiating surfaces are not formed on the slag covering. However, for some practices the flow rate of argon is increased to produce new heat radiating surface *e.g.*; to increase the rate of temperature loss for attaining the required temperature. Or to produce a fountain of steel into which ferroalloys and deoxidiser are added for close chemical composition control. The argon purging process is practiced for purposes listed below.

(*i*) To minimise the steep temperature gradients in a ladle of steel and the chill the steel by radiant heat losses. Temperature control is much more critical in continuous-casting operations and argon-purging is used for this purpose.

(*ii*) To improve steel cleanliness by promoting the agglomeration and flotation of non-metallic inclusions to the slag.

	QTM rating average inclusion area%	*Average oxygen ppm*
Non stirred heats	0.14	>200
Argon-stirred heats	0.09	109
(QTM : Quantitative television microscope)		

(*iii*) To minimize the fading of manganese and silicon contents in a heat during teeming.

(*iv*) To provide means for making additions of deoxidisers and ferroalloys to meet stringent chemical composition specifications. For this practice extreme care must be exercised to minimise or preferably eliminate the carryover of furnace slag during tapping.

(*v*) Minor reductions in hydrogen content are observed.

(*vi*) If reacting slags are added during purging, for instance mixtures of lime and fluorspar, specially while using basic lining in the ladles, a reduction in sulphur can be achieved.

Although argon gas stirring is effective in deoxidising molten steel, the total oxygen content will not be lowered beyond a certain level. This is due to the reoxidation of once deoxidised elements either

1. by contact with air on the surface of molten steel
2. by reaction with FeO and MnO contents of slag in ladle
3. by reaction with refractories, further continuing generation of oxidation products.

The deoxidising effect of argon gas purging is thus limited.

Further development on simple argon gas purging have been done by using non-oxidising synthetic slag with the specific purpose of manufacturing cleaner steel. In this process the melt is covered with a synthetic slag. There are two versions of this process.

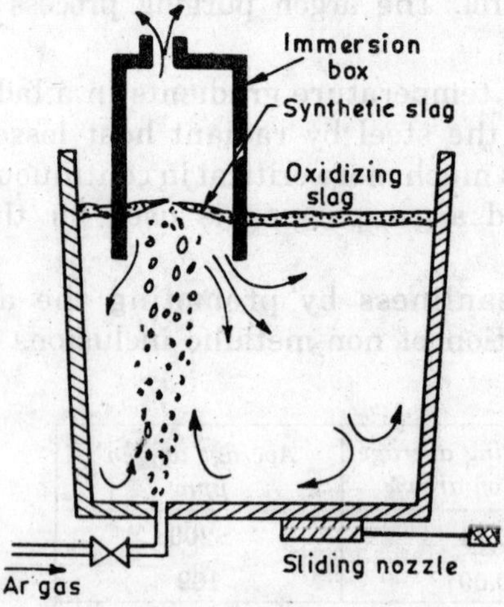

Fig. 3.13. Diagrammatic sketch of SAB process.

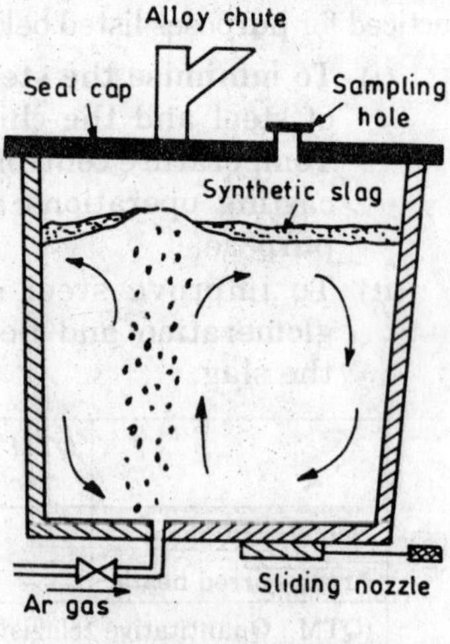

Fig. 3.14. Diagrammatic sketch of CAB process.

One process known as sealed argon bubbling (SAB) consists of placing a non-oxidizing synthetic slag in the immersion vessel made of castables. The vessel is immersed in the molten steel at a point where argon gas bubbles out of the molten steel, as shown in Fig. 3.13.

Another version, as shown in Fig. 3.14 requires, after sufficient flushing of slag during tapping, the addition of synthetic slag on to the molten steel surface in the ladle, and Ar gas purging is carried out after a cover is placed over the ladle. This method is called the capped argon bubbling (CAB) process. A typical composition of the slag used consists of $CaO : SiO_2 : Al_2O_3$ in 40 : 40 : 20 proportion, featuring a low melting point, high fluidity and large absorption capacity.

These processes are characterised by following advantages :

(*i*) Molten steel is not oxidised by air because the molten steel surface is exposed only to Ar gas.

(*ii*) Inclusions are caught by synthetic slag as they float up.

(*iii*) Under the SAB process, because the exposure of molten steel is limited within the immersion vessel, the decrease in the molten steel temperature is relatively small. In case CAB also temperature loss is low since ladle is covered completely.

The oxygen content of the melt depends on FeO content of the slag and it can be kept below 40 ppm if FeO content of slag is reduced to 1-2%.

The effects of CAB and SAB processes can be summarised as :

(*i*) The total oxygen content in steel can be reduced to 20-40 ppm.

(*ii*) Inclusions (especially Al_2O_3 decrease, particularly large inclusions of 20-40 μ and over decrease in number.

(*iii*) Control of composition is possible because ferroalloy and deoxidising agents can be added during treatment.

(*iv*) If deoxidising agent such as Al is added during argon gas purging instead of during tapping, nitrogen absorption during tapping can be avoided making it possible to produce low nitrogen steel.

3.2.6. Injection and Plunging Processes

Almost all the steel making reactions take place between metal and slag and therefore to speed up these reactions the surface area of contact between metal and slag should be increased. In 1933 Perrin introduced a process comprising of pouring liquid metal into a ladle containing molten reducing slag for effective desulphurization. The rigorous agitation of slag and steel not only desulphurised the steel to a great extent but also improved the degree of cleanliness of the steel. However, desulphurization by mere stirring of steel and slag can bring down sulphur content only to a limited extent. For extensive desulphurization it is desirable to maintain low oxygen potential of slag-metal system. This is also necessary to obtain oxidic cleanliness. Work soon began of developing special methods based on injection process guaranteeing maximum efficiency. However, the effectiveness of the process and its efficiency would ultimately depend on the use of those elements and reagents which form sulphides with lower free energy change than iron. Fig. 3.15 shows a section of the "Richardson Diagram" for sulphides, according to which desulphurization can best be achieved by using alkaline; alkaline earth metals or related compounds.

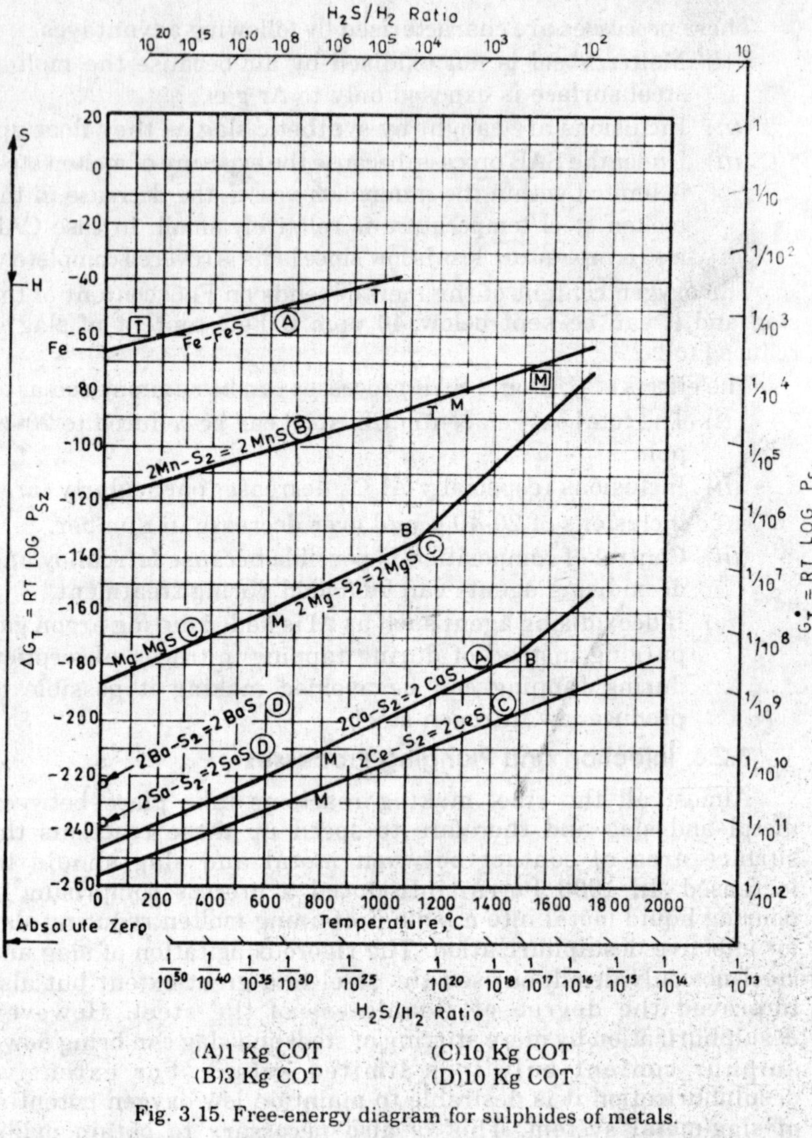

(A)1 Kg COT (C)10 Kg COT
(B)3 Kg COT (D)10 Kg COT

Fig. 3.15. Free-energy diagram for sulphides of metals.

For substantial desulphurization of steel heats, the oxygen potential of the slag/metal system should be as low possible. Reduction of oxygen in the system should also guarantee the minimization of oxide inclusions. The processes used for desulphurization reduce oxygen as well to a very limit of their detactability in basic lined ladle. Different metals like Ca, Mg, Ce and their compounds have been used for lowering down the sulphur level of the melt but Ca has given the excellent results along with desired economy. The clear-cut improvements in toughness are attributed to lower oxide and sulphide inclusion level and modification of these inclusions into less dangerous types. In case of calcium treatment, the inclusions which are otherwise present as MnS and alumino-silicates, transform into calcium aluminates which coagulate and separate well to the slag layer due to their lower melting point rendering the steel cleaner. The steel not only becomes cleaner but the remaining inclusions become less dangerous because they become globular and non-deformable at rolling temperatures. These inclusions do not elongate to form stringers which reduce the ductility seriously in the transverse and in short transverse direction in particular. This is specially true for flat products, where the phenomena of lamellar tearing is well known for heavy gauge plates.

A useful way of avoiding this problem is by injecting the desulphurizing agents in the steel melt in the ladle, known as ladle injection. Depending on the amount of desulphurizer injected a desulphurization degree between 60 to 90% is achieved.

Certain precautions must be taken for a successful desulphurization. The steel bath has to be well deoxidised prior to injection. Moreover, the furnace slag has to be removed as completely as possible. Meeting these requirements, it is possible to reach extremely low sulphur contents even below 0.005%.

Beside calcium which effectively changes the morphology of oxide and sulphide inclusions, there are elements like, titanium, zirconium, rare-earths and tellurium, which change the shape of sulphides, when present in small concentrations. However, since different elements differ on account of their impact on sulphide shape and mechanical properties, their selection should be dependent on the final steel properties required.

With rise in the demand of quality steels with particular toughness requirements, a number of processes have been developed in recent year for adding desulphurizing agents.

The lance-injection process known as Thyssen-Niederrhein (T-N) process is widely used for desulphurization and is shown in Fig. 3.16.

In this process a heat of steel is tapped into a preheated basic lined ladle, deoxidised and alloying addition are made. The carryover of the furnace slag is removed as far as possible and is replaced by synthetic slag, devoid of unstable oxides. The ladle is then brought to a desulphurizing station where a cover is placed over the ladle; a refractory lined lance is submerged to near the bottom of the ladle and powdered materials of the kind shown in table 1 are injected into the steel using a stream of inert gas. The amount of materials injected ranges between 1 to 2.5 Kg/ton depending upon the desulphurizing material and the amount of

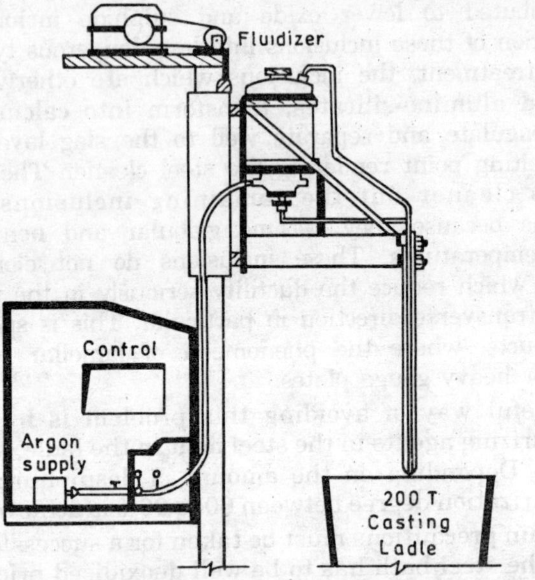

Fig. 3.16. T-N desulphurizing process.

sulphur to be removed. A 200t heat is readily desulphurized from 8 to 10 min to levels below 0.005% and as low as 0.002%.

Table 1. Desulphurization Agents for Powder Injection

Desulphuring agent	*Composition*	*Injected Kg/t*	*Amount of agent for* $(\Delta S/S_i) = 0.90$ *Kg/t*
SiCa	62% Si. 0.8% Al [30% Ca]	2-4.5	3.87
Calcium carbide	80% CaC_2. 15% CaO [50% Ca]	1-3	2.32
Calcium Cyanamide	55% $CaCN_2$ 33% CaO 12% C [20% Ca]	0.4-1	[4.22]
Mg [CaF_2]	5-20% Mg	1-3	22
CaO/CaF_2	90% CaO 10% CaF_2	3-6	3
CaO – Al_2O_3	50% CaO 50% Al_2O_3	1-5	4
CaO – Al_2O_3 – CaF_2	70% CaO 20% Al_2O_3 10% CaF_2	1-5	3

The reaction of the injected reagents is extremely efficient as a result of the transitoric phase contact and the permanent phase contact with basic top slag of the liquid bath. The relationship between the sulphur content and blowing time is shown in Fig. 3.17.

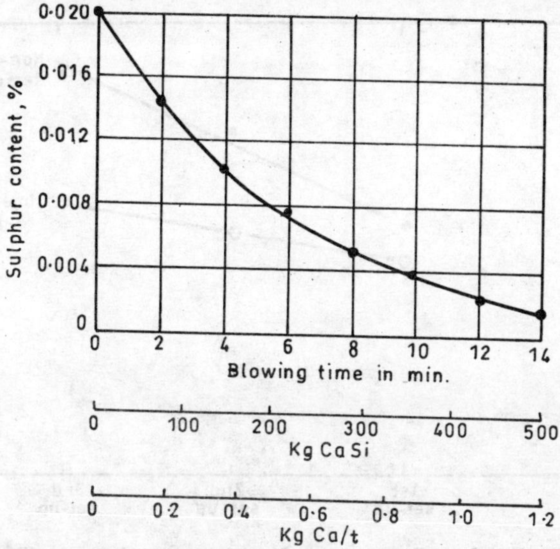

Fig. 3.17. Decrease of sulphur content during CaSi blowing into a 120 T-dolomite ladle.

The sulphur and oxygen content obtainable within few minutes are in the range of only few ppm. A temperature loss of approx. 35°C must be anticipated during the treatment. The hydrogen content tends to rise, therefore, care should be taken to see that dry materials are used. For steels susceptible to hydrogen cracking the injection treatment should be followed by vacuum/purging treatment.

The injection of powder has found widespread use in recent years, more than any other method used in ladle metallurgy. Apart from improving the quality of flat rolled products, it is being used today for nearly entire product range.

Ladle injection improves the fluidity of steel which is advantageous for continuous billet casting of high grade steels. So far continuous billet casting was predominantly used for making Mn/Si deoxidised basic grades. Aluminium-killed steels caused the tundish nozzle to be clogged due to alumina precipitation, having a high melting point. Feeding of aluminium wire in the continuous casting mould did not yield the desired quality improvement due to nonuniform distribution. It has been evidenced now that the nozzle blocking due to alumina precipitation in case of Al killed steels can be prevented by injecting Ca-bearing compounds and alloys. The injection of Ca-compounds into the steel reduces the dendritic alumina inclusions into spheroidal inclusions of calcium aluminate or CaO. The absence of alumina inclusions permits a lower teeming temperatures and virtually eliminates the nozzle blockage. The effect on teeming time for Ca-Si treated steel is shown in Fig. 3.18.

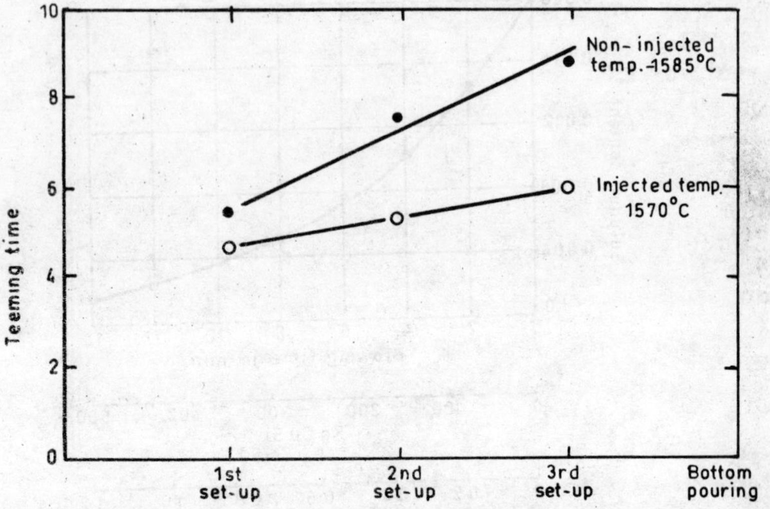

Fig. 3.18. Comparison of ingot teeming times, for SiCa treated heats and untreated heats.

The injection plant in conjunction with primary furnace offers further economic advantages by performing the deoxidation, desulphurization and alloying in ladle resulting in about 30% increase in the production capacity of the plant. The work is being done to further develop the injection method to control the formation of

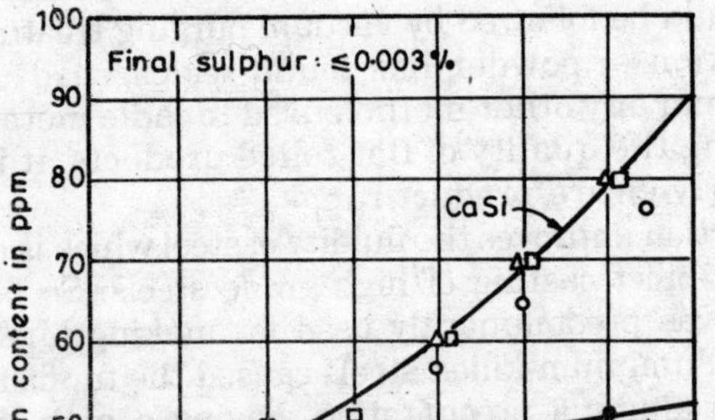

Fig. 3.19. Nitrogen pick up during injection of desulphurizing agents into a 120T dolomite ladle.

inclusions through the injection of specific kinds of elements (mag-coke, rare-earth metals). In this process nitrogen pick-up decreases the significantly by adding powdered slag with Ca-Si as shown in Fig. 3.19.

3.2.6.1. Other injection Methods

The encouraging results obtained by powder injection to desulphurize the steel melt to a level below .008% paved the way for other modified methods of injecting solids in the molten steel bath. The methods which have been successfully employed are :

1. Wire injection
2. bullet shooting and
3. Plunging

In wire injection, specially designed, sheathed calcium compound wires may be introduced continuously into the melt or the tundish during continuous casting. The feeding rate can be varied from 0.05 to 5.0 m/s. Wires may contain Ca and Al, or Ca-Si duly sheathed by sheet metal. During injection into ladle, inert gas purging is normally done for thorough mixing. Figs. 3.20 & 3.21 show the wire injection and bullet shooting processes.

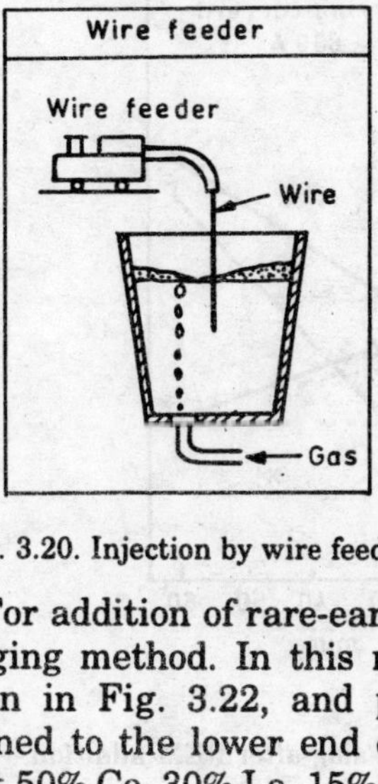

Fig. 3.20. Injection by wire feeding.

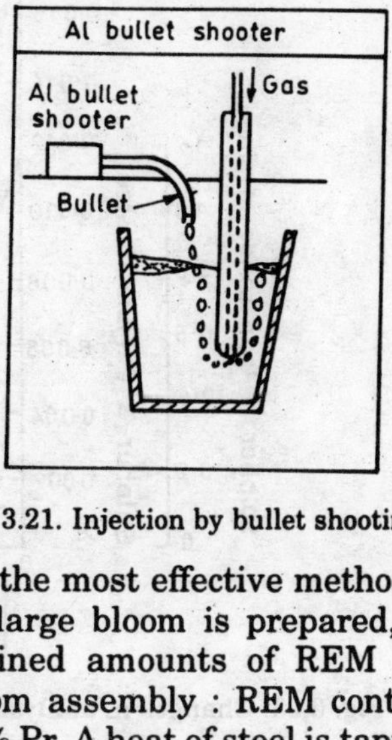

Fig. 3.21. Injection by bullet shooting.

For addition of rare-earth metals the most effective method is plunging method. In this method a large bloom is prepared, as shown in Fig. 3.22, and predetermined amounts of REM are fastened to the lower end of the bloom assembly : REM contain about 50% Ce, 30% La, 15% Nd and 5% Pr. A heat of steel is tapped into the ladle, alloyed and deoxidised followed by lowering the REM attached bloom into the molten steel. A small amount of magnesium or calcium is added to the REM to provide the agitation for uniform distribution of REM throughout the melt.

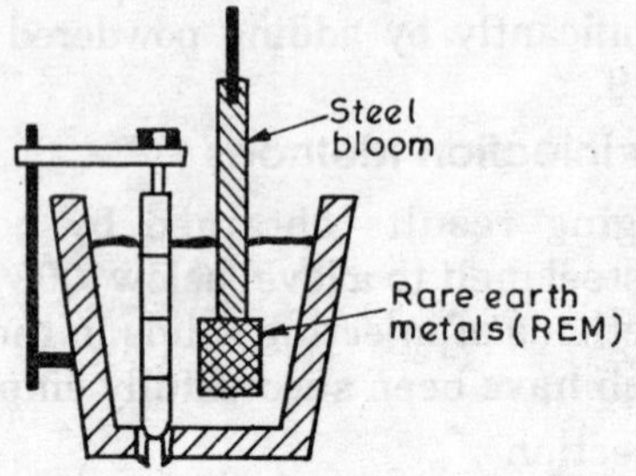

Fig. 3.22. Plunging method for adding REM into a ladle of steel.

Fig. 3.23 shows sulphur removal on account of REM addition. It has been calculated that consumption of REM reaches about 0.44 Kg/t, of which 0.24 Kg REM/t is needed to be in equilibrium with 0.004% S and remaining 0.02 Kg REM is consumed by reduction of slag. The removal of REM sulphides (RES) is a fairly slow process and takes about 15-30 min.

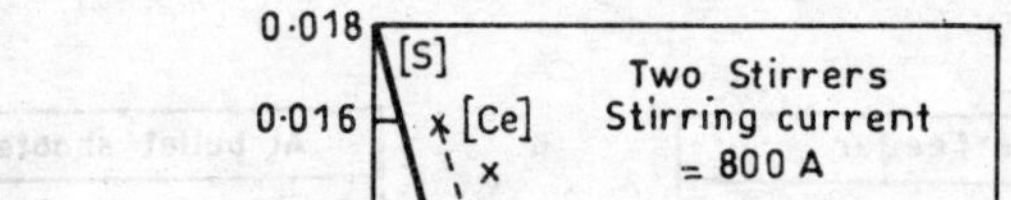

Fig. 3.23. Changes in analysis of steel and slag after REM addition.

Addition of REM favourably changes the morphology of sulphide inclusions from stringer type to globular sulphides resulting in more isotropic steel. This process is used on a limited scale only for such applications which require excellent ductility in the through-thickness direction and high charpy V-notch energy in the transverse direction in plate steels.

PART - II

3.3. MODERN PROCESSES OF STAINLESS STEEL MANUFACTURE

3.3.1. General

The manufacture of stainless steels has undergone tremendous change in the past few decades. In the recent years it has become a normal practice to carry out major operations in the vessel, outside the furnace. The drawbacks of the furnace practice emanate from the fact that while oxidising carbon to low levels the yields of chromium decrease and vary. Decarburisation at normal chromium contents of stainless steels at carbon levels above 1% results in fierceness of furnace flames. This is inhibited at carbon levels below 0.01% by nature of chemical equilibria. The use of significant quantities of expensive chromium units was unavoidable and indeed more than 20% of all added chromium had to be in the form of low carbon ferrochrome. On the other hand, due to very high temperatures (of the order of 2000°C) typically obtained during decarburising even at modest chromium levels, furnaces suffered increased risk of breakouts, and refractory consumption of hearths, roofs and sidewalls.

3.3.2. Principle

The austenitic stainless steels are specified to have very low carbon content (< .040% C) along with high chromium content to ensure absence of grain boundary corrosion. This requires selective oxidation of carbon from high chromium melts. The combined oxidation equilibrium of carbon and chromium may be expressed as,

$$3CO + 2Cr = Cr_2O_3 + 3C \qquad ...(12)$$

For which the equilibrium constant may be written as,

$$K = \frac{a_{cr_2O_3}\, a_C^3}{p_{CO}^3 \cdot a_{Cr}^2} \qquad ...(13)$$

At 1220°C the equilibrium constant, K, for the above reaction attains a value of 1. If the steels with very low carbon and high chromium contents are to be made, the value of K should be very very low, which can be attained at temperature much above 1220°C with $a_{Cr_2O_3}$ and p_{CO} equal to 1. However, selective oxidation of carbon at moderately high temperatures may be achieved if p_{CO} value may be reduced significantly either by creating vacuum or by purging or both. By lowering down the partial pressure of carbon monoxide the reaction equilibrium would be disturbed and carbon will react with Cr_2O_3 to form CO and Cr shall be transferred to the metallic phase. The relationship between Carbon and Chromium at different temperatures and p_{CO} values have been experimentally determined and are shown in Fig. 3.24. The figure shows the recovery of Cr in equilibrium with the Fe-Cr-C melt. This principle has been applied for production of stainless steel in VOD, AOD and CLU processes.

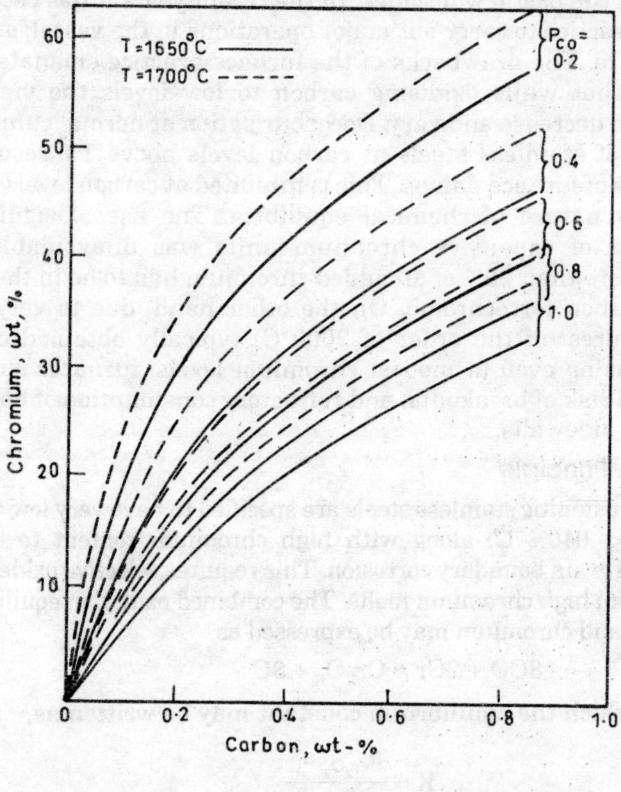

Fig. 3.24. Chromium-carbon equilibria at 1650°C and 1700°C for varying p_{CO}.

3.3.3. AOD (Argon Oxygen Decarburization) Process

The major feature of the AOD process is its ability to refine a wide compositional range of chromium bearing melts. This provides the chance of scanning the full range of available nickel and chromium bearing raw materials, and thereby minimising mixture costs through the selection and use, with the minimum technical constraints of an optimized steel making charge. Lean high-carbon ferrochromes with less than 60% Chromium and usually in the range 50-55% are presently, and are likely to remain, the cheapest form of chromium alloy units. In this ferrochrome carbon and silicon contents may be as high as, 7.5 and 4.5% respectively.

The amount of stainless steel scrap available for steel making has traditionally varied, reflecting the erratic nature, as far as the purchased scrap is concerned. For revert scrap, it is expected that the progressive move towards continuous casting and simultaneous improvements in hot and cold processing, will reduce the proportions which eventually may amount to little more than 20%. In total, the mean proportions of stainless steel scrap in charges are expected to be between 35 and 40% at which level alloy will supply nearly 2/3 of Cr and Ni units required. To obtain the Cr (about 12% of charge) through charge chrome would mean the simultaneous contribution to the burden of nearly 2% carbon and about 1% silicon. An additional carbon and silicon will be supplied in the form of various scrap and may be supplied in the form of other alloys, such as Fe-Ni, the most economical steel making burden is obviously out of the range of those which could be refined by hitherto, conventional furnace techniques. In practice, burdens will of course vary considerably, depending upon both the availability at that time and metallurgical factors which may prohibit scrap from or limit scrap in the charge, in order to satisfy low P or Co specifications, or to avoid risk of Ni contamination in staight chrome steels.

The AOD vessels have to tackle melts containing as much as 3-3. 5% carbon on occasions and generally within 1.5-2% carbon range. Only in exceptional circumstances is decarburization started in the furnace. Furnace oxygen blowing, under carefully controlled conditions is however used simultaneously to adjust the transfer of silicon content and to attain bath temperature and homogeneity as quickly as possible. Normal aim silicon content at metal transfer to the AOD is 0.20-0.30%, sufficient to protect the furnace chromium from oxidation at least to the extent that the chromic oxide in the final slag is seldom higher than about 0.55% without giving rise to unduly large slag volumes in the AOD which is necessary to preserve lining and to minimise chromium losses. As a further safeguard for the AOD lining, the final furnace slag is removed as completely as is practicable from the furnace prior to tap or additionally by decanting from the transfer ladle. The chromium loss is maximum upto the level of 0.93% in the typical through-chromium yield, from total charge to final liquid steel weight, but the lost chromium may be recovered. As shown in Fig. 3.25 an electric arc furnace is charged with high carbon

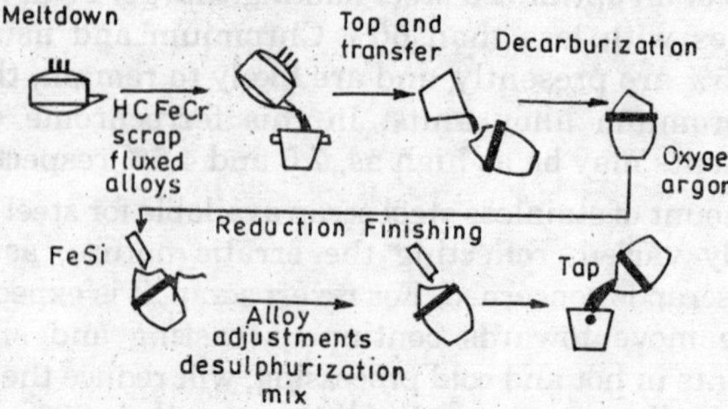

Fig. 3.25. Processing steps for refining stainless steel by the AOD process.

ferrochrome, ferrosilicon, stainless steel scrap, burnt lime and fluorspar. After the heat is melted and the desired temperature is obtained, the heat is tapped and deslagged, weighed and lip-poured into the AOD furnace.

Blowing in the AOD is aimed to start at about 1500°C *i.e.*, at a temperature just suffcient to get a clear and complete furnace melt, and not to impair clean metal transfer to the AOD, and for this, furnace tapping temperature in the range 1500-1600°C, depending on the vessel size, should be obtained. All the AODs are side blown through several tuyeres, at the most six. It may be two, three, five etc. The typical AOD vessel is shown in Fig. 3.26. Because the tuyeres, which are arranged horizontally away from the tapping side of the vessel and at about 20 cms above the vessel bottom, are a vital component in smooth operation and may otherwise cause severe localized wear of refractories. The location, number, design and control of tuyeres is very important for the AOD process.

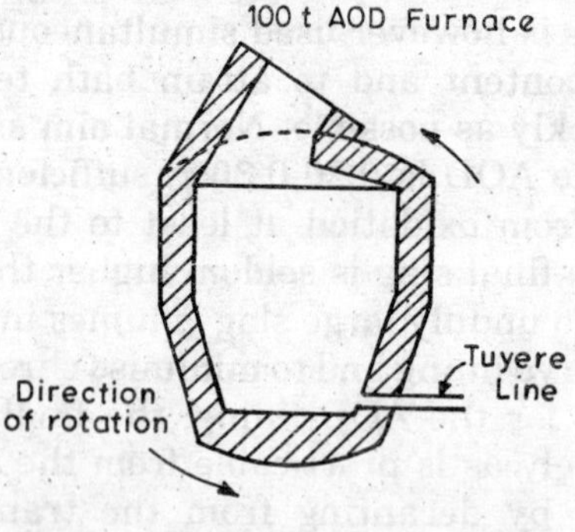

Fig. 3.26. A typical 100 t AOD furnace.

The gases emerging from the tuyeres give rise to only nominal penetration. However, under normal conditions the melt is subjected to quite vigorous action. This is due to the fact that gases after emerging from the submerged tuyeres rise and expand, due to the sudden rise of temperature and fall of pressure which is sufficient to generate extreme turbulence above the tuyere zone. This due to low friction, is immediately transmitted to the bath as a whole.

The gas blowing rates in AOD vessels are almost invariably in the range 0.7-1.0 m^3/min/t of vessel capacity which is a fraction of those used in BOF vessel. To blow at greater rates through side tuyeres and maintain bath stability would present difficulties but the proportion of time saved will be smaller. For example a 16t vessel is blown at about 11 m^3/min (0.7 m^3/min/t) and 110t vessel is blown at about 100 m^3/min (0.9 m^3/t). To increase the blowing rate by increasing the tuyere size caused premature damage to the hood by splashing. Blowing with small size tuyeres with their number increased, gives almost 50% increase in blowing rate with significantly less turbulance.

Blowing

In principle, the ratio of oxygen to the diluent gas should change continuously during decarburization in accordance with the carbon chromium equilibrium, Fig. 3.27, at that time and at that temperature. In practice this must be done in discrete steps. In the earlier practice the blowing procedures adopted were in following three stages.

Stage 1 : In this stage the blowing was carried out with oxygen to argon ratio as 3 : 1 during which the carbon level of the melt was brought down to about 0.30%.

Second Stage

The oxygen to argon ratio is decreased to 2 : 1 to blow down the carbon to 0.09-0.12%.

Final Stage

The oxygen to argon ratio is decreased further to 1 : 2 which takes carbon below specification to as low as 0.02%.

One could perhaps appreciate this if one remembers that fairly small pressure reductions are all that are necessary for effective decarburization and also that the system is self compensating in that, should conditions not favour decarburization, then the preferential oxidation of other elements notably chromium, manganese and iron will cause both the partial pressure of CO in the bubbles to fall through consequent argon enrichment and also the temperature to rise, so that conditions, for a given blown gas ratio, begin once more to favour the selective removal to carbon. The chromium and carbon contents in the melt in the three stages are shown in Fig. 3.27.

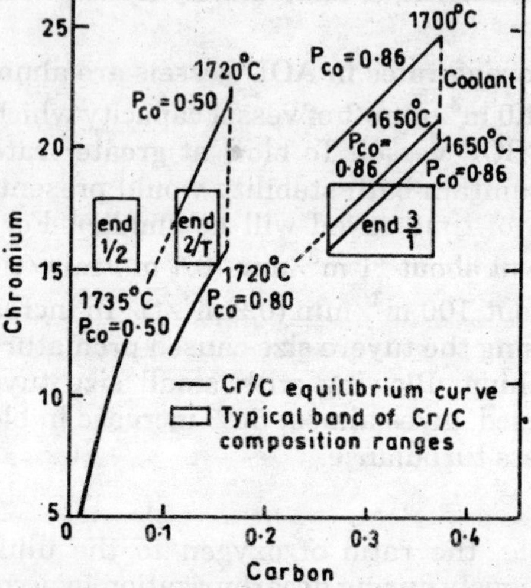

(*a*) Assumes 100% decarburization efficiency

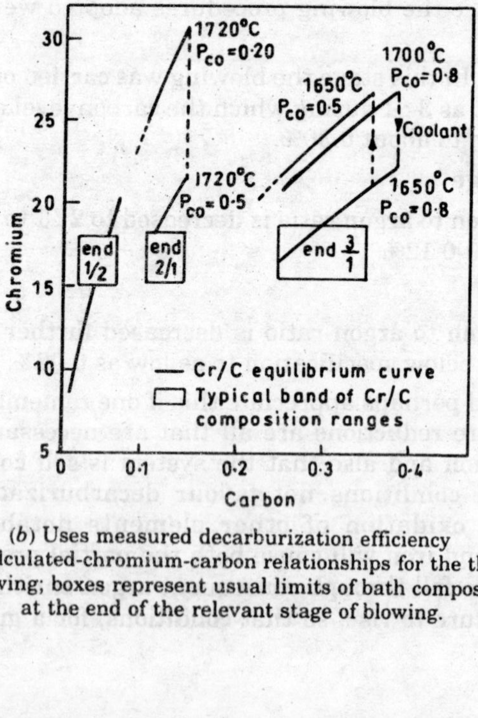

(*b*) Uses measured decarburization efficiency

Fig. 3.27. Calculated-chromium-carbon relationships for the three stages of blowing; boxes represent usual limits of bath composition at the end of the relevant stage of blowing.

It is now becoming accepted that 2 : 1 blow of stage 2 can be replaced by a 1 : 1 blow and that for the final 1 : 2 stage it may be wiser to carry this out at 1 : 3. Even lower O_2/Ar ratio have been used to achieve lower carbon levels. However, to compare one blowing schedule with another one needs to take into account the economic considerations; such as, argon consumption, chromium oxidation, cycle time and refractory performance. A major change in blowing practice has been the direct substitution of nitrogen for argon throughout stage 1, while retaining 3 : 1 ratio. Nitrogen has been used in case of ferritic steels also for which stricter limits on nitrogen control must be exercised. The nitrogen picked up in stage 1 is removed by the end of stage 3 upto 0.04% or 0.05% maximum. Control of decarburization in AOD depends on the start analysis and duration of blowing for a particular rate of oxygen injection being employed. These will in turn affect the final carbon content and temperature. For a 16t charge carbon removal values for stage 2 and 3 are .0020 and .0017% carbon per m^3 of oxygen blown respectively. In the stage 1 blow, the carbon contents of different heats vary and the decarburising efficiency in this stage is dependent upon the starting carbon level. The efficiency of decarburization may be defined either with respect to rate of oxygen blown or percent oxygen injected. The rate of carbon removal of 1.072 Kg/m^3 corresponds to 100% decarburising efficiency. Fig. 3.28 shows the decarburising efficiency in the first stage against the

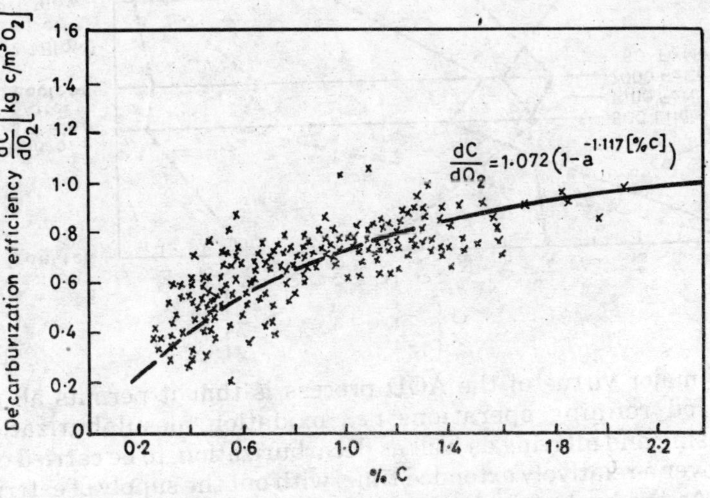

Fig. 3.28. Calculated decarburization efficiencies obtained from actual melts in phases of stage 1 blowing (3 : 1) plotted against mean carbon contents for 15t vessel.

mean carbon levels of starting melt. Time profiles of the main elements, Carbon, Silicon, chromium and manganese together with temperature are shown in Fig. 3.29 of a typical heat.

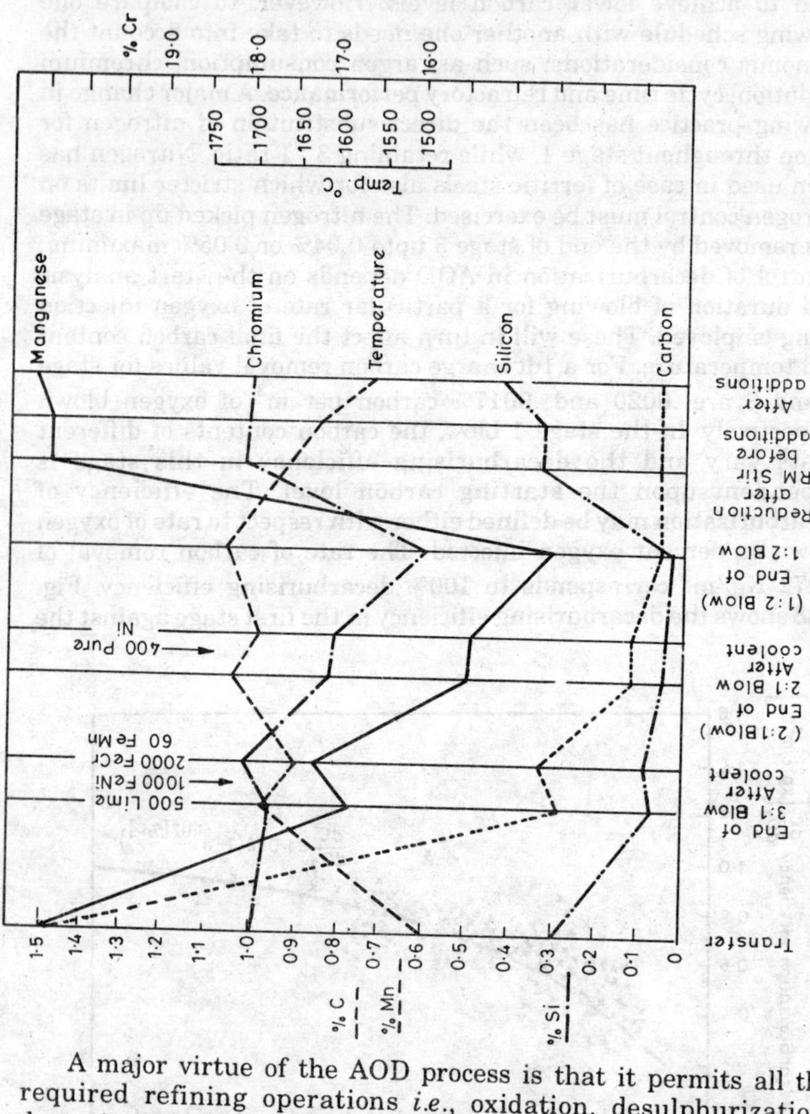

Fig. 3.29. Typical refining profile of a heat of 100t AOD vessel.

A major virtue of the AOD process is that it permits all the required refining operations *i.e.*, oxidation, desulphurization, degassing and alloying as well as decarburization, to be carried out, even over a relatively extended time, with out the supply of external heat. As the heat of oxidation of carbon is not always sufficient to permit this, the process depends upon the oxidation of small amount of chromium (1-2% of metal weight) to provide heat.

With high carbon melt the temperature of the bath may well rise above 1700°C in stage 1 and beyond this in stage 2 and stage 3 which is harmful for refractories. In such cases coolants are used to control the temperature of the bath during the AOD process.

Coolants

For any steel grade the process of refining is little different after the stage 1 (0.3%C). The first blow being variable, depending upon the charge, generates variable amounts of heat. Up to transfer carbon levels of about 0.7%, little is desired to be done to control temperature. Coolant addition begins at stage 1 when temperature reach to about 1690°C. The cooling effect desired depends on the transfer carbon *i.e.*, for 0.7-1.0% C it is 35°C and for 2%C or more it is 145°C. The exact amount of coolant depends on other factors like heat losses, cast temp. etc. Coolant quantity is usually little over 5% and excludes the reduction mix, fluxing addition and finishings.

Finishing additions

As the carbon is oxidised to the required level the bath is reduced by 75% Fe-Si. The amount of silicon added depends on the quantity of reducible oxides like Cr, Mn, Fe and the extent of deoxidation and alloying by silicon desired. However, with multistage process the silicon consumption for reduction is low. At the same time, lime is added to give lime silica ratio 1.5-2.0. This is flushed by the argon and stirring used throughout this period to desulphurise the bath to levels below 0.025 (<0.15%) but much depends upon the sulphur burden of the charge and how well the slag forms. If further desulphurization is required, the bulk of the reduction slag, after chromium recovery is removed and replaced by lime-spar mix, which after three minute stir lowers sulphur reliably to 0.007%. AOD offers a possibility of producing stainless steel with simultaneous low carbon and sulphur.

For production of ELC variety, care must be taken that lime is as fully burnt as possible (1% residual carbonate may lead to recarburization) as also Fe-Si should contain minimum fines. For very low nitrogen specification, nitrogen blowing at any stage of decarburization is forbidden (<.015%). For high proof stress steel however nitrogen is desired between 0.05-0.020%.

The addition in the final stage present no real problem. For titanium, selected titanium scrap is added in place of Fe-Ti and is added to vessel after slag removal before tapping, giving 70% yield consistently. AOD cycle time is 2h, half of which is consumed by decarburizing stages together with sampling, checking and addition of coolants.

The special alloys have proved to be well within the capability of AOD process. For instance, high nickel alloys decarburize so rapidly in presence of chromium (carbon removal starts well ahead of silicon removal and Cr losses are reduced) that it may seem undesirable to go for earlier practice of making 304 grade. In maraging stainless steels both silicon and manganese combined should not exceed 0.25%. Removal of manganese at such low levels even to an extent of 0.05% that too in presence of high chromium may be carried out in AOD vessel by appropriately modifying the different blowing stages.

High manganese nickel bearing 202 grade stainless steel is produced by modifying blowing technique which has reduced the problems of carbon removal at manganese levels of about 8%. All Mn and Cr unit are obtained from high carbon alloys. The nitrogen level of 202 grade (<0.20%) is again controlled by replacing argon by nitrogen in the final stage blowing of fully decarburized melt.

AOD Refractories

The main drawback of AOD process has been its poor refractory life. An AOD vessel gives 30-45 heats corresponding to 70-100h of treatment time. The principal refractories used have been direct bonded chrome-magesite bricks for the lower wear areas and high fired magnesite-chrome clinker bricks for high wear areas. In special wear areas *i.e.*, around tuyeres, rebonded fused grain magnesite chrome bricks have been found most resistant and direct bonded ones are least resistant to slag penetration. To minimise slag penetration it may be desirable to control the type and amount of slag. This may be achieved by replacing normal lime with dolomitic lime in various blowing stages. By this modification it has been reported that wear of lining at slag level has reduced considerably. The replacement of part of CaO by MgO has at the same time, shown no adverse effect on desulphurizing reactions.

In the tuyeres it has been observed that, besides erosion due to blowing, around the tuyere holes, knurdles are formed. These are nothing but out grown metal due to chilling effect of inert gas. These knurdles affect the direction of flow of the gases in side the convertor.

Advantages

1. Permit the refining of a larger quantity of melt and an increased gas blow rate because refining is carried out at atmospheric pressures and the vessel is pear shaped and side blown.
2. The decarburization rate is high and refining time is less for the given extent of decarburization.
3. There is no strict specification with regard to the carbon content of the melt taken for AOD refining.
4. Productivity is quite high.
5. Chromium yield is extremely satisfactory, 97-98%.

6. Guaranteed desulphurization can be carried out due to stirring under reducing atmosphere.

Limitations

1. Amount of gases (H, O, N) dissolved in the finished produced are slightly higher.
2. Short service life of the refractory lining.

This process is extensively used is USA for manufacturing high quality stainless and other chromium steels.

3.3.4. Vacuum - Oxygen - Decarburization (VOD)

This process has also been designed for the more economic production of stainless steels. The process is based on the fact that under vacuum and oxidising conditions decarburization is favoured against the oxidation of chromium. Therefore, this process is exclusively applied for production of stainless steel grades and other chromium-bearing steels. The reduced partial pressure of carbon-monoxide under vacuum shifts the Carbon-Oxygen equilibrium and kinetics favourably in the forward direction. The reaction of carbon with oxygen,

$$2\,[C] + \{O_2\} = 2\,\{CO\} \quad ...(14)$$

is more favourable in preference to,

$$4/3[Cr] + \{O_2\} = 2/3\,(Cr_2O_3) \quad ...(15)$$

and thus carbon content of the melt may be lowered to less than 0.02% in presence of about 15-18% Cr at about 1600°C. Since most of the stainless steels are even otherwise vacuum treated to bring down the amount of dissolved gases like H and N, such a process may be incorporated in the process design to produce extra low carbon stainless steels without excessive additional investment. The VOD process is carried out in following stages;

1. The melt charge for VOD is prepared either in LD or electric arc furnace.
2. The heat at required temperature is then poured into a preheated basic lined ladle equipped with porous plugs for blowing argon.
3. Oxygen preblowing at atmospheric pressure for preliminary decarburization and oxidation of silicon as well as Al, Ti and V if present.
4. Placing the ladle inside the vacuum chamber and evacuating the system to low pressures while oxygen is blown through lance above the bath of the steel and argon is blown through the porous plugs in the bottom of the ladle.
5. Further decarburization under additional vacuum time.

6. Deoxidation and vacuum treatment under reducing conditions.
7. Desulphurization.
8. Repressurizing to atmospheric pressure, taking samples for analysis.
9. If necessary, trimming additions in the form of ferroalloys.

A schematic diagram of the VOD unit is shown in Fig. 3.30.

One of the characteristics of this process is the independent vacuum chamber that may house various VOD ladles which are basic lined and equipped with slide gate, porous plug, ladle cover

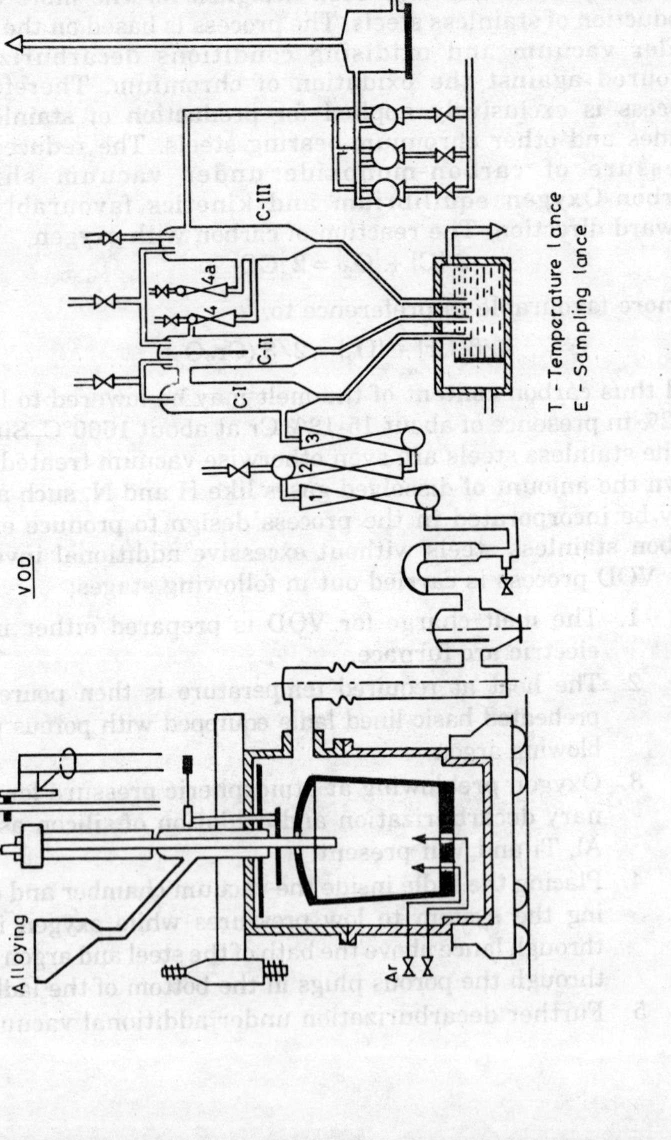

Fig. 3.30. A Schematic diagram of VOD unit.

and feature a ladle free-board of 1 metre. The freeboard is approximately 25-30% of the total capacity of the ladle to contain voilent stirring of bath during lancing.

The melting and preliminary refining of the charge of chromium and nickel to specification and carbon around 1% are carried out in a usual arc furnace. When carbon is reduced to about 0.6 to 0.4% the melt is transferred to preheated basic lined VOD ladle. A lance, fed through the cover, blows oxygen on to the bath surface, initially at atmospheric pressure and then under vacuum.

In step four the oxygen is blown through a lance under vacuum on to the metal surface. The blowing rates may vary between 500 and 2000 Nm^3/hr but most of the ladle VOD plants operate at 600 to 1000 Nm^3/h. The first and most important aim of the VOD process is decarburization with high yield of chromium. For this there has to be a precise control of the process by observing pressure, waste gas analysis and temperature of the bath.

Final carbon removal and simultaneous deoxidation take place in step four, during the formation of CO, at a vacuum in the range of 20 to 1 mbar. The oxygen dissolved in the melt at the end of blowing is consumed and some oxygen is recovered from the slag thus leading to some recovery of chromium. Obviously, such an inclusion free deoxidation and a chromium recovery can not be achieved in AOD. To the contrary, the chromium content in AOD decreases if low carbon and nitrogen are required. This decisive difference between various processes is shown in Fig. 3.31.

The removal of hydrogen, lead and zinc may be considered as a byproduct of the VOD process. Final deoxidation and desulphurization are achieved in step 5, 6 and 7 with best reducing condition prevailing. The vacuum is broken to add deoxidisers like Al and Fe-Si and desulphurization slag {$CaO-CaF_2$}. The vacuum is

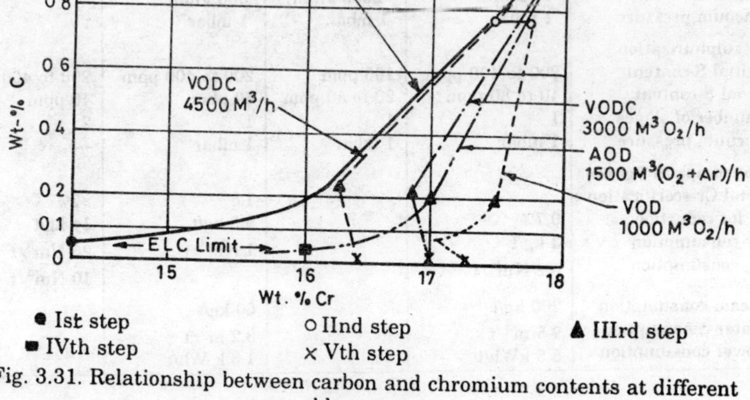

Fig. 3.31. Relationship between carbon and chromium contents at different blow rates.

further applied for about 20-35 min. The total VOD time varies between 2–2½ hrs. Oxygen content below 100 ppm and sulphur content below 50 ppm are results of standard VOD practice for stainless steel grades. For extra low nitrogen content, required in ferritic grades, special technique was developed in Japan, called strong stirred VOD process (SSVOD). In conventional VOD process, the treatment time of the melt increases with increase in charge size due to limited blow rate of 2500 $Nm^3 O_2/h$ because of reasonable free board. Application of VOD technique in a converter makes it possible to higher blowing rates of the order of 4500 $Nm^3 O_2/h$ which provides favourable condition for higher chromium recovery at more intensive argon stirring is termed as VODC.

Some characteristic operational data and results of VOD, SSVOD, VODC and also AOD are compiled in Table II, for 50 ton heats.

Table II : Comparison of characteristic operational data for the VOD. SS VOD, VODC and AOD processes for 50t heats

Example	*VOD 304L*	*SSVOD 18/Cr MO*	*VODC 304*	*AOD 304*
Preblow				
Initial C-content	0.8 to 1.5%	1.0 to 3%	1.0 to 3%	1 to 3%
Final C-content	0.6 to 0.8%	0.8 to 2%	0.2 to 0.3%	0.02 to 0.08%
Oxygen blow rate	2000 m^3/h	3000 m^3/h	4000 m^3/h	1500 m^3/h
Argon blow rate			500 1/Min	
Vacuum oxidising				
Initial C-content	0.6 to 0.8%	0.8-2.0%		
Final C-content.	0.04 to 0.08%	0.01 to 0.02%		
Oxygen blow rate	1000 m^3/h	1500 m^3/h		
Argon blow rate	30 1/min	1000 l/min		
Argon purity	99%	99.999%	99%	99%
Vacuum pressure	50 mbar	50 mbar		
Vacuum reducing				
Final C-content	100 to 300 ppm	3 to 10 ppm	150 to 200 ppm	
Final N-content	200 to 300 ppm	20 to 30 ppm	100 to 200 ppm	
Final O-content	100 ppm	100 ppm	100 ppm	
Argon blow rate	60 l/min	2000 l/min	500 l/min	
Vacuum pressure	1 mpar	1 mbar	1 mbar	
Desulphurization				
Initial S-content	200 to 400 ppm	100 ppm	200 to 400 ppm	200 to 400 ppm
Final S-content	40 to 80 ppm	20 to 50 ppm	40 ppm	40 ppm
Number of slages	1	1	1	2
Vacuum pressure	1 mbar	1 mbar	1 mbar	—
Consumption figures				
Total Cr-scorification			1.5%	>2%
Si for reduction	0.7%		9.5 kg/t	14 kg/t
Ar consumption	3 kg/t		1 Nm^3/t	20 Nm^3/t
N_2 consumption	0.2 Nm^3/t		—	10 Nm^3/t
Steam consumption	200 kg/t		60 kg/t	—
Water consumption	9.5 m^3/t		3.2 m^3/t	—
Power consumption	5.5 kWh/t		1.8 k Wh/t	—

Advantages

1. Gas content of the steel can be reduced to very low level (oxygen 30-80 ppm, Nitrogen 40-140 ppm and hydrogen 1-3 ppm).
2. Efficiency of carbon removal is high.

Limitation

1. Expansive vacuum system installation.
2. Charge material carbon content to be within 0.4-0.6%.

3.3.5. CLU Process

The process is similar to AOD process with a difference that hydrogen derived through dissociation of blown steam replaces argon as the inert constituent. The converter used in this process is similar to the one used in AOD process. However, the CLU converter tuyeres consist of concentric tubes. The gas mixtures are independently controlled in the inner tube and orifice. The CLU process was jointly developed by Creusot Loire in France and Uddeholm in Sweden and thus called CLU. In this process the refining gases are oxygen, and steam (super heated). The two gases are introduced together with nitrogen through tuyeres in the bottom of the vessel. The reaction between oxygen and carbon is;

$$\left\{\frac{1}{2}O_2\right\} + C = \{CO\} \qquad ...(16)$$

and the reaction between stream and carbon is;

$$\{H_2O\} + C \longrightarrow \{CO\} + \{H_2\} \qquad ...(17)$$

The hydrogen gas produced through the above reaction will together with nitrogen, act as a diluent for carbon-monoxide produced by above reaction. This will create a low partial pressure for CO which will favour above reaction and counteract oxidation of chromium through the reactions

$$\frac{3}{2}O_2\,(g) + 2\,Cr \longrightarrow Cr_2\,O_3 \qquad ...(18)$$

and

$$2H_2O + 2Cr \longrightarrow Cr_2O_3 + 3H_2 \qquad ...(19)$$

Throughout the decarburization period a balance exists between oxidation of carbon and oxidation of chromium. In order to avoid excessive chromium oxidation the gas mixture between O_2, H_2O and N_2 is changed in order to decrease the partial pressure of carbon oxidation. Superheated steam has another important function, in CLU process, as a temperature controller.

Reaction (17) is strongly endothermic and the temperature can therefore be controlled accurately by the gas mixture to balance reaction (17) against reactions 16, 18 and 19. This is done in a manner that temperature in the range 1660–1680°C is maintained during the carbon removal. At this temperature range no restriction on the initial concentration of carbon and silicon exists. When decarburization is complete, combined deoxidation and hydrogen removal period starts. Lime and ferrosilicon are added to reduce Cr_2O_3 and MnO and the melt is flushed with argon and nitrogen. During reduction period in CLU all the arc furnace slag is reduced which gives highest possible yield of chromium and manganese. Silicon or chromium at low silicon concentration is oxidised at the beginning of the cycle to raise the temperature to 1670°C and during this period between 5 to 15% of the charge weight is normally added as scrap and alloys. During the carbon refining no solid addition are made and the temperature is maintained by gas mixture alone. Finally, the reduction and hydrogen removal periods complete the heat cycle. Fig. 3.32 represents a typical CLU heat in which variation of temperature with progress of the process has been shown. The performance of CLU converter in terms of manganese and chromium recovery has been studied under following two conditions :

(*i*) Melt from EAF transferred to CLU converter with slag and melt Si-0.5%.

(*ii*) Melt transferred to CLU converter after removal of slag with melt Si-0.25%.

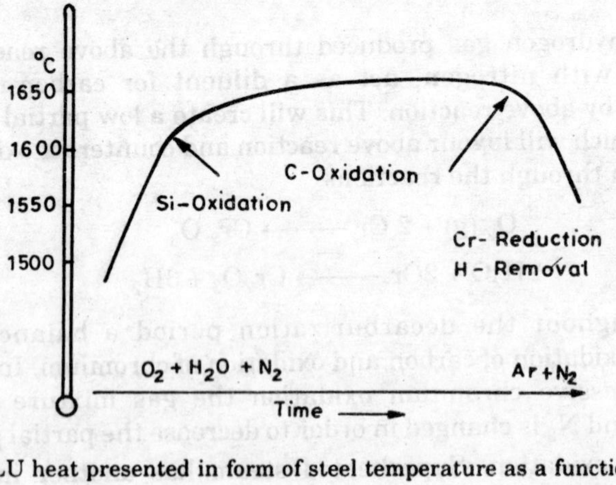

Fig. 3.32. CLU heat presented in form of steel temperature as a function of time.

The result obtained in case (*i*) with respect to Mn and Cr yield are shown in tables III and IV and in case (*ii*) are shown in Figs. 3.33 and 3.34.

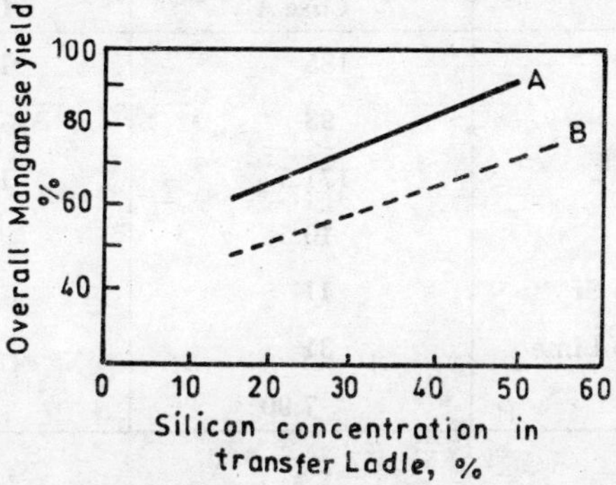

Fig. 3.33. Overall manganese yield as function of silicon concentration in transfer ladle.

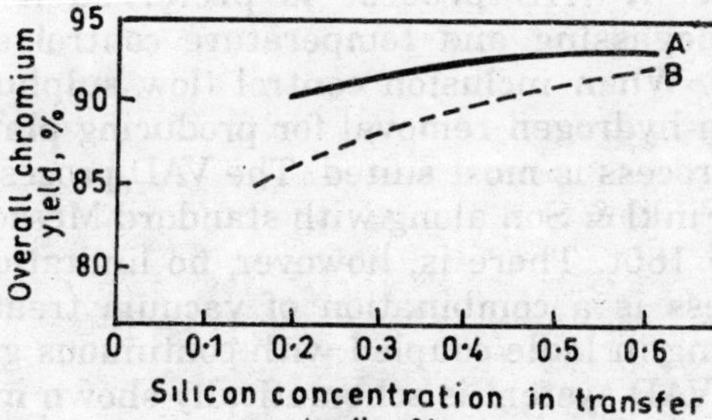

Fig. 3.34. Overall chromium yield as function of silicon concentration in transfer ladle.

Table III : Cost savings by increased manganese yield in case A for remelting of 1000 kg grade AISI

	Case A	*Case B*
Charged, Kg	16.5	16.5
Overall yield	90	55
To steel, Kg	14.9	9.1
Material saving, Kg	5.8	—
Extra consumption Lb Si 3.3		
Extra consumption, lb Lime 9		
Cost saving, USs	2.70	

Table IV : Cost savings by increased chromium yield in case A for remelting of 1000 kg grade 304

	Case A	*Case B*
Charged, kg	185	185
Overall yield, %	93	87
To steel, kg	171.5	161
Material saving, kg	10.5	—
Extra consumption, lb Si	11	
Extra consumption, lb Lime	31	
Cost saving USs	7.90	

3.3.6. Vacuum Arc Degassing (VAD) Process

Each secondary refining process has been developed for a specific purpose. A VAD process is preferred if the main requirement is degassing and temperature control along with sulphur removal. When inclusion control (low sulphur level) is needed alongwith hydrogen removal for producing plates, spring steels etc., this process is most suited. The VAD process was first developed by A. Finkl & Son along with standard Messo to handle heat size of upto 180t. There is, however, no limitation to large heats. The process is a combination of vacuum treatment and electric arc heating in ladle coupled with continuous gas stirring of the melt. The VAD system is schematically shown in Fig. 3.35.

The aims of VAD process are listed below ;

1. Arc heating under vacuum.
2. Vacuum treatment for dehydrogenization, denitriding and decarburization.
3. Alloying and composition control with high yield of alloying elements.
4. Desulphurization.
5. Improvement of micro-cleanliness.

Before the heat is taken to VAD ladle its temperature is usually about 1650°C. The steel is first tapped in the ladle equipped with a porous plug and then taken to a vacuum chamber which is equipped with electric arc heating. A slag-forming synthetic mixture is added at the top of the ladle and fused by the heat of arc, as the pressure within the chamber is lowered. After the slag is fused the heating is stopped and pressure inside the chamber in further lowered, preferably to less than 1 torr. The combination of vacuum and stirring gives rise to vigorous agitation, causing rapid mixing of the steel and slag resulting in

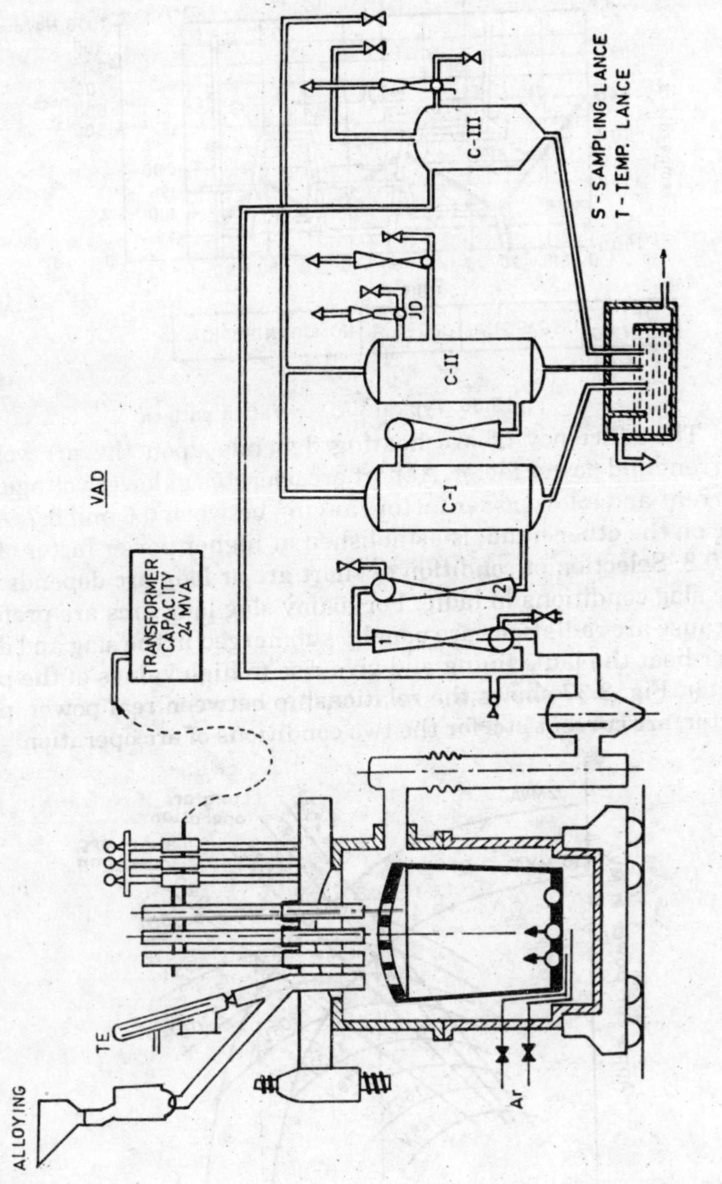

Fig. 3.35. A VAD unit system shown schematically.

effective desulphurization along with gas removal. This facility for arc heating enables perfect control of temperature all through the operation. A typical VAD operation pattern is shown in Fig. 3.36.

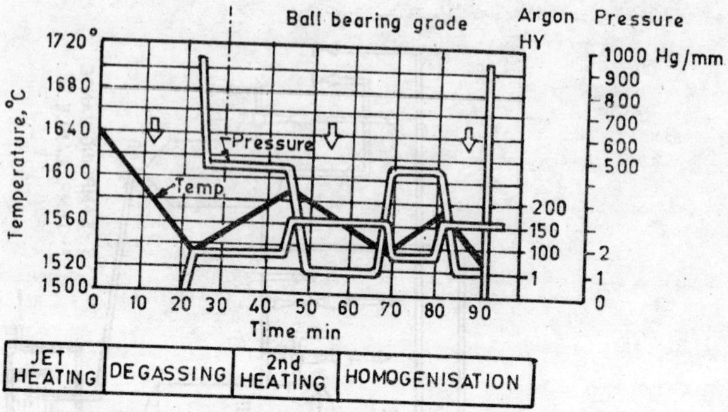

Fig. 3.36. Typical VAD operation pattern.

The efficiency of arc heating depends upon the arc voltage, current and power factor. A short arc operates at lower voltage, high current and a low power factor ranging between 0.6 and 0.7. A long arc on the other hand, is established at higher power factor of 0.75 to 0.8. Selection of condition of short arc or long arc depends upon the slag conditions in ladle. For foamy slag long arcs are preferred because arc radiations are usually submerged in the slag and do not over-heat the ladle lining and give rise to high values of the power factor. Fig. 3.37 shows the relationship between real power, power factor, arc current etc. for the two conditions of arc operation.

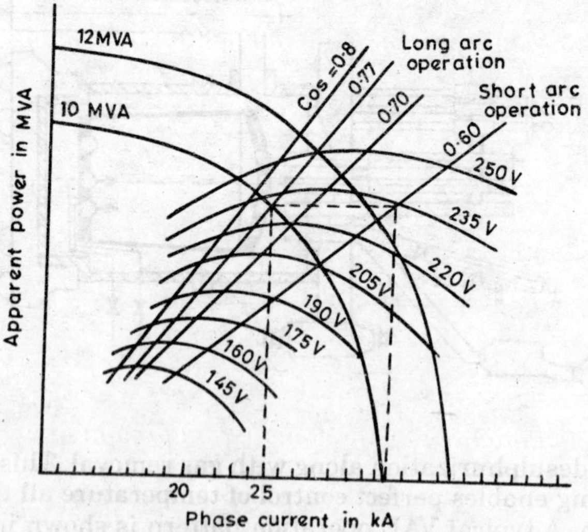

Fig. 3.37. Real electric power of VAD.

In general the long arc operations require low diameter electrodes where as for short arc operation large electrode diameter is must if high current values are to be used.

During the process campaign all through inert gas is used for bath circulation. Today, for purging, porous bricks are successfully used at the bottom of the ladle. The porous refractory used should have high uniformly distributed open porosity to allow a large volume of gas to flow at the lowest pressure. Argon stirring has a definite advantage over induction stirring because of the simplicity in controlling the stirring intensity. During heating period, stirring provides conditions of fast homogenization, prevents nitrogen, hydrogen and Carbon pickup and loss of alloying elements by oxidation. On the other hand it promotes deoxidation and decarburization reactions. Vigorous stirring is also necessary to achieve quick homogenization after alloying. The effectiveness of argon stirring depends upon the vacuum level. Fig. 3.38 shows the effect of pressure on the stirring energy of argon.

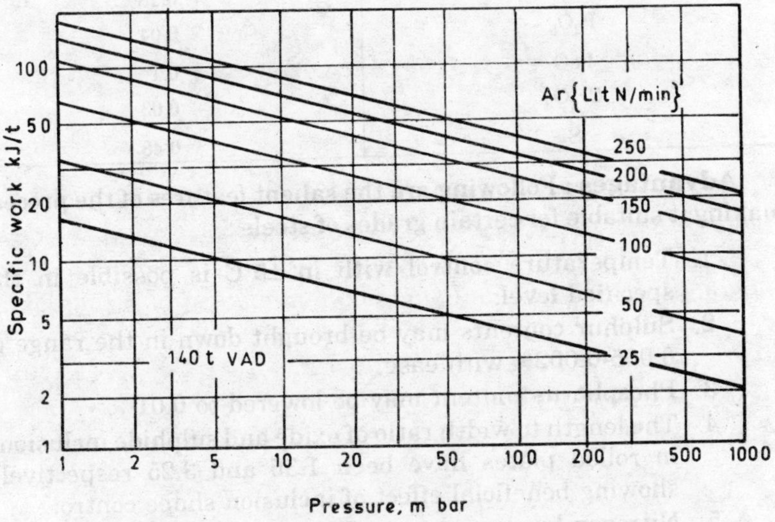

Fig. 3.38. Specific stirring work in a VAD ladle under vacuum.

To achieve desired results in the process it is necessary to exercise control on heating. High heating rates result in excessive fluid slag which causes significant loss of refractory lining along the slag line. On the other hand at low temperatures there are chances of carbon pickup by the melt from the electrodes due to short arc length. It has been reported that for improved performance of the ladle with intact heat shield and proper arc stability without glow discharge the pressure should be atleast 500 mbar.

The refractories employed in VAD vessel are selected for different regions on the basis of the conditions prevalent. For bottom and walls high alumina bricks and for slag line dolomite bricks are invariably used. For calcium treated steels bottom is of high alumina but walls are made of dolomite including slag line.

Fluxing agents used in this process are primarily lime-flourspar-bauxite. The typical slag analysis after treatment is given below in Table V.

Table V : Typical slag analysis

Constituent	*Wt%*
CaO	57.8
MgO	3.8
SiO_2	8.0
Al_2O_3	27.7
MnO	0.18
P_2O_5	0.04
FeO	0.1
Cr_2O_3	0.03
S	0.48

Advantages : Following are the salient features of the process making it suitable for certain grades of steels :

1. Temperature control with in ±5°C is possible in the specified level.
2. Sulphur contents may be brought down in the range of 0.005-0.002% with ease.
3. Phosphorus content may be lowered to 0.01%.
4. The length to width ratio of oxide and sulphide inclusions in rolled plates have been 1.35 and 3.25 respectively showing beneficial effect of inclusion shape control.
5. Nitrogen levels as low as 0.007% can be achieved.
6. Hydrogen levels below 2ppm can be achieved.

PART - III

3.4. LADLE FURNACE

The first ladle arc heating furnace was installed by Asea Metallurgy at SKF stahl in Hallefors, Sweden in 1965. The ladle furnace permits heating of melt in ladle after tapping from the electric arc. Ladle furnace may act as a holding vessel to be taken to continuous casting machine as well as further metallurgical treatments such as;

(*i*) Initial alloying
(*ii*) Desulphurization
(*iii*) Purging to improve the internal quality.

Ladle furnaces lead to decrease in cost of production, improve quality, product range and introduce process route flexibility.

Originally, a redundant EAF was converted into a ladle furnace to minimise capital costs and to salvage the equipments. It incorporated the upper part of the arc furnace and the shell was replaced by ladle with basic lining and slide gate valve. The ladle furnace used now is shown in Fig. 3.39. It features a new design taking into consideration optimum material flow and increased process flexibility as well as product range. The furnace roof is provided with openings for three electrodes, alloy additions, injection devices, fume evacuation, sampling and temperature measurement.

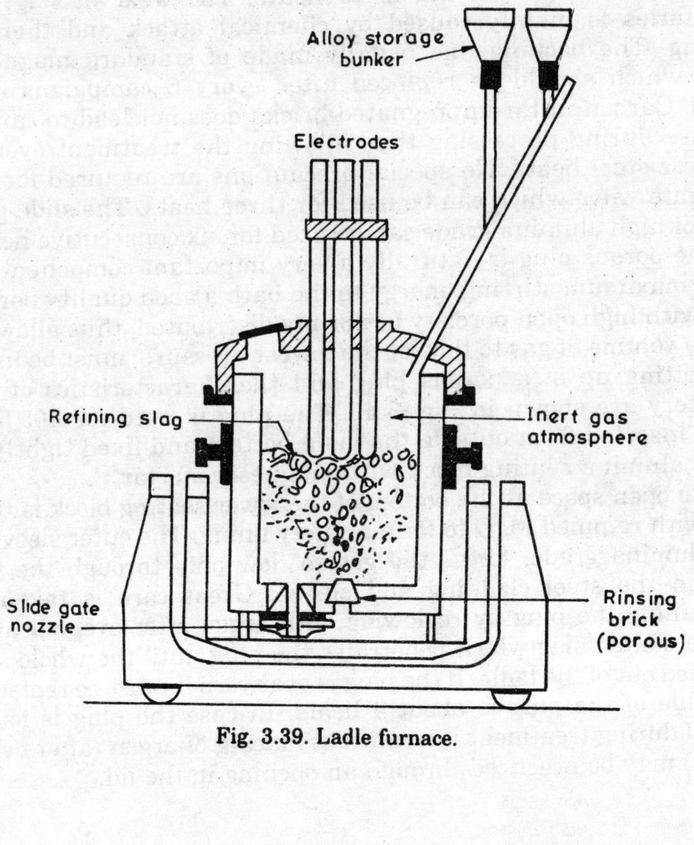

Fig. 3.39. Ladle furnace.

A supporting structure permits the furnace roof to be hydraulically lifted before the ladle is moved in. The roof also seals the ladle to retain the non-oxidising atmosphere. The traditional refractory roofs have been replaced by roof with tubular water cooled panels. Though the heat losses due to water cooled roof are 1.5 times more than the conventional one but it is more than compensated by the improved roof life. Ladle furnaces are normally heated with longer arcs with 100-140 mm thick layer of slag. This practice is generally called as submerged arc heating. The shorter arc length accompanies the risk of carbon pickup from the electrodes.

It has been observed that 70% of the operational costs are incurred by refractories (wear and safety lining), slide gate valve and porous plug. Generally several courses around the slag line are lined with basic materials and rest of the wear line with 87% alumina bricks. The mean life of the wear lining excluding slag line refractories ranges between 35 to 48 heats. The life of slag line refractories is comparatively shorter. For rebond Mg-Cr bricks the life is about 20 heats. For synthetic tar impregnated Mg-dolomite bricks the life is 15 heats. The wear of slag line refractories is mainly caused by chemical attack and thermal spalling. The backing line usually made of standard magnesia grade which should be replaced after every 6 campaigns (240 heats). Carbon in tar-impregnated bricks does not lead to carbon increase during processing thus allowing the treatment even of stainless steel heats. No special precautions are required for the slide gate valve which can be used for three heats. The slide-gate valve of high alumina grade can be used for six consecutive heats.

The porous plug is naturally a very important component. To give a maximum stirring energy to the bath a good quality porous plug, with high open porosity uniformly distributed, thus allowing a large volume of gas to flow at the lowest pressure, must be used. The setting up of a porous plug and the characteristics of the refractory are shown in Fig. 3.40. The plug is used for 40t heat and is inserted from outside the ladle bottom and fixed tightly in a high alumina seating block with magnesia mortar.

The open space at the bottom of the lower seating block is then filled with rammed MgO to form a safety lining. The outer sleeve of high-alumina grade, forces the gas to flow only through the gap between the steel cladding and sleeve. Great care is taken in maintaining the plug by reblowing with argon after every heat to remove molten slag which penetrates the plug until the whole slag is flushed out of the ladle. If the plug is overworn it must be replaced. Mean life of the plug is about 4 heats. In case the plug is found blocked during treatment intermittent lancing of argon (after every 10 min) may be practiced through an opening in the lid.

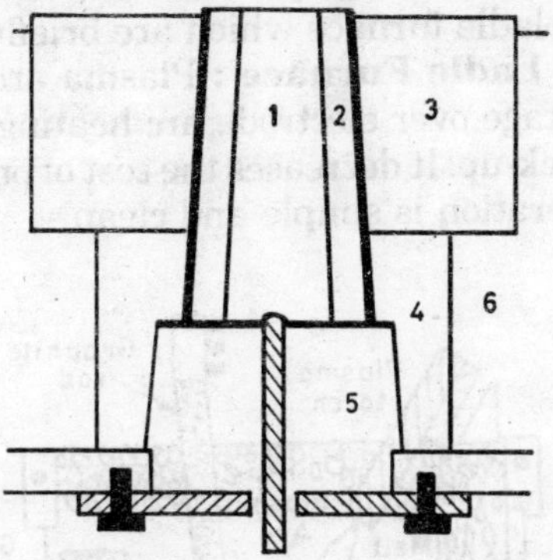

Fig. 3.40. Setting up of porous plug (DAIDO).

The energy efficiency of LF is higher than arc furnace as the arc is completely submerged in the slag. The slag is usually superheated by 30-50°C without causing any wear of slag line. Inspite of dipping electrodes in slag, carbon increase during heating is negligible, about 1 ppm/min for ELC stainless steel.

The reducing condition in LF obtained by complete sealing of the furnace and the oxygen left reacts with electrodes to form CO. Electrode carbon also reacts with oxides like MnO, FeO and Cr_2O_3 in slag and increase the recovery of elements. The amount of various elements pick-up, sulphur removal and metal losses are given below:

Carbon pick-up :	1-1.5 ppm/min of heating with pneumatic stirring and 0.6-1.0 ppm/min with electromagnetic stirring.
Desulphurization :	65-85% after 30 minutes from all initial sulphur content of 0.030%.
Total Oxygen content :	10-50 ppm, depending upon the steel grades, deoxidants used and ladle-roof airtightness.
Nitrogen pick-up :	Less than 5 ppm
Hydrogen pick-up :	Less than 1ppm, depending upon the slag deoxidation and initial hydrogen content.
Phosphorus pick-up :	5-50 ppm depending upon the slag type and amount in the ladle.
A1 and Si losses :	For reheating time of 15 min, 0.005% Al and 0.02% Si.

Besides electric arc heating system, other heating systems have also been tried with ladle furnace which are briefly discussed below :

(*i*) **Plasma arc Ladle Furnace :** Plasma arc heating shown in Fig. 3.41 has advantage over electrode arc heating in that it does not give rise to carbon pick-up. It decreases the cost of production, increases productivity and operation is simple and clean.

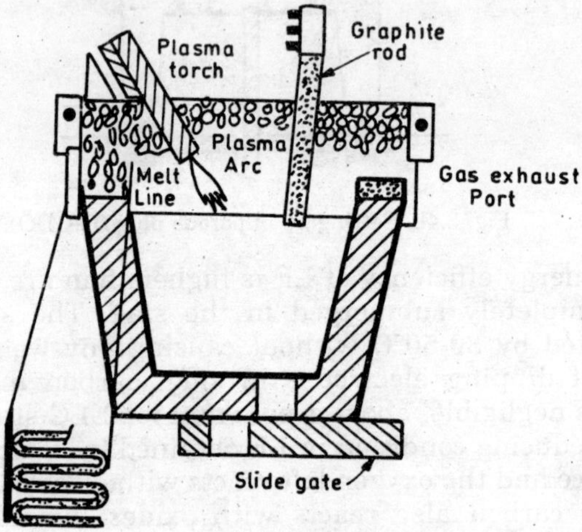

Fig. 3.41. Plasma ladle preheater.

(*ii*) **Oxy-fuel Ladle Furnace :** Combustion of fuel with pure oxygen may result in intense heat production and efficient heat transfer to metal at 1600 °C. This type of ladle furnace is shown in Fig. 3.42.

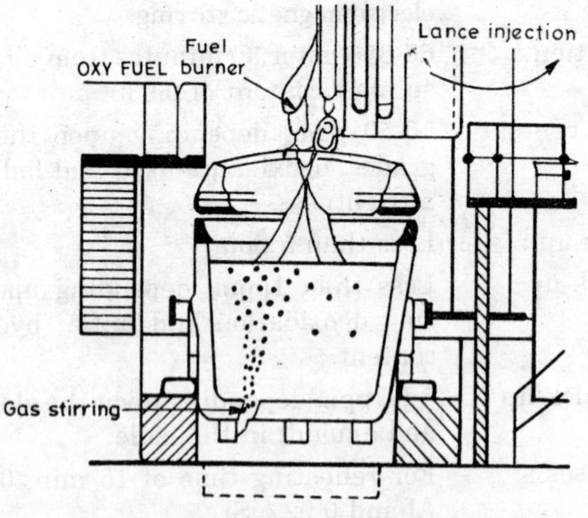

Fig. 3.42. Oxy-fuel ladle furnace.

(*iii*) **DC arc Furnace :** The DC arc furnace is shown in Fig. 3.43. The magnetic forces in the bath from the dc current flow cause the molten metal and slag to move downwards in the centre of the ladle below the dc arc and rise along the sidewalls. In these furnaces lower electrode consumption and refractory wear have been reported.

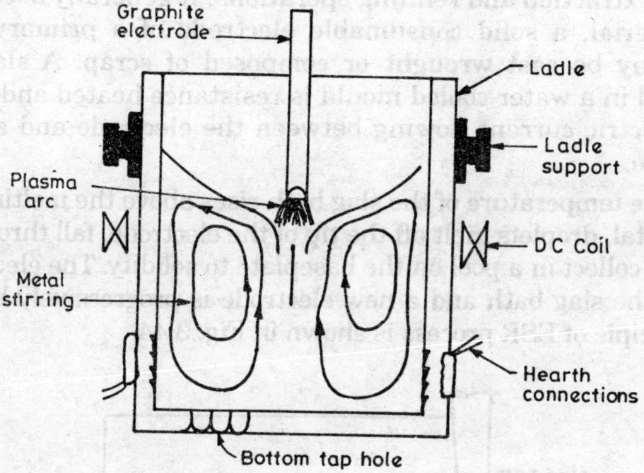

Fig. 3.43. ASEA dc arc ladle heater.

Some other systems incorporated in the ladle furnace are listed below :

(*i*) Oxygen blowing reheating
(*ii*) Induction ladle furnace
(*iii*) Argon injection through drilled graphite electrodes.

Advantages :

(*i*) Lower production cost, fewer rejects, increase in ferroalloy yield, savings in electrode and electric power for EAF.

(*ii*) Achieving steel analysis in narrow limits, good dissolution and homogenization of alloy additions and temperature equalization.

(*iii*) Quality improvement by modification of nonmetallic inclusions and enchanced steel cleanliness.

PART - IV

3.5. Electro-slag Refining Technology (ESRT)

3.5.1. General

Electroslag refining (ESR) is a secondary refining process for metals. It is used for further purification after completion of primary extraction and refining operations. It generally used, as its raw material, a solid consumable electrode of a primary metal which may be cast wrought or composed of scrap. A slag bath contained in a water cooled mould is resistance heated and melted by an electric current flowing between the electrode and a cooled base plate.

As the temperature of the slag bath rises above the melting point of the metal, droplets melt off the tip of the electrode, fall through the slag, and collect in a pool on the baseplate to solidity. The electrode is fed into the slag bath and a new electrode is progressively built up. The principle of ESR process is shown in Fig. 3.44.

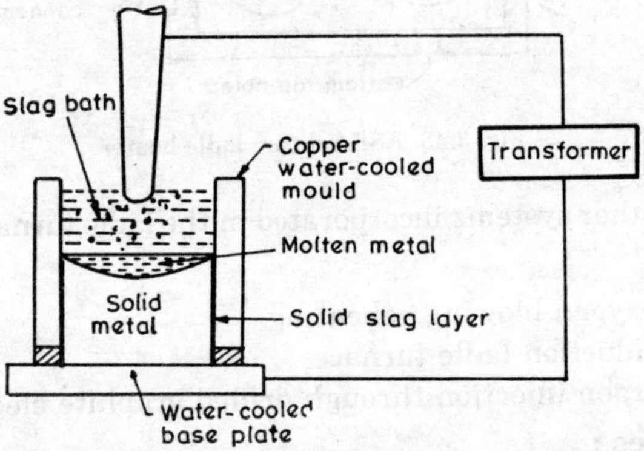

Fig. 3.44. Principle of ESR process (schematic).

In this process the molten slag is continuously displaced in the upward direction. Where the rising slag meets the cooled copper mould some solidification takes place, ensuring the presence of a continuous shell of solid slag between the mould and solidifying ingot. The shell remains intact after the steel is melted and forms a smooth lining to the mould. On stripping it flakes off and leaves an excellent ingot surface.

Refining of steel occurs because of the reaction between the metal and slag in three stages :

(*i*) During formation of a droplet at the electrode tip.

(ii) As the detached droplet falls through slag.

(iii) As the molten metal collects in the pool of metal below slag layer.

Refining metal and overall soundness of the ingot obtained may be ascertained by suitable choice of slags.

Advantages of ESR :

(i) Production of sound ingot with complete absence of pipe and porosity.

(ii) Much cleaner product with fewer and smaller inclusions.

(iii) Uniformity of structure and composition.

(iv) Elimination of zonal segregation.

(v) Controlled reduction of undesirable elements like S, O, N while retaining necessary alloying elements.

(vi) Ability to correct out-of specification electrodes by proper flux chemistry.

(vii) General improvement in ductility and impact transition characteristics.

(viii) Generally improved transverse characteristics.

(ix) Improvement in elevated temperature characteristics.

(x) Improvement in weldability.

(xi) Attainment of smooth surface of ingot thus eliminating the need for surface dressing before hotworking.

(xii) Superior hotworking characteristics.

(xiii) Control over direction and rate of solidification.

(xiv) Control of grain size.

(xv) Improved corrosion resistance.

(xvi) Protection of molten metal from atmospheric oxidation.

Electroslag refining has a very large number of degrees of freedom available in the process and this is particularly suitable as a modern secondary refining process. The principal reasons for the greater number of degrees of freedom of ESR are the large number of available combinations of slag chemistry, the more flexible power requirements and greater freedom of choice of electrode characteristics.

3.5.2. Slag Characteristics

The main requirements of a slag are that it should be molten at a temperature preferably slightly lower than that of the metal to be refined and that it should be stable at the temperature at which the process operates. It must, of course, be conductive electrically. Other important criteria are those of thermal conductivity, thermal capacity, viscosity and surface tension. The volatility of the slag should preferably be low. Reactivity of the slag must be appropriate for the degree of refining and control of chemical composition required.

The most usual choice of constituents is between CaF_2, lime, magnesia, alumina and Silica, but it is also possible to add MgF_2, BaF_3, TiO_2 and ZrO_2. By altering the amount of these constituents one can alter the melting point, the stability, the conductivity and the viscosity of the slag, and these in turn can alter the melting rate, the slag temperature, the droplet size and other factors affecting the refining rate.

The two more important functions of the slag layer are to generate heat and to control the chemical composition. The slag layer is the element of the electrical circuit between the transformer and the furnace which has the greatest resistivity and hence is the part of the circuit most responsible for conversion of electrical energy into heat. If the slag has insufficient resistivity at the required operating temperatures it will be very difficult to maintain satisfactory melting conditions.

The slag is the medium in which inclusions in the metal are removed by chemical reaction or dissolution and where the necessary adjustment of O, S and other elements is achieved.

If the slag is insufficiently reactive, then adequate cleanliness would not be obtained. If it is too reactive, elements which should be retained shall be removed. Other functions of slag :

1. By freezing on the mould wall it prevents direct contact between molten metal and the mould material thus isolating the metal thermally and electrically from the mould and also ensures smooth ingot surfaces.
2. It protects the molten metal from direct exposure to atmosphere.
3. It provides a heat reservoir at the top of the molten metal and thus prevents piping and internal shrinkage.
4. It helps achieving required ingot structure by controlling the condition of solidification.
5. Helps controlling grain size and shape.

For melting to occur in the ESR process it is necessary that the slag be wholly liquid and be at a temperature in excess of the melting point of the metal being refined.

The temperature attained in the slag depends upon :

1. Resistance of slag.
2. Current passing through the slag bath.

The voltage (20-50 V) is required to derive the current. Too high a voltage may cause arcing and prove detrimental to successful refining by promoting high oxidation losses from the metal.

The temperature attained in the slag bath, however, is not only dependent on the rate at which the heat is generated but also on the rate at which the heat is extracted. The heat is extracted by

1. Conduction 2. Convection
3. Radiation 4. Latent heat of fusion of metal

Of these, conduction losses, to the cooling water via the mould walls and the base plate, are the most important.

The slag constituents must be stable at the process operating temperature and they are therefore selected primarily on account of the low vapour pressure. They should refine the metal and at the same time they should not contaminate the metal in any way. Slags having melting point at, below or above the melting point of the metal can be used. The choice directly affects the economy of the process. Other characteristics considered for final selection are the physical properties; vapour pressure, electrical conductivity, sp. heat, latent heat of fusion, heat of solution, viscosity, surface tension and density. For proper selection of slag constituents it is necessary that the chemical and physical characteristics of the slag be known correctly and for this various phases present in the slag should be known.

The most important slag constituents are calcium fluoride, lime, magnesia and alumina. Other important constituents include the fluorides of Ba, Mg and oxides of Zr and Ti. Chlorides have also been used in case of low melting metals and alloys. The most important slag system comprise of fluoride-oxide mixtures. These slags contain CaF_2, CaO- & Al_2O_3 and CaF_2, MgO and Al_2O_3.

Flourides alone or in combination can be used in electroslag refining process. Fluoride slags are oxygen free, have low m.p. and have low electrical resistance thus give rise to low operating temperatures. CaF_2, MgF_2, and BaF_2 because of their low vapour pressure and high temperature stability are the primary components of fluoride slag. Of these CaF_2 is most widely used since it has high resistivity, less hygroscopic and cheaper of the three. Pyrohydrolysis of the three yields their respective oxides in varying magnitude, maximum being in case of BaF_2 (giving BaO).

NaF, AlF_3 and cryolite (Na_3AlF_6) are generally unsatisfactory, as they boil and evolve copious fumes at the temperatures commonly experienced during the process.

CaF_2 : Calcium fluoride or fluorspar is the most important constituent of fluoride slags. Calcium fluoride used commercially is usually acid grade, low sulphur fluorspar concentrate. A typical analysis of which is given below;

CaF_2 98% SiO_2 11, CaO 0.3%, Al_2O_3 0.2% S 0.1% and Pb 0.1%.

This can be purified of H, S, Pb and some silica by prefusion. The removal of hydrogen is associated with an increase in lime (CaO) to about 2% which can in certain cases be as high as 5%. The reaction between CaF_2 and moisture takes place as per the following reaction;

$$CaF_2 + H_2O \rightleftharpoons 2HF + CaO.$$

The lime formed by pyrohydrolysis is sufficient to give desulphurisation when metal or alloy is electroslag refined through prefused fluorspar. M.P. of pure fluorspar is 1418°C but commercial grade has M.P. of about 1380°C. As little as 0.5 Wt% lime (CaO) is sufficient to lower the liquidus temperature to 1380°C.

Fluorspar when used alone devoids heating the slag bath to a high enough temperature as it has a low electrical resistance (0.23 ohm. Cm. at 1650°C). However, when considering the fluoride slags only, CaF_2 slags are easiest to work. Metal, refined through CaF_2 slag are generally the following;

1. Pure Iron
2. Mild steel
3. Ball bearing steel
4. 18/8, Ti stabilised stainless steel

The characteristics of different fluorides are briefly discussed below ;

BaF_2 : High vapour pressure, low M.P. (1290°C)

It acts as an arc stabiliser in arc melting.

It is electrically stable in ESR.

It is Toxic, hygroscopic and soluble in metal.

MgF_2 : High vapour pressure, low M.P. (1260)

Lower thermal stability

It is difficult to work in ESR

MgF_2 content should not exceed 20-30% in fluoride slags.

It is hygroscopic (water of crystallization up to 15-10%).

Pyrohydrolysis will yield MgO which is less effective than lime for removing sulphur.

Binary Fluorides :

CaF_2–MgF_2

Eutectic composition contain 51% MgF_2 and eutectic temperature is 945°C. Slags of eutectic composition have been used for melting copper and its alloys with limited success which has been attributed to high MgF_2 content and it has therefore, been recommended that slags should not contain more than 20-30% MgF_2. The binary system of CaF_2 and MgF_2 is shown in Fig. 3.45.

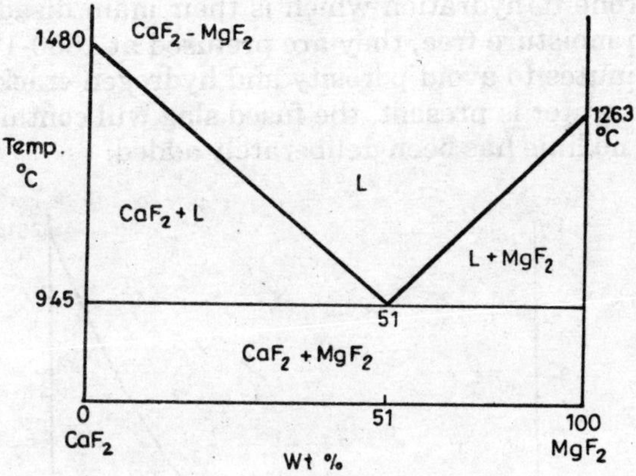

Fig. 3.45

Ternary Fluorides :

CaF_2–BaF_2–MgF_2

The pseudo-ternary phase diagram of this system is shown in Fig. 3.46.

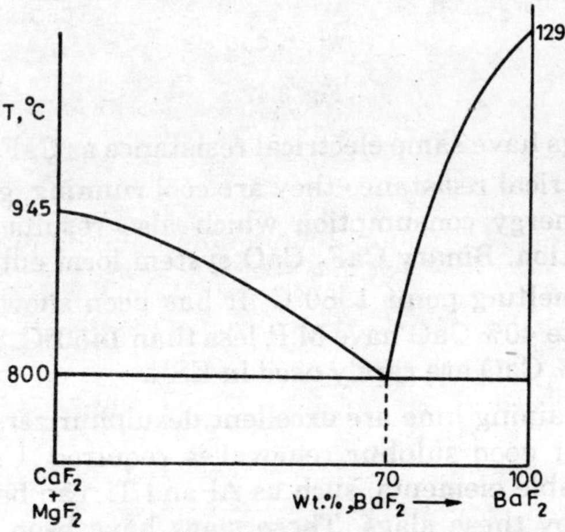

Fig. 3.46

Composition consisting of 70% BaF_2, 15% MgF_2, and 15% CaF_2 with a m.p. of about 800°C have been used successfully for melting of copper and its alloys *e.g.* manganese bronze.

Fluoride-Oxide systems

CaF_2–CaO

The phase diagram the system CaF_2-CaO is shown in Fig. 3.47. These slags are mostly used in ESR having their major components as CaF_2 and CaO. These slags are cheaper. When having free lime they are prone to hydration which is their main disadvantage. To make them moisture free, they are prefused at 1650-1750°C for at least 10 minutes to avoid porosity and hydrogen cracking. During prefusion if water is present, the fused slag will contain 2-5% CaO even when no lime has been deliberately added.

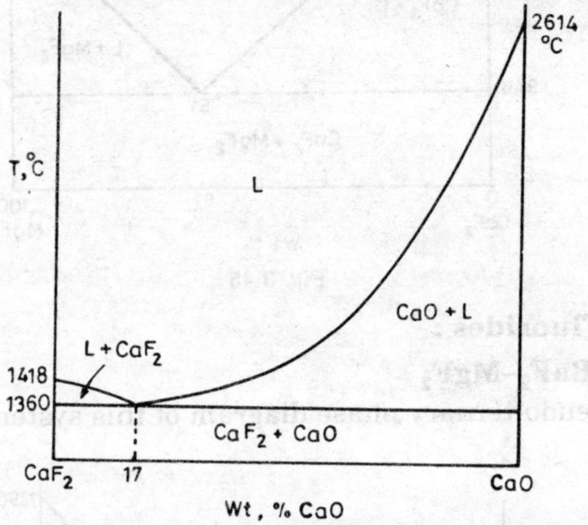

Fig. 3.47

These slags have same electrical resistance as CaF_2. Because of their low electrical resistance they are cool running, give low melt rates, high energy consumption which also results in unstable melting condition. Binary CaF_2-CaO system form eutectic at 17% CaO having melting point 1360°C. It has been shown that slags containing upto 40% CaO have M.P. less than 1450°C. Slags having more than 40% CaO are rarely used in ESR.

Slags containing lime are excellent desulphurizers and can be used whenever good sulphur removal is required. Unfortunately readily oxidisable elements, such as Al and Ti, can be removed to some degree by these slags. These slags have been successfully used for; ball bearing steels - 1% Cr, 1% C (oxygen level -12 ppm), pure iron, mild steel, and 18/8 stainless steel.

Chemical analysis of ball-bearing steel (EA 31) before and after ESR through a (70F/30) slag is shown in Table VI.

Table VI : Chemistry of steel after ESR treatment

Element	*Electrode Wt%*	*ESR ingot Wt%*
C	1.2	1.18
Si	0.35	0.19
Mn	0.54	0.52
S	0.012	<0.007
P	0.008	0.008
Cr	1.55	1.49
Ni	0.14	0.14
Al	0.006	0.005
O_2ppm	20	0.6

CaF_2–Al_2O_3 :

This system shows eutectic at 9 or 27 Wt% Al_2O_3 having melting point 1290°C. These slags constitute probably the most important range of slags in ESR.

A typical analysis of calcined alumina is given below;

Al_2O_3 98%, K_2 0.07%, Fe_2O_3 0.3%, SiO_2 0.2%.

Addition of Al_2O_3 markedly increases the electrical resistance of the slag. The alumina-bearing slags, on account of their high electrical resistance (electrical resistivity at 1650 °C, of $CaF_2/CaO/MgO/Al_2O_3$ (90F/0/0/10) is 0.26 ohm. cm. and of (65F/5/0/30) is 0.57 ohm. cm.), are hot working, give fast melting rates and low energy consumption per unit weight of metal refined. These slags are not susceptible to hydration. However, if the slag temperature is too high the metal pool will deepen, resulting in solidification pattern being radial rather than longitudinal. The Al_2O_3 content of $CaF_2 - Al_2O_3$ slags rarely exceeds 40% by weight, as above this level the melting point becomes too high.

Ternary Fluoride-oxide systems

(*i*) **CaF_2–CaO–Al_2O_3 :** These slags find the most universal application in the field of electroslag refining. The slag composition falls into three groups.

Acidic-no free lime, $Al_2O_3 : CaO \simeq 1 : 1$

Acidic-no free lime, $Al_2O_3 > CaO$

Basic-free lime, little or no alumina.

The electrical resistivity and M.P. of 60F/20/0/20 slag are 0.5 ohm. cm and 1250°C respectively. These slags have high fluidity. Melt rates would be low to moderate but degree of unidirectional solidification, freedom from pipe, and cleanliness will be excellent.

If it is required to increase the melt rate for a given power input, to increase the temperature of operation, or to make more economic use of energy, slag resistance may be increased by increasing the alumina content. However care be taken while increasing the alumina content so that the liquidus temperature of the slag does not increase excessively. *e.g.*: 50F/25/0/25, 40F/30/0/30. 50F/20/0/30.

If the resistance is too much for (20F/40/0/40) or (0F/50/0/50), then although the melt rate is high and economic use of energy (low KWh kg^{-1}) is made, the metal pool deepens, the solidification pattern becomes more radial, centre looseness results. 65F/5/0/30 - at 1650°C has 0.57 ohm-resistivity. Slag composition containing more than 40% Al_2O_3 are not generally used as liquidus temperatures increase. Also with Al_2O_3 more than 5% in slag the aluminium and O_2 content may increase in ESR melt.

(*ii*) **CaF_2–MgO–Al_2O_3 :** These slags have high liquidus temperature (1450–1800 °C) and can be operated with much difficulty. For 40F/0/10/50 flux approximate melting temperature falls around 1600°C.

Vapour Pressure : The process operating temperatures, measured in the slag bath between the electrode tip and the molten metal pool, usually lie in the range, 1000–2500°C. 1700°C is the temperature commonly observed when melting ferrous alloys.

CaF_2 has lowest vapour pressure and least volatile of the fluorides. Metal oxides other than those of alkali metals have even lower vapour pressure. However, only those oxides may be used which have lower V.P. e.g.; Lime, Al_2O_3, MgO and ZrO_2.

Other important characteristics of the slag which play an important role in ESR process are :

1. Electrical conductivity
2. Viscosity
3. Density
4. Heat capacity

3.5.3. Metal-Slag reactions

In most cases very close control over chemical composition is desirable and in order to achieve a fine degree of control, it is necessary to understand the significant reaction occurring in the slag which control the properties of these elements is slag and metal.

Oxidation reactions

During ESR process there is usually loss of alloying elements which have high affinity with oxygen such as Al, Ti and Si. Reaction of alloying elements is of great importance, as even small losses of certain elements may adversely affect the metallurgical properties of the alloy being refined. Certain additions are made to achieve required results. Addition of aluminium and Titanium is generally done to deoxidise the melt and for grain size control and as carbide stabilizer respectively.

It is also important to keep low non-metallic inclusions (oxides and silicates) and hence oxygen content to the minimum. Non metallic inclusions which are large and angular like alumina, can cause metal defects which lead to fracture initiation and ultimately failure of the component in service.

Oxidation losses are expressed as

(*i*) Wt% element in electrode – Wt% element in ingot

(*ii*) $$\frac{\text{Wt\% element in electrode} - \text{Wt\% element in ingot}}{\text{Wt\% element in electrode}} \times 100$$

Generally little or no oxidation loss is observed in case of W, Co, Sn, Zn, Ni and Cu, where as elements like Al, Mg, Ti, Si, Mn, Cr, and Iron are lost. The loss of these elements decreases with decrease in their reactivity which falls as one goes from Al to iron.

Sulphur Reactions

It is generally desired to keep S content of steels and other alloys at low level and normally in the range. ·005–·010 wt%. The degree of S removal in ESR is usually between 50-75%. In case of lime rich slags S content as low as .0025wt% may be achieved.

(*i*) slag-metal reaction

$$[S] + (O^{2-}) \rightleftharpoons (S^{2-}) + [O] \quad ...(20)$$

$$K = \frac{(S^{2-}) \cdot [O]}{[S] \cdot (O^{2-})} \quad ...(21)$$

hence $$\frac{(a_s^{2-})}{[a_s]} = K \frac{(a_o^{2-})}{[a_o]} \quad ...(22)$$

or $$\frac{(\text{Wt\% } S^{2-})}{[\text{Wt\% } S]} = K \frac{(N_o^{2-})}{[N_o]} \quad ...(23)$$

(*ii*) Slag-gas reaction.

$$(S^{2-}) + \frac{3}{2}\{O_2\} \rightleftharpoons \{SO_2\} + (O^{2-}) \qquad ...(24)$$

$$K = \frac{\{p_{SO_2}\} \cdot (a_o^{2-})}{\{p_{O_2}\}^{3/2} \cdot (a_s^{2-})} \qquad ...(25)$$

$$\frac{\{p_{SO_2}\}}{(a_s^{2-})} = K \frac{\{po_2\}^{3/2}}{(a_o^{2-})} \qquad ...(26)$$

or

$$\frac{\{Wt\%\ S\}}{(Wt\%\ S)} = Const * \frac{(po_2)^{3/2}}{N_o^{2-}} \qquad ...(27)$$

Electrolytic reaction

ESR may be operated using ac, dc direct polarity (electrode negative) or dc reverse polarity (electrode positive). In dc reverse polarity the following reactions occur.

$$[S] + 2e \longrightarrow (S^{2-}) \qquad ...(28)$$

At the surface of the slag the S in slag reacts with atmospheric oxygen

$$(S^{2-}) + \frac{3}{2}\{O_2\} \rightleftharpoons \{SO_2\} + (O^{2-}) \qquad ...(29)$$

dc reverse polarity gives better de-S than direct polarity.

However, ac gives much better de-S.

Phosphorus reaction

During ESR removal of phosphorus occurs according to the following reaction :

$$2[P] + 5[O] + 3\,(O^{2-}) \rightleftharpoons 2(PO_4^{3-}) \qquad ...(30)$$

$$K = \frac{(N_{PO_4^{3-}})^2}{[a_P]^2\ [a_O]^5\ (N_{O^{2-}})^3} \qquad ...(31)$$

Distribution of phosphorus in slag and metal may thus be represented as,

$$\frac{(Wt\%P)}{[WT\%\ P]} = const.\ [Wt\%\ O]^{5/2} \times (Wt\%\ CaO)^{3/2} \qquad ...(32)$$

and therefore, effective removal of phosphorus will occur under basic, oxidising conditions of slag and at moderate temperatures.

3.5.4. Inclusions Control

During electroslag refining, cleanliness is invariably improved. The only inclusion present are those which precipitate from the combination of oxygen dissolved in melts with certain alloying elements as it cools. These inclusion are usually small in size and are uniformly distributed. The interfacial surface tension between slag and metal should be kept moderate to avoid any slag entrapment and to favour solution of inclusions in slag. Large exogenous inclusions are usually eliminated in ESR as they are removed into the slag.

FOUR

Non-metallic Inclusions in Steel

4.1. INTRODUCTION

Whenever an etched polished steel is examined under a microscope (generally at a magnification of 100) particles of various sizes and shapes are observed. These particles which are referred to as non-metallic inclusions, occur in a wide range of compositions and distribution depending upon a number of parameters that are determined by the steel making process. The total number of inclusions per ton of steel may be of the order of 10^{12}–10^{19}. They form an inseparable part of steel.

Inclusions affect most of the chemical and mechanical properties of steel such as machinability, fatigue, cold and hot shortness, corrosion resistance, tensile strength, wear resistance, impact strength and so on. Except for the non-metallic inclusion of globular sulphides for improving the machinability of steel, all other inclusions affect the properties of steel adversely. Therefore in recent times there has been an increasing demand to control the inclusion rating of steels. This is especially true in case of steels, which are required to undergo more stringent applications.

This underlines the importance of various aspects of inclusions in steel. The present work deals briefly with the following aspects of inclusions in steel :

(*i*) Identification of inclusions, their chemical composition and distribution
(*ii*) Origin of inclusions
(*iii*) Inclusion control

Further how the information gathered by inclusion identification can be applied to trace its origin and to suggest means of their control shall be illustrated by means of a case study.

4.2. IDENTIFICATION OF INCLUSIONS

The first step in the study of inclusions is their identification for which the following techniques can be applied :

(*i*) Metallographic Examination
(*ii*) Microhardness Testing
(*iii*) X-ray diffraction testing
(*iv*) Microprobe Analysis

Apart from the above the oxygen analysis of the metal is done at various stages, as oxygen is responsible for most of the inclusions present in steel.

4.2.1. Metallographic Examination

Metallographic examination is usually the only test conducted on finished or semifinished steel. To conduct this test a small sample is cut out of the parent material at a suitable position. It is ground and then finely polished and subjected to metallographic examination at different magnification with the help of metallurgical microscope. The study is conducted either under ordinary light or plane-polarized light. This test has gained so much importance that while mentioning the specifications of steel the inclusion rating is also mentioned. This inclusion rating is generally mentioned according to JK charts. Various categories of inclusions according to these charts are shown in Fig. 4.1. As seen from the figure, the inclusions are classified into following four categories, each of which is represented by two columns called fine and coarse series. These four categories are :

(*i*) Category A which represents sulphide type inclusions
(*ii*) Category B which represents alumina type inclusions
(*iii*) Category C which represents silicate type inclusions and
(*iv*) Category D which represents globular oxide type inclusions.

Each of these categories is further subdivided into five classes numbering 1 to 5. As the number of inclusions seen under a microscope at a magnification of 100 increases, the inclusion class increases. For example in case of D1 thick series one will see 3-4 inclusions under a microscope each having a size of 1-2 mm and above (*i.e.* above 12μm actual size).

Further the inclusion of A type *i.e.* sulphide inclusion are subdivided into three types (Fig. 4.1 *a*) according to the shape. These types are as follows :

(*i*) **Type I :** These are randomly distributed globular type inclusions. These are the least harmful among all sulphide inclusions.

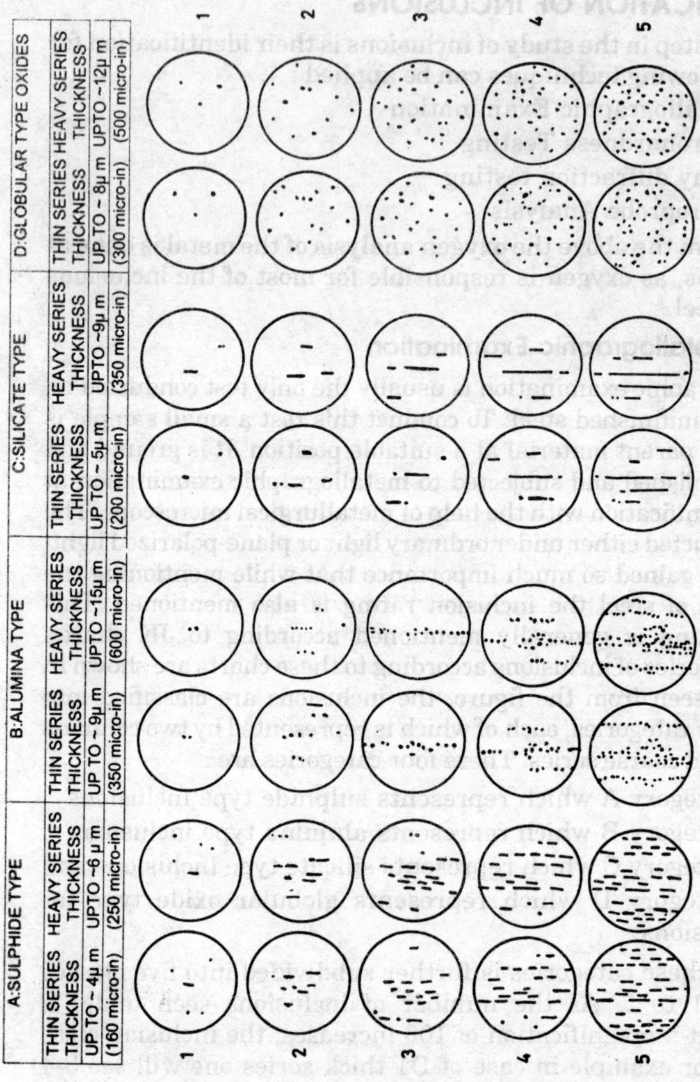

JK chart is among standards used to classify inclusions by size, number, and distribution within a given field. This information is used by specifiers. However, it does not amount to a rating system which indicates, for example, that a given size inclusion is detrimental to steel properties. JK standards shown do not include those for vacuum melted steels, for which there are seven more categories. Also, round fields shown duplicate shape of those seen in microscope. In other standards for newer inclusion counting systems, the field may be square to show the areas covered.

Fig. 4.1 Classification of inclusions.

(*ii*) **Type II** : These are film or chain type eutectic sulphide inclusions. These are the most harmful among the sulphide inclusions.

(*iii*) **Type III** : These inclusions are large in size and are of irregular shapes. These appear to form on grain boundaries. This type of inclusion is less harmful than the Type II inclusions.

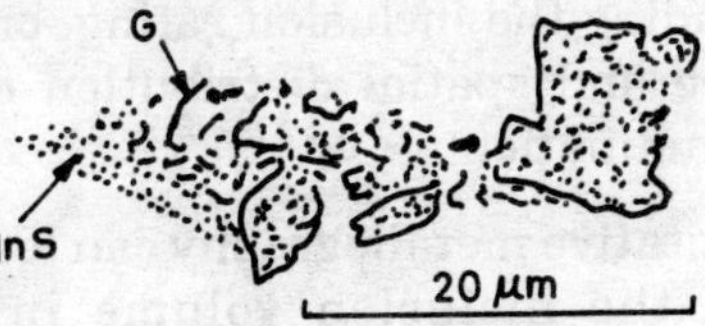

Sulphide inclusions in steel : MnS of type I with Cr in solid solution and Cr-Galaxite

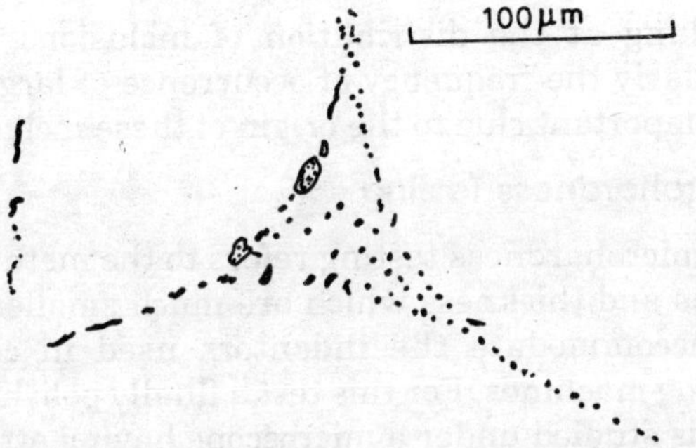

Sulphide inclusions in steel : MnS of type II

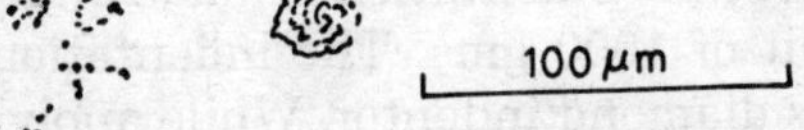

Sulphide inclusions in steel : MnS of type III

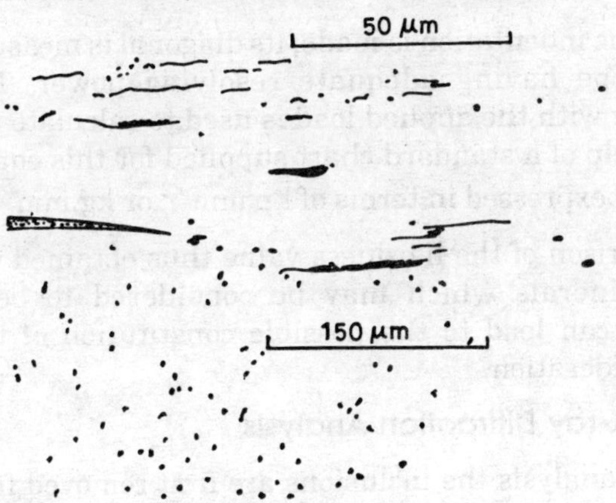

Fig. 4.1. (*a*) **Sulphide inclusions in steel : MnS stringers in a rolled bar.**

Apart from finding the inclusion rating, one takes note of size distribution, shape and spatial distribution of inclusions during metallographic examination.

Further quantitative metallography can lead to finding the size distribution and the inclusion volume in the sample. It is interesting to note that such studies have revealed that 98% of the total number of the inclusions are below the size of 0.2 μm, but they form only 1-2% of the total volume of inclusions.

While looking at the distribution of inclusions, one should record particularly the frequency of occurrence of large inclusions as it gives an important clue to the origin of these inclusions.

4.2.2. Microhardness Testing

The term microhardness testing refers to the method of testing specimen areas and thickness which are much smaller than those required to accommodate the indentors used in conventional hardness testing machines. For this test a finally polished sample is prepared and is studied under a microscope having attachment for microhardness measurement. Generally the sample is seen at a magnification of 100. Then the inclusion whose hardness is to be measured is spotted. The indentor of the tester is pointed to this inclusion and a static indentation is made under a load which has a maximum limit of 1000 gms. The indentation is generally made with a Vicker's diamond indentor. While applying load care should be taken that it is applied without impact or friction. As the size of indentor is extremely shallow, it must be accurately shaped.

After the indentation is made, its diagonal is measured by using a microscope having adequate resolving power. This data in conjunction with the applied load is used to calculate the hardness with the help of a standard chart supplied for this equipment. The hardness is expressed in terms of kp mm^{-2} or kg mm^{-2}.

Comparison of the hardness value thus obtained with those of various minerals which may be considered to be present in inclusions, can lead to the possible constitution of the inclusion under consideration.

4.2.3. X-ray Diffraction Analysis

In this analysis the inclusions are first removed from the steel sample. This can be done by either of the following three methods :

(*a*) Picking out the inclusions from the surface of a polished sample with the help of a pointed tool while the inclusion is observed under a microscope.

(*b*) Chemically dissolving steel in some medium which does not react with the inclusion. The common agent used for this purpose are dilute acids, iodine or chlorine gas. Once steel is dissolved in one of these mediums, inclusions precipitate in the form of a powder. These are collected for further investigation.

(*c*) Electrochemical dissolution of steel. In this case the sample is made the anode in an electrolytic cell having a neutral electrolyte. Iron from steel dissolve at the anode and gets deposited at the cathode whereas inclusions get precipitated at the bottom of the electrolyte. These inclusions are collected for diffraction studies.

Once the powdered sample of inclusion is obtained, it is pasted to a fibre glass capillary with the help of a resin binder or filled in it. Then the capillary is mounted at the centre of an x-ray diffraction camera and properly centred as shown in Fig. 4.2. A film is mounted on the periphery of the camera on the side facing the sample. Monochromatic x-rays are directed towards the sample from one of the windows provided for the purpose. The scattered x-rays fall on the film and thus it gets exposed to specific wavelengths depending upon the structural characteristics of the phases present in the sample. After some time the film is removed and developed. The wavelength of various lines appearing on the film are calculated.

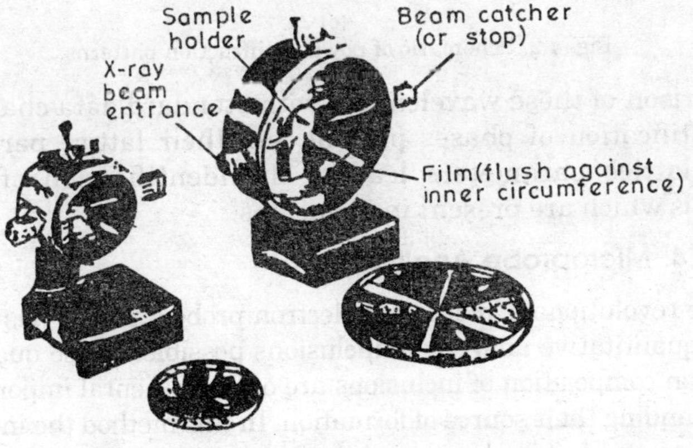

(*a*)

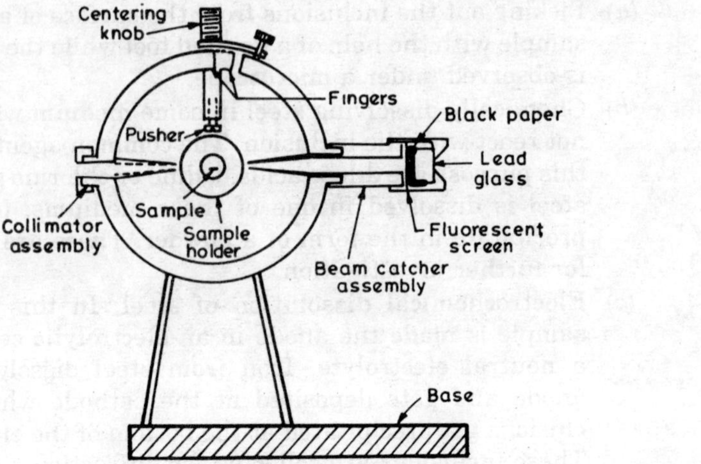

(*b*) X-Ray Powder Diffraction Cameras.

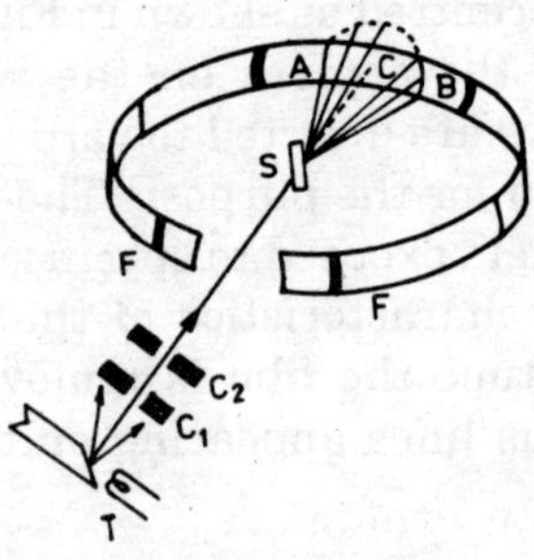

(*c*)

Fig. 4.2. Schematic of powder diffraction patterns.

Comparison of these wavelengths with standard data charts helps in identification of phases present and their lattice parameters. These lattice parameters lead to the identification of various minerals which are present in inclusions.

4.2.4. Microprobe Analysis

The revolutionary method of electron probe analysis has made an in-situ quantitative analysis of inclusions possible. These quantitative results on composition of inclusions are of fundamental importance for understanding their source of formation. In this method the inclusion is exposed to an electron beam and the intensity of reflected electrons is measured. From these measurements the concentration of various elements present in various phases of inclusions is calculated. Further the mean composition of inclusion can also be calculated by using planimetric method.

Thus with the help of above four techniques one is able to arrive at the type, shape, size, distribution, phase and the composition of various inclusions present in steel. In order to help identification of phases present in inclusion, the commonly occurring phases are given in Table 1 and 2 alongwith their physico-chemical properties. As oxygen is most important element leading to the formation of inclusions so knowledge regarding its solubility at various stages of steel processing is of great help in locating the source of inclusions. So apart from the tests conducted directly on inclusions, oxygen analysis forms an important part of studies on inlcusions. Hence its method of analysis shall be considered here.

4.2.5. Oxygen Analysis

Oxygen present in steel can be determined either by oxygen probe or with the help of oxygen analyser.

An oxygen probe consists generally of a stabilised zirconia tube which is an oxygen ion conductor. On the inner side of this tube a standard electrode in the form of a metal/metal oxide mixture (generally $Cr–Cr_2O_3$ or $Mo–MoO_3$) is provided. The outer side is provided with a contact wire. When this tube is dipped in the metal E.M.F. develops because of the difference in oxygen partial pressure on the two sides of the tube. This E.M.F. is a function of temperature and oxygen dissolved in metal and from its value the amount of dissolved oxygen can be calculated making use of standard thermodynamic relationships.

In case of oxygen determination by an analyser, samples are drawn from the metal at stages at which the oxygen content is to be found out. A small amount of metal (generally less than 1 gm.) is placed in the graphite crucible of the analyser. The crucible is heated in a furnace. It leads to the reaction between oxygen in the sample and carbon in the crucible. The evolved gas is analysed by conductivity measurement and from the amount of gas evolved the amount of oxygen present in the simple is calculated.

Table 1. Properties of Various Phases Present in Inclusions

Name	*Formula*	*Microscopic Examination*	*Microhardness KP/MM²*	*Struct, X_Ray Diffraction*	*Composition (Wt. %) Microprobe Analysis*						*M.P.(°C)*	*Remarks*
					MnO	*FeO*	*SiO_2*	*Al_2O_3*	*Cr_2O_3*	*MgO*		
Corundum	Al_2O_3	Spinel Type (Needles)	3000-4500	Hexagonal				100			2050	
Cristobalite	SiO_2	Rossets	1600	Tetragonal			100				1713	
Tridymite	SiO_2	Laths (Dark thick plates slightly split at ends)	1600				100					
Maganosite	MnO		400		100						1850	Generally not found in pure state
Quartz	SiO_2	Regular plates (Rectangular or Triangular)	1600	Trigonal			100					
Wustite	$Fe_{1-x}O$	Dark Plates		Cubic							1370	Generally found in combination with MnO
Galaxite	$MnO.M_2O_3$	Rectra Triangular	1500-1700	Cubic	41			59			1560	
Mullite	$3Al_2O_3.2SiO_2$	Rhombic or Rhombohedral Crystals with Glassy Residue in the Centre	1500	Orthorhombic			28	72				

Name	Formula	Microscopic Examination	Microhardness KP/MM^2	Struct, X_Ray Diffraction	Composition (Wt. %) Microprobe Analysis						M.P.(°C)	Remarks
					MnO	FeO	SiO_2	Al_2O_3	Cr_2O_3	MgO		
Rhodonite	$MnO.SiO_2$	Glassy Silicate Like Silica	750	Triclinic	54		46				1291	
Spessarite	$3MnO.Al_2O_3.3SiO_2$	Feather Like Branch Structure	1000-1100	Cubic	43		36	21			1195	
Tephroite	$2MnO.SiO_2$	Banded Structure	950	Orthorhombic	70		30				1345	
Mn-Anorthite	$MnO.Al_2O_3.2SiO_2$				24		41	35				Phase coexist when stabilised with silica
Mn-Cordierite	$2MnO.2Al_2O_3 5SiO_2$				24		46	32				
Manganese Sulphide	MnS	Light, Grey plastic slightly transparent	170	Cubic							1610	
Fayalite	$2FeO.SiO_2$	Similar Appearance to tephroite		Orthorhombic							1205	
Escolite	Cr_2O_3	White regular Idiomorphic Crystal		Hexagonal							2265	
Chromite Galaxite	$MnO.Cr_2O_3$	- Do -		Cubic								
Chromite	$FeOCr_2O_3$	Needles		Cubic							1850	

Name	Formula	Microscopic Examination	Microhardness KP/MM^2	Struct, X_Ray Diffraction	Composition (Wt. %) Microprobe Analysis						M.P.(°C)	Remarks
					MnO	FeO	SiO_2	Al_2O_3	Cr_2O_3	MgO		
Spinel	$MgO.Al_2O_3$	Idiomorphic Hard Regular Crystals	2100-2400	Cubic							2135	Composition given in Table-2
Eseantite Protoenstatite Clinoenstatite	$MgOSiO_2$	Crystals Embedded in Silicate Matrix	1000	Orthorhombic							1557	
Calcium Aluminates		Spherical Shapes	930-2200									
Calcium Silicates		Glassy										
Calcium Alumino Silicates		Glassy with embedded Crystals										
Iron Sulphide	FeS	Spread along grain boundaries										Generally mixed with Mn S and/or Fe

Table 2. Composition (Wt. %) of the spinel phase in different inclusions with MgO

Inclusion type	*MgO*	*MnO*	*FeO*	*CaO*	*Al_2O_3*	*Cr_2O_3*	*Steel type*
Spinel	30	—	—	—	70	—	Rolled stainless
Spinel in Ca-aluminate	33	—	—	—	68	—	Low alloyed
Spinel in Ca-aluminate	31	—	—	—	67	—	Low alloyed
Spinel in Ca-aluminate	31	+	+	2	68	—	Low alloyed
Spinel and forsterite in silicate	25	7	—	—	70	—	Vacuum steel
Spinel in Ca-Al silicate	25	6	—	—	67	—	Low alloyed
Spinel in silicate	25	12	—	2	62	—	Vacuum steel
Spinel	12	23	—	—	59	—	Low alloyed
Spinel in silicate	5	34	1	—	54	—	Low carbon steel
Spinel in silicate	5	25	1	—	46	—	Rolled stainless
Spinel in silicate	3	34	2	—	61	—	Steel

4.3. SOURCES OF INCLUSIONS

Once the inclusions are identified, one proceeds to find out their origin. From the point of view of sources, inclusions can be classified into two categories, namely, indigenous or endogenous and exogenous inclusions. In the context of present knowledge this classification is oversimplified. It is now well known that indigenous precipitation occurs on exogenous nuclei during the different stages of steel making process. Indigenously formed inclusions may even precipitate at some point on the refractories and corrode them leading to inclusions which are neither pure and indigenous or exogenous. These may react further with the metal leading to modification in their composition. Thus from the moment an inclusion nuclei appears (either indigenously or exogenously) in the melt, its composition goes on changing continuously till precipitation on it becomes negligible. But still depending upon the source of its nucleus an inclusion can be termed indigenous or exogenous and we shall stick to this terminology for further discussions.

First indications whether an inclusion is indigenous or exogenous, comes from its size, shape, frequency of occurrence and distribution. Indigenous inclusions are generally small in size, large in number and are usually uniformly distributed in a sample. They are also typical of steel in which they occur. In Table 3 the common inclusions occurring in various steels are tabulated. This table also serves as a good guide for identification of inclusions found in steel from various processes.

Table 3

The $MnO-SiO_2-Al_2O_3$ system : Inclusion classified according to steel analysis. The steels are grouped according to main alloying elements and within each group arranged with increasing carbon content. The general type of inclusion is given, but the reader is referred to the captions in the different figures given for each steel

Steel type	*Steel analysis Wt. %*						*Inclusion type*
	C	*Si*	*Mn*	*Cr*	*Ni*	*Mo*	
Carbon steel							
Rimmed deep drawing steel from sheet	0.05	0.29	0.07	—	—	—	Manganosite
Basic open hearth, Al-killed from plate	0.11	0.25	1.10	0.06	0.05	—	Corundum
Basic open hearth, Al-killed from plate	0.11	0.25	1.10	0.06	0.05	—	Corundum and galaxite
Bottom cast ingot from sheet	0.12	0.20	0.70	0.10	—	—	Tridymite, rhodonite, and sulphide
Basic Kaldo semi-killed from ingot	0.14	0.07	1.00	—	—	—	Rhodonite
Basic Kaldo semi-killed from ingot	0.16	0.06	1.00	—	—	—	Tephroite
Basic electric, Si-killed isolate	0.16	0.35	0.67	—	—	—	Cristobalite and thodonite
Basic electric, Si-killed isolate	0.16	0.35	0.67	—	—	—	Cristobalite, thodonite, and sulphide
Basic electric, Si-killed isolate	0.16	0.35	0.67	—	—	—	Cristobalite
Carbon steel, killed	0.22	0.24	0.50	—	—	—	Scale of iron oxides
Bottom cast ingot, from rolled bar	0.24	0.30	1.39	—	—	—	Galaxite
Carbon steel from forged ingot	0.33	0.25	0.60	—	—	—	Galaxite and Corundum
Basic electric, surface crack in ingot	0.33	0.25	0.60	—	—	—	Wilistite and magnetite
Basic electric, Si-killed from billet	0.45	0.22	0.70	—	—	—	Mullite and corundum
Basic electric, Si-killed from billet	0.45	0.22	0.70	—	—	—	Spessartite
Carbon steel	0.46	0.22	0.61	—	—	—	Wilstite and magnetite

Steel type	Steel analysis Wt. %						Inclusion type
	C	Si	Mn	Cr	Ni	Mo	
Carbon steel, surface crack in ingot	0.46	0.22	0.61	—	—	—	Wilstite, magnetite and faylite (?)
Lancashite iron	0.55	0.07	0.04	—	—	—	Fayalite and iron oxides
Lancashite iron	0.55	0.07	0.04	—	—	—	Iron oxides
Lancashite iron	0.55	0.01	0.04	—	—	—	Haematite
Carbon steel from billet	0.60	0.14	1.11	—	—	—	Galaxite
Low-alloy steel							
Basic electric killed from billet	0.10	0.32	0.46	2.20	0.9	1.01	Corrundum and spinel
From bottom fatigue test sample	0.15	0.29	0.75	0.86	1.45	0.08	Spinel
Basic open hearth, killed, from ingot	0.33	0.34	0.60	2.43	0.12	0.49	Corrundum
Basic open hearth, killed, from ingot	0.33	0.34	0.60	2.43	0.12	0.49	Anorthite, carbon and Cr-galaxite
Basic open hearth, killed, from forging	0.62	0.26	0.79	0.65	0.01	—	Corrundum, spessartite and sulphide
Basic open hearth, killed, from forging	0.62	0.26	0.79	0.65	0.01	—	Mullite, spessartite and sulphide
Acid open hearth, killed, from tube	1.0	0.30	0.30	1.5	—	—	Quartz, cristobalite and Cr-galaxite
Alloy steel							
From vacuum cast ingot	0.04	0.40	1.34	18.5	10.1	0.50	Cr-galaxite
From hot rolled sheet	0.04	0.40	1.34	18.5	10.1	0.50	Cr-galaxite
From rolled billet	0.05	0.43	0.96	18.0	9.8	0.50	Tridymite, Cr-galaxite, and sulphide
From hot-rolled strip	0.05	0.54	1.27	18.3	9.1	0.25	Cr-galaxite
Basic electric, killed, from sheet	0.05	0.54	1.30	18.6	9.2	—	Chromic
From cold-rolled sheet	0.05	0.66	1.03	17.8	9.0	—	Cr-galaxite crushed
Basic electric, from billet	0.18	0.13	0.39	13.6	0.66	—	Cr-galaxite sulphide
Basic electric, from forging	0.8	0.35	—	11.5	0.60	0.50	Tridymite, quartz and Cr-galaxite

Steel type	Steel analysis Wt. %						Inclusion type
	C	Si	Mn	Cr	Ni	Mo	
High speed steel							
From billet	0.01	0.34	0.35	4.30	—	4.90	Mullite
Cast steel							
Basic open hearth from forging	0.15	0.44	0.84	0.21	—	—	Spessartite
Ferroalloys							
Ferrochromium	0.03	0.08	—	73	—	—	Corrundum with Cr_2O_3
Ferrochromium	0.03	0.09	—	73	—	—	Cristobalite
Ferrochromium	0.03	0.09	—	73	—	—	Cr_2O_3 and cristobalite
Ferrochromium	0.03	0.08	—	73	—	—	Cr_2O_3
Ferrochromium	0.03	0.08	—	73	—	—	Cr-galaxite
Ferrochromium	0.03	0.08	—	73	—	—	Cr_2O_3 and two galaxites
Ferrochromium	0.03	0.08	—	73	—	—	Two galaxites
Ferrochromium	0.03	0.20	—	66.0	—	—	Cr_2O_3
Ferrochromium	0.03	0.57	—	68.6	—	—	Cr_2O_3 and cristobalite
Ferrochromium (81.2%V)	0.15	0.50	—	—	—	—	Corrundum

Exogenous inclusions are generally much larger, rare in occurrence and have a haphazard distribution. Further these inclusions in general have a very irregular shape.

Let us now deal with the sources of these two types of inclusions in more detail.

4.3.1. Indigenous Inclusions

These inclusions are formed inside the metal as a result of the chemical or physical change. These inclusions are formed by one or more of the following modes :

(*i*) By deoxidation of steel.
(*ii*) By decrease in solubility during cooling and solidification.
(*iii*) By increased concentration of solute from dendritic segregation during solidification.
(*iv*) By solubility changes during cooling and phase changes below solidus line.

The inclusions formed by the first mode are referred to as primary inclusions whereas those formed by subsequent modes are referred to as secondary inclusions. Now we shall deal with these modes in further detail.

Inclusions formed by deoxidation : Deoxidation is the main source of the indigenous inclusions. The number of inclusions formed during deoxidation depends upon the initial oxygen content of the metal, deoxidising power of the deoxidiser and its amount. The deoxidation power of a deoxidiser depends upon the solubility product of the oxide which is formed as a result of deoxidation. This solubility product value for various oxides can be calculated from thermodynamic data. Higher is this value weeker is the deoxidiser. The deoxidation power of a deoxidiser can also be gauged from Fig. 4.3 in which the equilibrium oxygen concentration of the melt is plotted against the dissolved concentration of deoxidiser in the melt. Higher is the concentration of oxygen for a given deoxidiser content in the melt, weaker is the deoxidiser. In other words higher is the position of curve in Fig. 4.3 weaker is the deoxidiser. From this figure it is seen that the deoxidisers in order of ascending deoxidation are Mn, Cr, B, Ti, Al and Zr. Apart from various metals used in pure or in alloyed form with iron, one makes use of alloys of two or more deoxidisers. Such deoxidisers are termed as complex deoxidisers. The properties and behaviour of various deoxidisers shall now be described briefly.

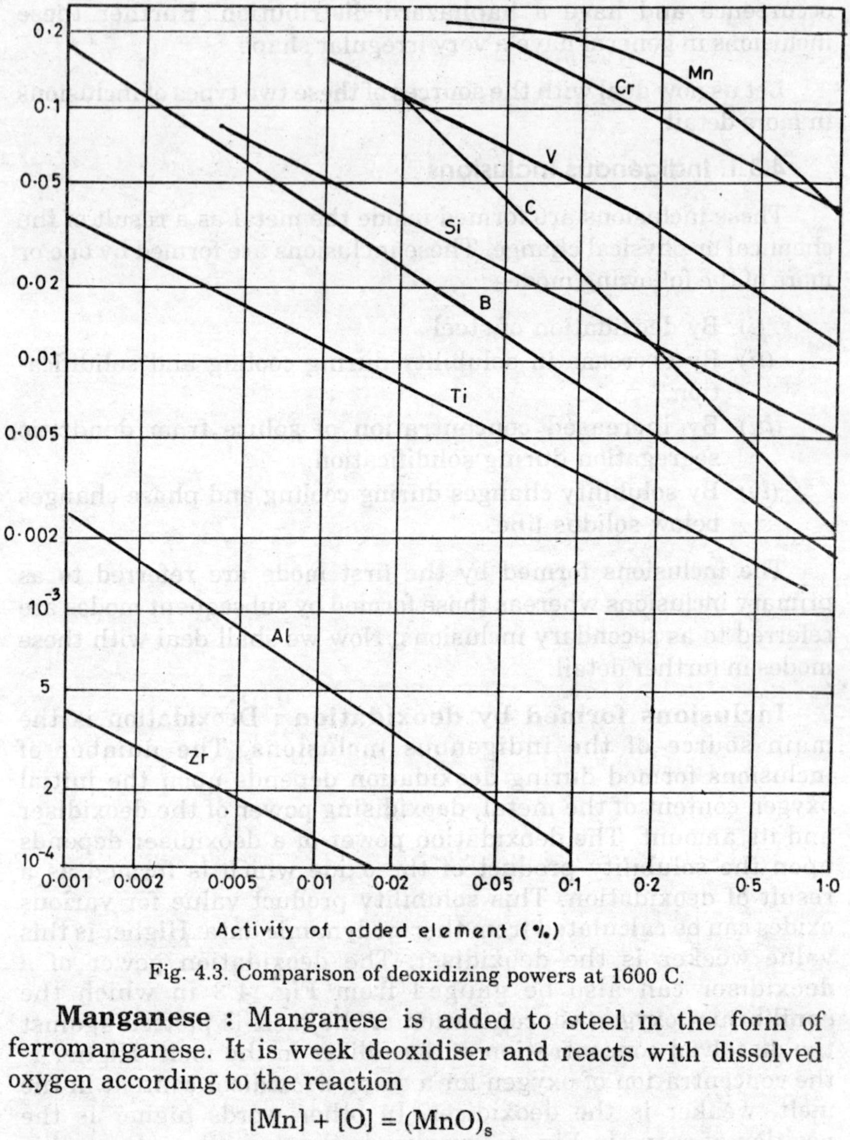

Fig. 4.3. Comparison of deoxidizing powers at 1600 C.

Manganese : Manganese is added to steel in the form of ferromanganese. It is week deoxidiser and reacts with dissolved oxygen according to the reaction

$$[Mn] + [O] = (MnO)_s$$

Another reaction which generally occurs alongwith this reaction is

$$[Fe] + [O] = (FeO)_s$$

Thus the product of deoxidation in this case is a solid solution of FeO and MnO.

The amount of each of the oxides in the deoxidation product is decided by the reaction

$$(FeO) + [Mn] = (MnO) + [Fe]$$

If K_{Mn} is the equilibrium constant for the above reaction, one can write the following expression for the ratio of MnO to FeO.

$$\frac{MnO}{FeO} = K_{Mn}\,[Mn]$$

This equation shows that with increase in Manganese content of the metal the MnO content of the deoxidation product increases. In case solid MnO is formed as a deoxidation product, the solubility product for this oxide is a constant depending upon temperature. (0.02 approx.)

Apart from reacting with oxygen, manganese reacts with sulphur and forms sulphide or oxysulphide inclusion of type I. As sulphide inclusions of type I are least harmful, one tries to get sulphur precipitated as MnS as far as possible.

Silicon : Silicon is used in the form of ferro-silicon. It is a stronger deoxidiser than manganese. Silicon reacts with dissolved oxygen according to the reaction.

$$[Si] + 2\,[O] = (SiO_2)_s$$

leading to the formation of solid silica. In case oxygen content of the metal is high, fayalite ($2FeO \cdot SiO_2$) may be the product of deoxidation.

Aluminium : It is used either as aluminium shots or as ferro-aluminium. Aluminium is a much stronger deoxidiser than either manganese or silicon. Its deoxidation action can be described by the equation.

$$2[Al] + 3[O] = (Al_2O_3)$$

The deoxidation product in this case is Al_2O_3 and its solubility product is 1.0×10^{-14} at 1600°C. Aluminium shots have a very low efficiency of utilisation which lies around 25%. On the other hand ferro-aluminium has an aluminium utilisation efficiency of around 85% but it suffers from the drawback that it gets disintegrated in the presence of moisture and therefore, it can not be stored for long durations of time.

Aluminium has the advantage of improving the hardenability of steel, but simultaneously suffers from the disadvantage of changing the sulphide inclusions to more harmful classes namely type II and type III.

Chromium : It is rarely used as a deoxidiser, but when added to steel it leads to a deoxidising action. Chromium is generally added in the form of ferro-chrome. It is a stronger deoxidiser than manganese and forms either Cr_2O_3 or $FeO–Cr_2O_3$ as deoxidation product.

Titanium : It is added to a steel bath in the form of ferro-titanium and reacts with oxygen according to the equation

$$[Ti] + 2\,[O] = (TiO_2)$$

The product of deoxidation is this case is TiO_2, which has a solubility product of 0.3×10^{-6} at 1600°C. This value and Fig. 4.3 shows that though titanium is a strong deoxidiser, it is weaker than aluminium. Titanium affects sulphide inclusions in a similar manner as aluminium. It also combines with dissolved nitrogen and carbon and leads to the formation of nitride and carbide. In such cases of special steels in which boron is to be kept in the dissolved form, Titanium is used as a deoxidiser to fix nitrogen as well so that boron nitride is not formed. In some case of titanium deoxidation carbo-nitrides may as well be formed.

Zirconium : It is a slightly stronger deoxidiser than aluminium, but is rarely used as deoxidising agent. The product of deoxidation in this case is zirconia (ZrO). Small additions of Zirconium changes the sulphide inclusions from type I to more harmful type II which changes to type III on further zirconium addition.

Silico-manganese : Although deoxidation by silicon is more effective than that by manganese, simultaneous deoxidation by both these elements becomes still more effective with both silicon and manganese reactions taking place simultaneously. This leads to the formation of liquid mangano-silicates as the product of deoxidation. These inclusions are globular in shape and are more deformable at higher temperature and are thus least harmful. MnO and SiO_2 content of these inclusions is determined by the equation

$$2\,(MnO) + [Si] = (SiO_2) + 2[Mn]$$

If K_{si} is the equilibrium constant of this reaction, one can write for MnO and SiO relationship the expression

$$\frac{(SiO)}{(MnO)^2} = K\,\frac{[Mn]^2}{[Si]}$$

Calcium Silicide : It is a very strong deoxidiser and leads to the formation of CaO and $2CaO \cdot SiO_2$ as products of deoxidation. When it is added alongwith aluminium, product of deoxidation is Calcium Aluminate. It has advantage of leading to the formation of globular type of inclusions. Even sulphide inclusion are converted to type I when deoxidation is carried out with Calcium Silicide. Only problem with this oxide is that it is to be injected into the melt with the help of a gas otherwise it shows a very poor efficiency.

Rare Earths : Commonly used rare earths for addition to steel are cerium and lanthanum. These are added in the form of misch-metal to steel. Both of these rare earths have similar deoxidising power. They also form very stable sulphides and lead to the formation of type I inclusions in the bath to bring down the sulphur content of the metal.

Apart from the above deoxidisers, another very important deoxidiser is carbon. As it leads to a gaseous deoxidation product, so it can not be considered in the present context. A brief summary of the characteristics of deoxidisers is given in Table 4.

Table 4 : Properties of Deoxidisers

Deoxidiser	*Used from*	*Deoxidation product*	*Solubility Product*	*Remarks*
Manganese	Ferro-Manganese	(FeO-MnO) & (MnO)s		Forms MnS (Type-I Inc.) With sulphur
Silicon	Ferro-silicon	SiO_2 or 2 $FeO.SiO_2$		
Aluminium	Al-shots or Ferro-Aluminium	Al_2O_3		Changes sulphides to Type II & III Inc
Chromium	Ferro-chrome	$FeO\text{-}Cr_2O_3$ or Cr_2O_3		
Titanium	Ferro-Titanium	TiO_2		Also forms carbides, Nitrides and Carbo-Nitrides
Zirconium		ZrO_2		
Silico-Manganese	Silico-Manganese	XMnO $YSiO_2$		Liquid deoxidation product
Calcium Silicide	Ca-Si	CaO+2CaO.SiO_2		Leads to globular inc. and also modifies sulphides to Type-I Inc.
Rare Earth	Misch Metal			Combines with sulphur to form Type-I Inc.

Secondary Inclusions : These are formed by any of the modes (*ii*) to (*iv*) stated under indigenous inclusions. They are formed because the solubility product of oxides goes on decreasing with decrease in temperature. It also falls sharply on solidification. Hence, further precipitation of deoxidation products takes place. This precipitation is generally referred to as secondary precipitation. In fact, it rarely leads to the formation of new inclusions, but the precipitation takes place on existing ones. Thus the existing inclusions grow in size on cooling as well as upon solidification of the melt.

4.3.2. Exogenous Inclusions

These inclusions are formed as a result of entrapment of foreign non-metallic material from external sources such as slags and refractories with which the metal comes in contact during its processing. Following materials may act as sources of exogenous inclusions :

(*i*) Furnace and ladle slag
(*ii*) Ferro-alloys
(*iii*) Refractories
(*iv*) Reoxidation of the melt
(*v*) Ingot scum
(*vi*) Teeming and antipiping compounds

Each of the above source shall now be discussed briefly.

Furnace and Ladle Slag : Furnace slag may be carried into the ladle and the ingot while pouring and teeming. This slag may partly get trapped in the metal and act as a source of inclusion. The inclusions from this source are generally rich in calcium and shall contain very little of alumina. They shall have a strong resemblence in composition to the slag last made in the furnace.

Sometimes a flux is added to the metal in the ladle to facilitate further refining of the metal by slag metal reaction or by purging. The slag formed by this flux may get trapped in the metal leading to the formation of exogenous inclusions. Further this slag may even react with ladle lining to give rise to a new series of inclusions. This ladle slag inclusions may contain the ingredients of the flux as well as that of the ladle lining. If deoxidation is partly or completely conducted in the ladle, the deoxidation products may precipitate on those inclusions leading to further modification in their composition. Hence ladle slag inclusions may vary over a wide range of composition but they are known to contain invariably silica, alumina alongwith small amount of K_2O, TiO_2 and MnO.

Ferroalloys : Ferroalloys rarely act as sources of inclusions as the inclusions present in ferroalloys either float up when the alloy is added to the bath or they may even dissolve. But in exceptional cases a direct inheritance of these inclusions may occur. For example if the ferroalloys are not completely molten to dissolved in case an unusually low temperature is used during melting, some inclusions may originate from them. These inclusions are generally rich in the oxide of the principal alloying element present in the ferro-alloy. For example, if ferro-chrome is used as a ferro-alloy, any inclusion originating from it will have Cr_2O_3 as its main constituent and will have a characteristic and often complex structure.

Refractories : Refractories form a major source of exogenous non-metallic inclusions. During the whole steel making process molten metal comes in contact with the refractories. This leads to the possibility of their erosion or corrosion and thus become a source of inclusion. Inclusions from this source must not appear merely as refractory pieces in the metal, but they may act as nuclei for indigenous precipitation from the moment they mix with the metal. Thus these can have a continuous variation in their composition as compared to the original refractory. Inclusions formed by corrosion may have a structure which often has no direct relationship to the refractory structure and their composition may also be considerably different. This makes the fixing of refractories as source of exogenous inclusions quite difficult. In such cases amounts of refractory oxides such as TiO_2, ZrO_2 and K_2O provide valuable hints regarding the establishment of origin of the inclusions. Hence inclusions should be thoroughly checked for the presence of these elements before making a comment on their origin.

As different types of refractories are used at different stages in processing of steel, so the inclusions arising from these refractories shall have a vast variation in composition. Thus inclusions which come from furnace refractories may contain traces of Ca along with Mg. Inclusions from ladle, launder and teeming refractories are likely to contain Mg, Ti and K and so on. The best way to confirm whether an inclusion has originated from refractories is to make simultaneously microprobe analysis of inclusions as well as that of samples from various refractories and compare the ratio of various oxides present in both the cases.

Reoxidation of the melt : Exogenous inclusions may be formed through chemical reaction of the melt with air during pouring and teeming. The surface of the metal stream during pouring and teeming is in constant contact with air and can get oxidised through the reaction.

$$2\,Fe + O_2 = 2FeO$$

FeO thus formed can react with other components of the melt or with the refractory lining to form inclusions. For example in case of Silicon killed steel following reaction may take place

$$4FeO + Si = 2FeO \cdot SiO_2 + 2\ Fe$$

In case of reaction with refractories a general reaction scheme which may involve corrosion can be written as

$$XFeO + XMn + YAl_2O_3 \cdot ZSiO_2$$
$$= XMnO \cdot YAl_2O_3 \cdot Z \cdot SiO_2 + XFe. \quad \text{(from refractories)}$$

Oxides formed as a result of components dissolved in the melt generally precipitate on the inclusions already existing in the melt and hence modify their composition. But the inclusions formed as a result of reaction with refractories occur sparingly and maintain their own identity. These inclusions are generally large in size.

Ingot Scum : During teeming of the metal in a mould, a pool of scum generally forms on the surface of the melt. It grows as the ingot gets filled up. This scum is more pronounced in case of top poured ingots than in case of bottom poured ones. Sometimes this scum gets partly trapped in the metal as a result of stirring and it leads in general to the formation of subsurface inclusions. In some rare cases of top poured ingots this scum may be carried deep into the melt by the falling stream and may be trapped in steel before floating out and hence may lead to the formation of inclusion.

Teeming and Antipiping Compounds : These compounds which are used during teeming are generally applied in the form of fine powders. They may get trapped into the metal during natural stirring of the bath and hence may become a source of inclusions. Though such occurrence is extremely rare but the possibility of their entrapment exists. As these compounds contain sodium in one form or the other, so any inclusion originating from them will invariably contain this element.

In order to identify the source of an inclusion quickly, the elements coming to inclusions from various sources are given in Table 5. This table is not exhaustive. In fact, before starting to trace the source of an inclusion one should first prepare a similar table and then proceed to conduct the identification tests for the inclusion and then find its source with the help of the prepared table.

Once the sources of inclusions are identified one tries to control the inclusions.

Table 5 : Possible Inclusion sources discussed in the present text

	Possible Inclusion Source	*Key Element*
Furnace	Furnace slags	Ca
	Furnace - Refractories	Ca
	Ferroalloys	Cr, Al, Si
Tapping	Launder-Refractories	Mg, Ti, K
	Oxidation	FeO
Ladle	Deoxidation	.
	Ladle slag	Ca, Mg
	Ladle refractories	Mg, Ti, K
Teeming	Stopper and nozzle Refractories	Mg, Ti, K
	Oxidation	FeO
	Deoxidation of mould Refractories	Mg, Ti, K.
Heat Treatment and rolling	Surface Oxidation	FeO
	Surface Sulphurisation	FeS
	Inner Oxidation	SiO_2
	Hot-shortness	FeS
Welding	Welding Slags	Ca, Ti
	Electrode Coatings	Ti, V
	Steel inclusions	
	Hot tearing	S

4.4. INCLUSION CONTROL

It is not possible to altogether avoid the presence of inclusions in steel, but their number, size, shape and distribution can be controlled to a considerable extent. The subject of inclusion control implies not only the control of their number but also their size, shape and distribution.

4.4.1. Indigenous Inclusions

The major sources of inclusions, predominantly indigenous is oxygen. Hence the first principle of inclusion control involves the control of oxygen present in steel. To accomplish this a number of secondary steel treatments also referred to as clean steel technologies have been developed. But these technologies require a high capacity and recurring expenditure. Hence they can only be applied to steels which are required for specialised applications. For common steels only the practice is to be suitably modified to accomplish this aim to an optimum value.

The first step in reducing the number of inclusions is to make them float out of the metal after deoxidation. The number of inclusions floating out of the melt is a function of their size and retention time of the ladle before teeming. Assuming the inclusions follow Stoke's law, the velocity of their rise in liquid metal will be given by the equation.

$$v = \frac{2gr^2 (\rho m - \rho i)}{a\eta}$$

where v is the velocity of the inclusion in cm sec^{-1}

g is the acceleration due to gravity in cm sec^{-2}

r is the radius of inclusion in cm.

ρm and ρi are the densities of metal and inclusion respectively in gm cm^{-3} and η is the viscosity of the metal in poise. If one substitutes mean value of plain carbon steels in the above equation, the above equation assumes a simplified form

$$V = 2 \times 10^4 (7 - \rho i) r^2 \text{ cm sec}^{-1}$$

This equation shows that velocity of rise is a quadratic function of radius of inclusion. It will increase sharply with increase in the radius of inclusion. Thus, for example, an alumina inclusion of 10 μm radius will rise at a velocity of 0.07 cm sec^{-1} and shall take about 35 minutes to rise through a 150 cm height of the metal whereas an inclusion of 100 μm radius will rise at the rate of 7 cm sec^{-1} and shall take only 20 secs to rise through the same height of the bath. This leads to the conclusion that indigenous inclusions formed during deoxidation should have as big a size as possible so that they are able to float out of the melt during the retention time of the ladle.

Though a strict control of inclusions size is not possible, but some guidelines are available on the basis of laboratory experiments conducted by some research workers. These experiments have shown that one can control the size by controlling the number of nuclei formed per cubic cm of steel during deoxidation. Fig. 4.4 shows a relationship

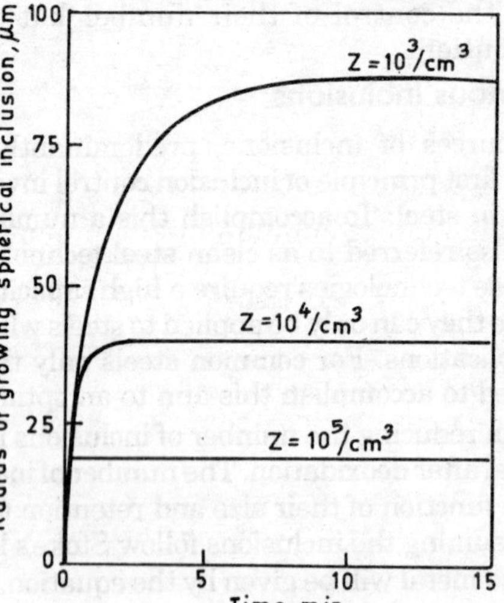

Fig. 4.4. Rate of growth of inclusions.

between the number of nuclei per cm^3 and their rate of growth for a steel having an initial oxygen content of 500 ppm. It shows that with the decrease in the number of nuclei the ultimate inclusion size increases, but the inclusion takes a longer time to grow. During its growth inclusion will keep on rising in the bath and the liquid gets constantly depleted of inclusions from the bottom. Fig. 4.5 shows the total time taken for inclusions formed during deoxidation at different nuclei rating to travel certain distance in a given time. It is seen from this figure that lower is the number of nuclei per cm^3 faster they travel. But at low nuclei rating the diffusional path of oxygen will increase leading to higher residual oxygen in the bath. Thus there exists a critical value, Z_c of number of inclusions which should be present per unit volume of the bath. If the number of nuclei is less than this value the inclusion rating of the resulting steel will be lower, but there will be higher residual oxygen in the melt. But if the number of nuclei is more than Z_c, the bath will be effectively deoxidised but smaller inclusions will be formed which may not separate out during the retention period of the ladle and hence inclusion rating of the metal will be high. The control of nuclei rating is a matter of practice and can be developed only through experience.

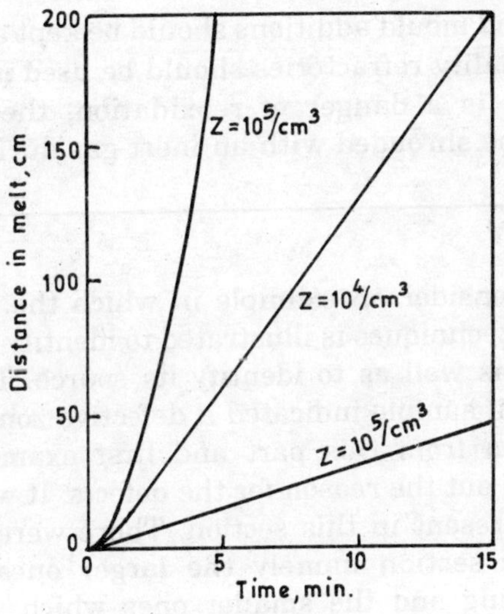

Fig. 4.5. Rate of rise of inclusions.

Whatever may be the distribution of the deoxidiser in the bath, the above discussion leads to the conclusion that inclusions cannot be altogether removed. So if they have to be present in steel, they should be present in a form which is least harmful. Such a form is globular form of inclusions. Further, inclusions should be plastically deformable at working temperature. This can be achieved by careful selection of deoxidisers keeping in view the properties of various deoxidisers as explained in the previous section.

4.4.2. Exogenous Inclusions

The above discussion was confined mainly to indigenous inclusions but more harmful are the seldom occurring exogenous inclusions which are generally larger in size. To avoid inclusions from these sources following necessary precautions depending upon the quality of steel to be produced, should be taken during steel making practice.

(*i*) The steel making operation should start with a clean furnace bottom.
(*ii*) Late ore additions should be avoided.
(*iii*) Somewhat higher residual manganese content should be maintained.
(*iv*) Slightly prolonged ore boil should be adopted.
(*v*) FeO content of slag during reducing period should be kept as low as possible.
(*vi*) Ladle and mould additions should be kept at a minimum.
(*vii*) Good quality refractories should be used in spouts.
(*viii*) If there is a danger of reoxidation, the steel stream should be shrouded with an inert gas during teeming.

4.5. CASE STUDY

Let us now consider an example in which the application of above mentioned techniques is illustrated to identify the inclusions in a rail section as well as to identify its source. The ultrasonic inspection of a rail sample indicated a defective zone in its rib. A sample was taken from this part and first examined under a microscope to find out the reason for the defects. It was found that inclusions were present in this section. There were two types of inclusions in the section namely the larger ones which were separately occurring and the smaller ones which were more in number. Microhardness, X-ray diffraction and microprobe analysis of these inclusions revealed that the following phases were present in these inclusions (as shown in Fig. 4.6).

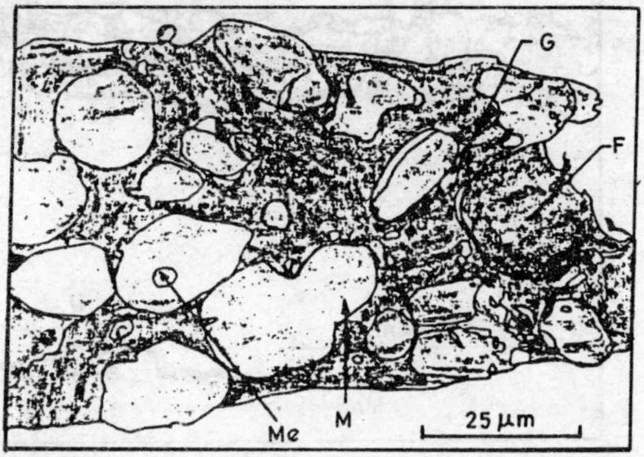

Inclusions :

(*F*) Forsterite with 8 MnO, 3 FeO, 55 MgO 37 SiO_2

(*G*) Spinel with 7 MnO, 25 MgO, 70 Al_2O_3

(M) Matrix with 7 MnO, 3 CaO, 6 MgO, 57 SiO_2, 22 Al_2O_3

(Me) Steel : Mean Analysis (calc) 8 MnO, 1 FeO

Table 6 : Spectrochemical analysis of the composition of different possible exogenous inclusion sources as compared with inclusions in situ

Exogenous Source	*MnO*	*FeO*	*CaO*	*MgO*	Al_2O_3	SiO_2	TiO_2	*Fig.*
Furnace Slag Ladle Slag :	10	7	45	10	13.5	15	—	
Before Degassing	11	2.5	30	8	9.5	34	—	
After Degassing	10	+	27	6	14.5	37	—	
Ladle Refractories :								
Top Lining	—	+	+	+	75	22	+	
Main Lining	—	2	+	+	33	62	+	
Ingot Mould Refractories	—	2	—	20	38	40	—	
Ingot Scum	22	6	+	12.5	12.5	40	—	
Inclusion I Mean Analysis	13	—	23	7	20	41	—	207
Inclusion II Mean Analysis	8	+	+	34	10	46	+	208 a
Inclusion III Mean Analysis	31	—	4	14	18	43	+	208 b

(*a*) Large Inclusions

1. Forsterite - 8 MnO, 55MgO, 37 SiO_2 designated as (F)
2. Spinel - 7 MnO, 25 MgO, 70 Al_2O_3 designated as (G)
3. Matrix - 7 MnO, 3CaO, 6MgO, 57 SiO_2, 22 Al_2O_3 designated as (M)

Average Composition : 8 MnO, 1FeO, 1CaO, 34MgO, 46 SiO_2, 10Al_2O_3

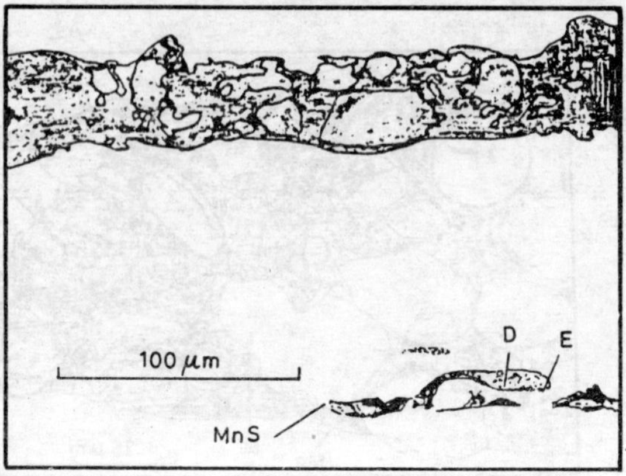

Inclusions :

(E) Enstatite with 34 MnO, 26 MgO, 39 SiO_2

(D) Matrix With 27 MnO, 7 CaO, 4 MgO, 45 SiO_2, 14 Al_2O_3 (Sd) MnS

Mean Analysis (calc) 31 MnO, 4 CaO, 14 MgO, 43 SiO_2 - 8 Al_2O_3

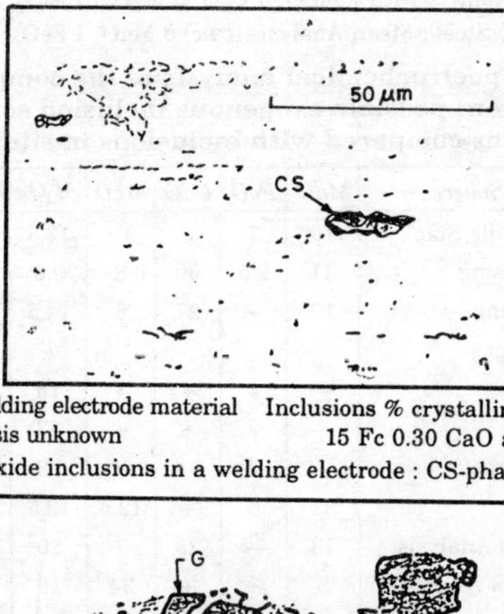

Material type welding electrode material analysis unknown

Inclusions % crystalline CS-phase with 15 Fc 0.30 CaO and 50 SiO_2

Oxide inclusions in a welding electrode : CS-phase

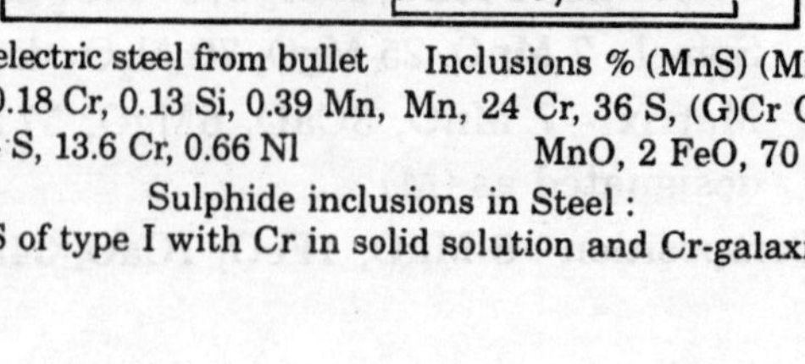

Steel type basic electric steel from bullet steel analysis, % 0.18 Cr, 0.13 Si, 0.39 Mn, 0.023 P, 0.14 S, 13.6 Cr, 0.66 Nl

Inclusions % (MnS) (Mn Cr)S with 34 Mn, 24 Cr, 36 S, (G)Cr Galaxite with 30 MnO, 2 FeO, 70 Cr_2, O_3.

Sulphide inclusions in Steel :
MnS of type I with Cr in solid solution and Cr-galaxite.

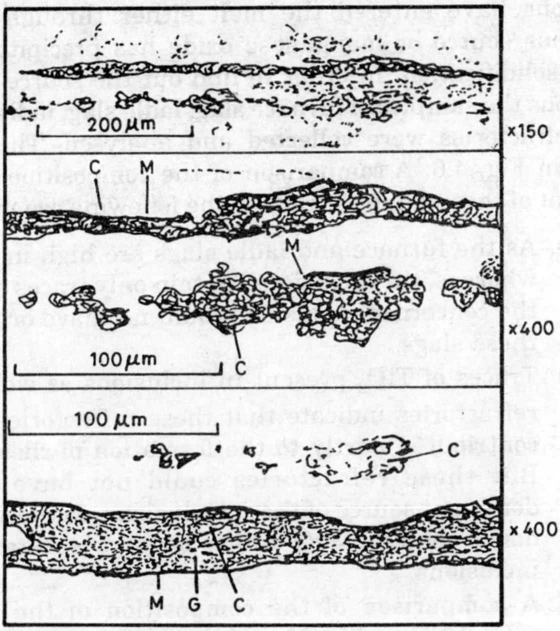

Fig. 4.6. Details of case study.

(*b*) Small Inclusions

1. Enstatite - 34 MnO, 26 MgO, 39 SiO_2 designated as (E)
2. Matrix - 27 MnO, 7CaO, 4MgO, 45SiO_2, 14Al_2O_3 designated as (D)
3. MnS streaks

Average Composition : 31 MnO, 4CaO, 14 MgO, 43 SiO_2, 8Al_2O_3.

The steel from which these rails were made was silicon killed, vacuum degassed in the ladle and was top poured in a big end down ingot. The steel had a composition of 0.59% C, 0.39% Si and 1.53% Mn.

As the inclusions were quite large it looks improbable that they were deoxidation products. Further another point against the inclusions being indigenous was that the steel was vacuum degassed. Further the general microscopic appearance of the large inclusions showed them to be of exogenous origin. As the inclusions are large in size they could have easily separated had they got only a few minutes time to float out. Thus it can be inferred that these appear in the melt at a late stage of steel making process but are preset before solidification.

Presence of MnO in large inclusions indicates that these inclusions have entered the melt either through corrosion of exogenous source or manganese oxide has precipitated on them during solidification. In order to find out the source of exogenous inclusions the samples of furnace slag, ladle slag, ladle and teeming path refractories were collected and analysed. This analysis is shown in Fig. 4.6. A comparison of the composition of inclusions with that of these samples leads to the following conclusions :

(*i*) As the furnace and ladle slags are high in CaO content whereas, the inclusions contain only traces of CaO, hence the concerned inclusions could not have originated from these slags.

(*ii*) Traces of TiO_2 present in inclusions as well as in ladle refractories indicate that these refractories might have contributed partly to the formation of these inclusions. But these refractories could not have become the dominant source of these inclusions, as these refractories donot contain MgO which forms a substantial part of inclusions.

(*iii*) A comparison of the composition of the ingot mould refractories with that of the inclusion shows that high MgO in the inclusions might have come from ceramic bottom ramming mass.

Hence it is inferred that the large inclusions resulted mainly from the ramming mass which got mixed with molten metal as a result of stirring action caused during pouring. This mass got partly separated from the melt and floated to the top as shown by the composition of ingot scum. Manganese oxide in these inclusions came from precipitation of deoxidation products on inclusions.

The main component of smaller inclusions are the same as that for the large ones except that MnO content of these inclusions is much higher. This shows that the exogenous source in this case also is the same as that for large inclusions, but the effect of indigenous precipitation in this case is far greater than that in case of larger ones.

FIVE

Techno-economics of Stainless Steel Making : A Thermodynamic Approach

5.1. INTRODUCTION

Stainless steels offer an excellent combination of mechanical, heat resisting and corrosion resistance properties and hence find extensive applications in house hold, chemical, petrochemical, fertiliser and nuclear industries. Today over 200 different alloys of this class have been developed which can be classified into five categories depending upon their microstructure, namely, austenitic, ferritic, martensitic, duplex and precipitation hardening stainless steels. All these contain chromium which is an essential alloying element in all these steels in excess of 12%. Hence the technoeconomics of stainless steel making depends predominantly on chromium retention at low carbon contents.

In early days of stainless steel industry *i.e.* before 1930 the manufacturers followed a simple and uniform arc melting process. But as the demand and production increased and hence the incentive for lowering the production cost became more attractive a number of technologies were developed. These technologies can be categorised into following four processes :

(*i*) Dilution process
(*ii*) Rustless process
(*iii*) Vacuum processes (VOD + VCD + VD)
(*iv*) AOD and CLU processes

The chronological development of these processes is shown in Fig. 5.1.

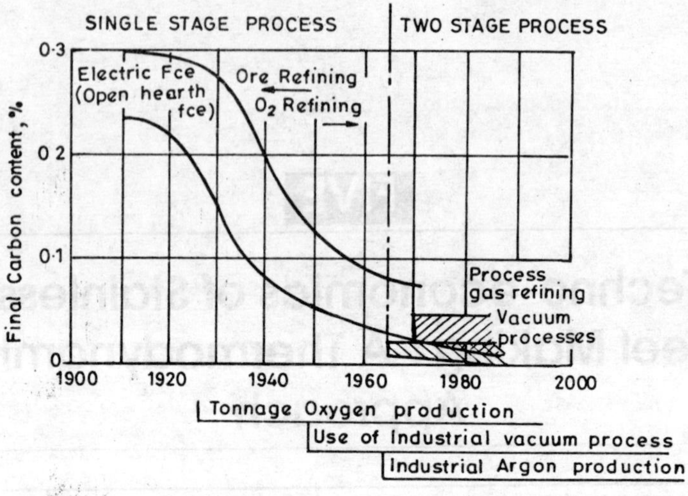

Fig. 5.1. The chronological development of various steel making processes with respect to final carbon content achieved in single stage and two stage processes.

Dilution process was the first processes by which stainless steel was produced in an electric arc furnace. The process consists of making a low carbon steel in a conventional electric arc furnace and then adjustment of its composition by the addition of costly low carbon ferro-alloys and nickel.

In rustless process a molten bath prepared in an electric arc furnace using raw materials including stainless steel scrap and high carbon ferro-chrome is oxidised to bring down the carbon to a desired level. The accompanying oxidised chromium is partially reduced by the addition of ferro-silicon and finally the composition of the metal is made up by the addition of low carbon ferro-alloys. The relative oxidation of carbon and chromium is controlled by the reaction.

$$3C + Cr_2O_3 = 2Cr + 3\ CO \quad ...(1)$$

Hence the efficient working can only be achieved by the control of reaction temperature and slag composition so that the loss of chromium is reduced to minimum.

Eq. (1) shows that loss or recovery of chromium is a strong function of partial pressure of carbon mono-oxide. This aspect led to a phase of two stage processes which began with the refining of high chromium bath prepared in general in an electric arc furnace, under vacuum.

The availability of argon on industrial scale led to simultaneous development of another set of two stage processes commonly known as AOD and CLU processes. In these processes the first oxidation is affected by a mixture of argon and oxygen, finally by reducing the oxygen content of the gas chromium is recovered. In case of CLU process a mixture CO_2 and H_2O is used alongwith oxygen and argon.

The ultimate composition of the bath after a period, oxidation or reduction, in all the above processes shall be controlled by the underlying physico-chemical principles namely, the thermodynamics and kinetics of the reactions. Though kinetics plays a very important role in predicting the end point of various stages of the processes, it is the thermodynamics which will decide the ultimate achievable limits. The present work is devoted to the analysis of these achievable limits vis a vis the cost in case of each of the above processes. So a thermodynamic analysis of the reactions taking place during stainless steel making shall now be undertaken.

5.2. THERMODYNAMIC ANALYSIS

Stainless steel making starts with the treatment of a molten metal bath of predetermined compositions. Such a bath may contain apart from Fe, Cr, C a number of other elements such as Mn, Si, P and S. During oxidation and/or reduction stages these elements shall also be reacting, so the overall thermodynamic analysis will require considering these reactions alongwith those involving Cr, C and Fe. In Table 5.1 the thermodynamic data of the various chemical reactions taking place during stainless steel making is given. This data can only lead to the calculation of ratio of the activities of various components in equilibrated phases. In order to calculate the composition of the metal phase under given conditions, one requires data on the solution thermodynamics of various components as well as that on interaction parameters. This data is given in Table 5.2 and 5.3 respectively.

Table 1 : Thermodynamic Data of Chemical Reactions

$$[C]_{1\,wt\%} + [O]_{1\,wt\%} = [CO]$$

$$\Delta G^\circ = -\,22470 - 39.8T \text{ Jules}$$

$$[Mn]_{1\,wt\%} + [O]_{1\,wt\%} = (MnO)$$

$$\Delta G^\circ = -\,245280 + 109.1\,T \text{ Jules}$$

$$[Si]_{1\,wt\%} + 2[O]_{1\,wt\%} = (SiO_2)$$

$$\Delta G^\circ = -\,596400 + 231T \text{ Jules}$$

$$2[Cr] + 3[O]_{1\,wt\%} = (Cr_2O_3)$$

$$\Delta G^\circ = -\,802032 + 271.35\,T \text{ Jules}$$

$$4/3\,Al + [O_2]_{1\,wt\%} = 2/3\,(Al_2O_3)$$

$$\Delta G^\circ = -\,1008840 + 154.56\,T \text{ Jules}$$

$$2(CaO) + (SiO_2) = (Ca_2SiO_4)$$

$$\Delta H^\circ = -\,125000 \text{ Jules}$$

Table 2 : Solution Reactions

$1/2\ (O_2) = [O]_{1\ wt\%}$

$\Delta G^\circ = -118510 - 2.39\ T$ Jules

$C_{gr} = [C]_{1\ wt\%}$

$\Delta G^\circ = 22680 - 42.42\ T$ Jules

$Cr = [Cr]_{1\ wt\%}$

$\Delta G^\circ = 19320 - 46.97\ T$ Jules

$Mn = [Mn]_{1\ wt\%}$

$\Delta G^\circ = 14647 - 47.91\ T$ Jules

$Si = [Si]_{1\ wt\%}$

$\Delta G^\circ = -117600 + 23.27\ T$ Jules

$N_2 = [N]_{1\ wt\%}$

$\Delta G^\circ = 3612 + 23.98\ T$ Jules

Table 3 : Interaction Coefficients

Element	*O*	*C*	*Mn*	*Cr*	*Si*	*Ni*
O	-20×10^{-2}	-13×10^{-2}	0	-4.1×10^{-2}	-2×10^{-2}	0.6×10^{-2}
C	-9.7×10^{-2}	22×10^{-2}	—	-2.4×10^{-2}	10×10^{-2}	1.2×10^{-2}
Mn	0	—	—	—	—	—
Cr	-13×10^{-2}	-10×10^{-2}	—	—	—	—
Si	-25×10^{-2}	24×10^{-2}	—	—	32×10^{-2}	0.5×10^{-2}

As an application of this data let us consider the Cr and C contents of the metal in relation to temperature at one atmosphere pressure of carbon-mono-oxide. In Fig. 5.2 metal chromium content is plotted against carbon content. This figure shows that with increase in temperature chromium retention in the metal increases for the same amount of carbon present. Hence it can be inferred that during decarburisation highest admissible temperature should be maintained to keep chromium loss to minimum.

The above mentioned data leads to the calculation of metal composition provided slag composition is known. In order to fix slag composition let us first consider Fig. 5.3. In this figure chromium distribution is plotted against slag basicity. This figure suggests that for minimum loss of chromium the slag basicity should be above 2. Such high basicity, if maintained, shall also help in

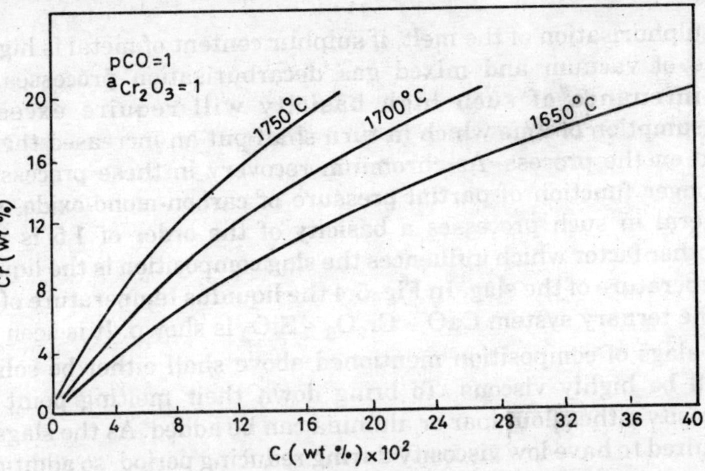

Fig. 5.2. Carbon - chromium relationship in molten steel.

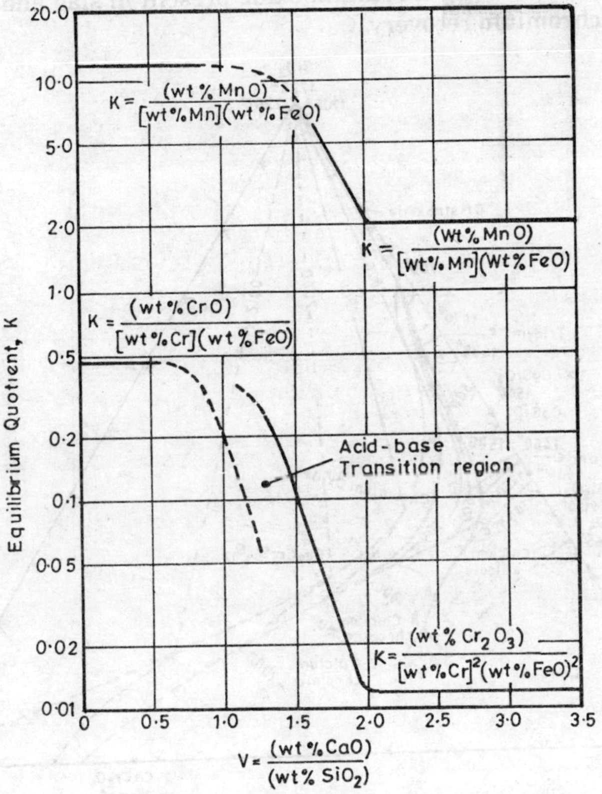

Fig. 5.3. Chromium distribution shown as a function of basicity under different steel making conditions.

desulphurisation of the melt, if sulphur content of metal is high. In case of vacuum and mixed gas decarburisation processes, the maintenance of such high basicity will require excessive consumption of lime which in turn shall put an increased thermal load on the process. As chromium recovery in these process is a stronger function of partial pressure of carbon-mono-oxide, so in general in such processes a basicity of the order of 1.5 is used. Another factor which influences the slag composition is the liquidus temperature of the slag. In Fig. 5.4 the liquidus temperature of slag of the ternary system $CaO - Cr_2O_3 - SiO_2$ is shown. It is seen that the slags of composition mentioned above shall either be solid or shall be highly viscous. To bring down their melting point and viscosity either flourspar or alumina can be added. As the slags are required to have low viscosity during reducing period, so addition of aluminium instead of alumina can also be done. Aluminium if added will help in reducing chromium oxide present in slag and will thus help in chromium recovery.

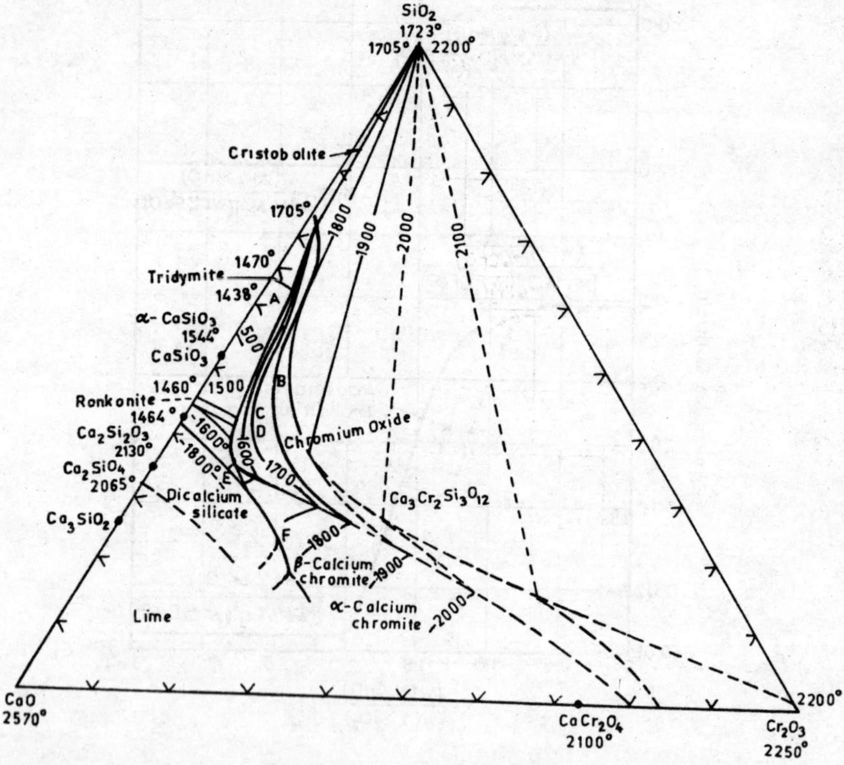

Fig. 5.4. Liquidus temperature in the system $CaO–Cr_2O_3–SiO_2$.

Combination of above requirements, the heat capacity data for various input materials is given in Table 4.

Table 4 : Thermodynamic Data

Material	*Specific Heat KJ/Kg/°K*	*Latent Heat KJ/Kg*
Molten Metal	0.84	—
Lime	0.77	1425
Argon	0.52	—
Oxygen	0.91	—
Cr_2O_3	0.69	940
Slag	2.0	—
Ni	0.44	300
Fe	0.46	274
Mn	0.48	266
Si	0.68	1660
Al	0.9	395
C	0.7	—
Cr	0.46	265
MgO	0.95	1942
SiO_2	0.7	182
Fe_2O_3	0.95	866
CaF_2	0.92	915
Al_2O_3	0.78	1070

Thus with the help of this data one is able to calculate the total material and energy requirements for various processes. Following this procedure the requirements for various case studies have been computed as given under case studies.

5.3. ECONOMICS

The cost of production of stainless steel can be split under following heads :

(*i*) Raw Material Cost
(*ii*) Energy Cost
(*iii*) Maintenance
(*iv*) Depreciation and
(*v*) Interest

As the present work concerns the effect of process parameters on the cost, so only first two aspects of production cost shall be considered. Thus the contribution of these two factors to the cost per ton of steel is equal to the cost of individual raw materials including power per unit mass multiplied by the amount of raw material consumed per ton of steel. The rates of various raw materials per ton under Indian conditions is given in Table 5 alongwith their composition.

Table 5 : Raw Materials

Material	*Composition*								*Cost $/Kg*
	C	*Mn*	*Si*	*P*	*S*	*Fe*	*Cr*	*Others*	
Molten metal (1500°C)	0.10	0.15	0.06	0.04	0.04	99.5	—	—	0.34
HC Ferro-Mn (25°C)	7.0	75.0	—	0.3	—	17.8	—	—	0.57
LC Ferro-Mn (25°C)	0.1	85.0	6.0	—	—	8.9	—	—	2.28
HC Ferro-Cr (25°C)	7.0	—	3.0	—	—	30.0	60.0	—	1.14
LC Ferro-Cr	0.05	—	5.0	—	—	29.7	65.0	—	3.14
Aluminium	—	—	—	—	—	—	—	Al = 100	2.29
Nickel	—	—	—	—	—	—	—	Ni = 100	20.00
Ferro Silicon	—	—	70.0	—	—	30.0	—	—	1.43
Fluorspar	—	—	—	—	—	—	—	CaF_2 = 100	0.40
Lime	—	—	—	—	—	—	—	CaO = 95 SiO_2 = 5	0.06
Chrome ore	—	—	—	—	—	23.0	33.0	MgO = 13 Al_2O_3 = 13 SiO_2 = 2.5	0.11
Oxygen	—	—	—	—	—	—	—	—	0.40
Argon	—	—	—	—	—	—	—	—	4.57
Energy	—	—	—	—	—	—	—	—	8.29

5.4. CASE STUDIES

In order to illustrate the application of above mentioned principles and compare the processing cost, the following target composition of commonly produced austenitic stainless steel is considered as an example.

C-0.08, Mn-1.1, Si-0.7, Cr-18, Ni-8, S-0.03 max. and P-0.045 max.

To calculate the material and energy requirements, the following assumptions are made :

(*i*) The raw material used in all cases shall be the ones from the list given in Table 5. This does not restrict the applicability of the principles outlined above as these can be applied to any set of starting materials.

(*ii*) The activities in the slag phase can be calculated using Temkin Model except for activity of Cr_2O_3 which is assumed to follow a linear relationship below the saturation limit. On the basis of these assumptions, the following cases involving different processes have been considered.

Case-1 : The material and energy requirements for dilution process are given in Table 6. In this case a low carbon heat is made in an electric arc furnace and the composition is adjusted by the addition of low carbon ferroalloys.

Case-2 : Initial use of high carbon ferrochrome and scrap followed by oxidation of carbon which allows the maximum chromium content of the bath to be around 6%. Finally the composition of the bath is adjusted by low carbon ferroalloys. The material and energy balance is given in Table 7.

Case-3 : Some as case 2 followed by a reduction step which leads to higher chromium recovery. The heat and material balance for this case is presented in Table 8.

Case-4 : As the next case, two stage process involving argon purging is considered. The material and energy balance for this case is given in Table 9.

Case-5 : Same as above except that the reduction has partly been done by the addition of aluminium to form a fluid slag (Table 10).

Case-6 : As next case, a vacuum process is considered. For this a material and heat balance is shown in Table 11.

Case-7 : Finally as last case, vacuum with partial chromium recovery from chrome ore is considered. The balance sheet for this case is presented in Table 12.

5.5. DISCUSSION

A summary of thermodynamic processing cost based on material requirement is given in Table 13. It is seen from this table that the savings are mainly due to progressive substitution of costly low carbon ferro-alloys by high carbon ones and the oxidation of excess carbon by modifications in technology. This table also shows that nickel contributes substantially to the overall cost. Further Nickel is an imported material in our country. Therefore, a further reduction can be affected by the substitution of nickel by other alloying elements in stainless steels. In our country, nickel is replaced by a combination of Manganese, nitrogen and copper. The amount of Nitrogen in steel for making it completely austenitic is given below :

For Mn contents 6.5 - 7.8%

$$\%N_2 = 0.04\ (1.3\%\ Cr - \%\ Ni - 30\%\ C - 11.3)$$

For Mn contents 7.8-8.5%

$$\%N_2 = 0.04\ (1.3\%\ Cr - \%Ni - 30\%\ C - 10.8)$$

With the addition of Nitrogen, the nickel content has been reduced to about 2%, which has considerably reduced the cost of production.

Table 6 : Energy and Material Balance (Basis 1 Ton of S.S. Produced) Dilution Process

Target Comp.; C = 0.08, Mn = 1.1, Si = 0.7, S = 0.03, P = 0.045, Cr = 18, Ni = 8.0

Material		*C*	*Mn*	*Si*	*Cr*	*Ni*	*Others*	*Cost $*	*Energy required KJ*
Item	*Amount kg.*								
Input									
Hot Metal	625	0.625	0.94	0.38	—	—	P=0.25 S=0.25	210.86	—
Low Carbon Ferrochrome	280	0.14	—	14.0	182	—	—	960.00	2.48×10^5
Nickel	80	—	—	—	—	80	—	1600.00	0.97×10^5
Low Carbon Ferro-manganese	13	0.01	11.05	0.78	—	—	—	29.71	0.1×10^5
Ferrosilicon	10	—	—	6.50	—	—	—	14.29	-0.07×10^5
Fluorspar	8	—	—	—	—	—	—	3.20	0.18×10^5
Lime	40	—	—	—	—	—	—	2.29	1.04×10^5
Products									
Stainless Steel :	1000								
Slag :	68								
							Total Material Cost	2820.35	4.30×10^5
							Energy Cost	1.6	= 20 KWH

*For this process the hot metal carbon is required to be brought down to about 0.1%.

Table 7 : Energy and Material Balance (Basis 1 Ton of S.S. Produced) Rustless Process without reduction

Target Comp.; C = 0.08, Mn = 1.2, Si = 0.7, S = 0.03, P = 0.045, Cr = 18, Ni = 8.0

Material		C	Mn	Si	Cr	Ni	Others	Energy required KJ	Cost $
Item	*Amount Kg.*								
Input Hot Metal	630	0.63	0.95	0.38	—	—	P=0.25 S=0.25	79380	216.00
High carbon Ferrochrome	82	5.7	—	2.50	49.2	—	—	91348	93.71
Low carbon Ferrochrome	208	0.62	—	10.40	135.2	—	—	189280	713.14
Nickel	80	—	—	—	—	80	—	103700	1600.00
Low Carbon Ferro-manganese	12	0.12	10.20	0.72	—	—	—	—	27.43
Lime	50	—	—	—	—	—	—	135750	2.86
Fluorspar	5	—	—	—	—	—	—	12280	2.00
Oxygen	18.50 (=13 cum)	8.50	—	8.0	2.00	—	—	—	5.20
Products								600000	2660.34
Stainless steel :	1005 kgs							170 KWH	14.29
Slag :	75 kgs								2674.63

Table 8 : Energy and Material Balance (Basis 1 Ton of S.S. Produced) Rustless Process with Reduction

Target Comp.; C = 0.08, Mn = 1.1, Si = 0.07, S = 0.03, P = 0.045, Cr = 18, Ni = 8.0

Material									
Item	*Amount Kg.*	*C*	*Mn*	*Si*	*Cr*	*Ni*	*Others*	*Energy required KJ*	*Cost $*
Input Hot Metal	625	0.64	0.95	0.38	—	—	P=0.25 S=0.25 Fe=631.8	80010	217.71
High carbon Ferrochrome	130	8.89	—	3.90	78.0	—	Fe=38.1	141478	148.57
Ferrosilicon	17	—	—	11.9	—	—	Fe=5.1	—	24.29
Low carbon Ferrochrome	165	0.48	—	8.0	107.5	—	Fe=47.52	—	565.71
Low Carbon Ferro-manganese	12	0.01	10.20	0.72	—	—	Fe=1.07	—	27.43
Nickel	80	—	—	—	—	80	—	—	1600.00
Lime	75	—	—	—	—	—	CaO=75	203625	4.29
Fluorspar	12	—	—	—	—	—	CaF_2=12	29472	4.80
Oxygen	35.15 (=25 cum)	12.3	0.05	20.4	2.4	—	—	—	10.86
Products							Energy :	150 KWH	12.57
Stainless steel :	1000 kgs								
Slag :	130 kgs								<u>2616.23</u>

Table 9 : Energy and Material Balance (Basis 1 Ton of S.S. Produced) A.O.D. Process

Target Comp.; C = 0.08, Mn = 1.2, Si = 0.7, S = 0.03, P = 0.045, Cr = 18, Ni = 8.0

Material		*C*	*Mn*	*Si*	*Cr*	*Ni*	*Others*	*Energy required KJ*	*Cost $*
Item	*Amount Kg.*								
Input Hot Metal	625	0.63	0.94	0.38	—	—	P=0.25 S=0.25	78750	214.29
High carbon Ferrochrome	320	22.4	—	9.6	192	—	—	356480	365.71
Ferrosilicon	32	—	—	22.4	—	—	—	—	45.71
Low Carbon Ferro-manganese	12	0.60	10.20	0.72	—	—	—	9840	27.43
Nickel	80	—	—	—	—	80	—	103700	1600.00
Lime	100	—	—	—	—	—	—	271500	5.71
Oxygen	64.56 (45 cum)	29.29	—	29.75	5.52	—	—	6000	18.00
Argon	25.00 cum								114.29
Products : Stainless steel : 1000 kgs Slag : 170 kgs						Energy requirement is balanced by heat of Oxidation			2391.14

Table 10 : Energy and Material Balance A.O.D. Process with Aluminium

Target Comp.; C = 0.08, Mn = 1.2, Si = 0.7, S = 0.03, P = 0.045, Cr = 18, Ni = 8.0

Material									
Item	*Amount Kg.*	*C*	*Mn*	*Si*	*Cr*	*Ni*	*Others*	*Energy required KJ*	*Cost $*
Input Hot Metal	625	0.63	0.94	0.38	—	—	P=0.25 S=0.25	78750	214.29
High carbon Ferrochrome	320	22.4	—	9.6	192	—	—	356480	365.71
Ferrosilicon	20	—	—	14	—	—	—	(-) 13180	28.57
Low Carbon Ferro-manganese	12	0.01	10.2	0.72	—	—	—	9840	27.43
Nickel	80	—	—	—	—	80	—	103700	1600.00
Lime	45	—	—	—	—	—	—	122175	2.57
Aluminium	5.5	—	—	—	—	—	—	10450	12.57
Oxygen	53 = 36.9 cum	29.2	0.04	20.3	3.50	—	—	—	
Argon	22.00 cum					—	—	—	105.7
Products : Stainless Steel : 1000 kgs Slag : 78 kgs							Energy requirement is balanced by heat of Oxidation		2368.65

Table 11 : Energy and Material Balance (Basis 1 T of S.S. Produced) V.O.D. Process, Normal Process

Target Comp.; C = 0.08, Mn = 1.2, Si = 0.7, S = 0.03, P = 0.045, Cr = 18, Ni = 10

Material		C	Mn	Si	Cr	Ni	Others	Energy required KJ	Cost $
Item	*Amount Kg.*								
Input Hot Metal	625	0.63	0.94	0.38	—	—	P=0.25 S=0.25	78750	214.29
High carbon Ferrochrome	300	21	—	9	180	—	—	334200	342.86
Nickel	80	—	—	—	—	80	—	103700	1600.00
Ferrosilicon	15	—	—	10.5	—	—	—	—	21.43
Low Carbon Ferro-manganese	12	0.01	10.2	0.72	—	—	—	—	27.43
Low Carbon Ferrochrome	20	0.06	—	1.0	13	—	—	—	68.57
Lime	80	—	—	—	—	—	—	—	—
Fluorspar	10	—	—	—	—	—	—	—	4.57
Oxygen	50.59 (35.2 cum)	27.86	0.04	16.69	6.00	—	—	—	14.00
Argon	45 litres								0.29
Products : Stainless steel : 1000 kgs Slag : 125 kgs						Energy requirement is balanced by heat of oxidation			2297.43

Table 12 : Energy and Material Balance VOD Process with Chromite Ore

Target Comp.; C = 0.08, Mn = 1.1, Si = 0.7, S = 0.03, P = 0.045, Cr = 18, Ni = 10

Material: Item	Amount Kg.	C	Mn	Si	Cr	Ni	Others	Energy required KJ	Cost $
Input Hot Metal	610	0.61	0.95	0.37	—	—	P=0.24 S=0.24	76860	209.14
High carbon Ferrochrome	260	1.82	—	0.78	156	—	—	289640	297.14
Low Carbon Ferro-manganese	12	0.01	10.20	0.72	—	—	—	9840	27.43
Low carbon Ferrochrome	10	0.3	—	0.5	6.5	—	—	9100	34.29
Ferrosilicon	22	—	—	15.4	—	—	—	14498	31.43
Chrome Ore	80	—	—	—	26.4 as Cr_2O_3	—	as Fe_2O_3	210800	—
Nickel	80	—	—	—	—	80	—	103700	1600.00
Lime	70	—	—	—	—	—	—	190050	4.00
CaF_2	7	—	—	—	—	—	—	—	2.80
Oxygen	19.03 (=13.2 cum)	2.59	0.04	12.3	4.1	—	—	—	5.32
Argon	45 litres						—		0.29
Products : Stainless Steel : 1000 kgs Slag : 120 kg							Energy requirement is Balanced by heat of Oxidation		2211.84

Table 13 : Comparison of Costs of Production by Various Processes

	Dilution		*Rustless*				*AOD*				*VOD*			
			With Reduction		*Without Reduction*		*Normal*		*With Aluminium*		*Normal*		*With Chromite Ore*	
Material	*Amount Kg*	*Cost $*	*Amount Kg*	*Cost $*	*Amount Kg*	*Cost $*	*Amount Kg*	*Cost $*	*Amount Kg*	*Cost $*	*Amount Kg*	*Cost $*	*Amount Kg*	*Cost $*
Molten Metal (1550°C)	625	210.86	625	216.00	625	217.71	625	214.29	625	214.29	625	214.29	610	209.14
HC Ferro-Mn (25°C)	—	—	—	—	—	—	—	—	—	—	—	—	—	—
LC Ferro-Mn (25°C)	13	29.71	12	27.43	12	27.43	12	27.43	12	27.43	12	27.43	12	27.43
HC-Ferro-Cr (25°C)	—	—	130	148.57	82	93.71	320	365.71	320	365.71	300	342.86	260	297.14
LC Ferro-Cr	280	960.00	165	565.71	208	713.14	—	—	—	—	20	68.57	10	34.29
Aluminium	—	—	—	—	—	—	—	—	5.5	12.57	—	—	—	—
Nickel	80	1600.00	80	1600.00	80	1600.00	80	1600.00	80	1600.00	80	1600.00	80	1600.00
Ferro-silicon	10	14.29	17	24.29	—	—	32	45.71	20	28.57	15	21.43	22	31.43
Fluospar	8	3.20	12	4.80	5	2.86	—	—	—	—	10	4.00	7	2.80
Lime	40	2.29	75	4.29	50	2.86	100	5.71	45	2.57	80	4.57	70	4.00

	Dilution		Rustless				AOD				VOD			
			With Reduction		Without Reduction		Normal		With Aluminium		Normal		With Chromite Ore	
Material	Amount Kg	Cost $	Amount Kg	Cost $	Amount Kg	Cost $	Amount Kg	Cost $	Amount Kg	Cost $	Amount Kg	Cost $	Amount Kg	Cost $
Chrome ore	—	—	—	—	—	—	—	—	—	—	—	—	80	—
Oxygen	—	—	35.15 (25.0 cum)	10.86	18.5 (13 cum)	5.20	64.56 (45 cum)	18.00	53	16.94	50.59 cum	14.00	19.03	5.32
Argon	—	—	—	—	—	—	25 cum	114.29	22 cum	100.57	45 liters	0.29	45 lit	0.29
Energy	20 kwh	—	150 kwh	—	170 kwh	—	—	—	—	—	—	—	—	—
Total cost	—	2820.35	—	2503.66	—	2616.2	—	2391.14	—	2368.65	—	2297.44	—	2211.84
(Total Cost - Ni, cost)	—	1220.66	—	1003.66	—	1061.2	—	791.14	—	768.65	—	697.44	—	611.84

REFERENCES

1. K. Strauss (Edi) - "Applied Science in the Casting of Metals", Pergamon Press, (1970), p. 133.
2. V. Kudrin - "Steel Making", MIR Publishers, Moscow, (1985), p. 115, 257.
3. Amit Chatterjee and S. Govindarajan (Edi) - "Monograph on Continuous Casting at Tata Steel", Tata Steel, Jamshedpur, Feb. (1991), p. 36, 56, 76, 86, 104.
4. S.K. Dutta, P.J. Roychowdhury and V.L. Gadgeel - "Steel making through DR-EAF route under Indian Conditions", IE(I) Journal MM (Spl.), Vol. 68, March (1988), p. 52-62.
5. B.P. Singh and L. Prasad - ""Graphite Electrodes– Technology Upgradation", Metal News, Vol. 9, No. 1, March (1987), p. 4.
6. Brahma Deo and R.K. Sandhir - "Materials Optimisation in Arc Furnace Steel Making", Trans IIM, Volume 40, No. 3, June (1987), p. 281-283.
7. Raman Datta and Sanak Mishra - "Continuous Casting Technology - the present Scenario and future prospects", Trans. IIM, Vol. 48, No. 4, August (1995), p. 339-358.
8. Ashok Kumar and L.K. Singhal - "Continuous Casting Technology - the present Scenario and future prospects", IE (I) Journal - MM, Vol. 75, May (1994), p. 17-36.
9. Naveen Nakra - "Operational Experience at Continuous Casting Plants" Usha Alloy Steels Ltd., Jamshedpur, p. 22-35.
10. "Thin Slab Continuous Casting Process", SDI Year Book, Tata Steel, (1987), p. 50-53.
11. R.H. Tupkary - "Introduction to Modern Steel Making", Khanna Publishers, (1983).
12. C.E. Sims (Edi) - "Electric Furnace Steel Making", Vol. I & II, Phys. Chemistry of steel making committee, AIME (1967).
13. Rolland Kiessling and Nils Lange - "Non Metallic Inclusions in Steel", The Metals Society, London, Second Edition, (1978).
14. Ladle Furnace, SDI Year Book, Tata Steel (1990).
15. Secondary steel making - VAD Process, SDI Year Book, Tata Steel (1990).
16. Secondary Steel Making - RH Process, SDI Year Book, Tata Steel (1990).

17. Subharwal, SPS & Ganguly A., Iron making Steel making, (1992), Vol. 19, No. 3, p. 233.
18. Subba Rao, M.V., Minerals & Metals review, 24, July (1991).
19. Subba Rao, M.V., Minerals & Metals review, 32, Feb. (1992).
20. Bodsworth, C., and Zetouni, A.W., Iron making Steel making, (1991), Vol. 18 No. 2, p. 107.
21. "The outlook for electric steel making worldwide", Minerals and Metals review, Annual (1984), p. 143.
22. "Small Scale Steel making" Ed. R.D. Walker.
23. Proceedings, Seminar on "Electric Steel Making", Dec. 21 & 22 (1979), New Delhi.
24. Sidhu, S.S., "Steel Industry in India, Problems and Perspective".
25. "Secondary Steel making", Proc. Conference, Oct. 23-24, (1984), The Metal Society, The Institute of Metals, London (1985).
26. Proc. 3rd Indo-Soviet Bilateral Symposium, Nov. 26 & 27 (1984), Ranchi, Organised by SAIL and I.I.M.
27. National Seminar on Secondary Steel Refining, Dec. 8-10 (1994), BHEL, Hardwar.

Index

A

B

C

D

E

F

G

H